*Design of Modern Steel Highway Bridges*

# Design of Modern Steel Highway Bridges

**C. P. HEINS**
*University of Maryland*

**D. A. FIRMAGE**
*Brigham Young University*

**A Wiley-Interscience Publication**

**JOHN WILEY & SONS,** New York · Chichester · Brisbane · Toronto

*Library of Congress Cataloging in Publication Data:*

Heins, Conrad P.
  Design of modern steel highway bridges.

  "A Wiley-Interscience publication."
  Includes index.
  1. Bridges, Iron and steel—Design and construction.
I. Firmage, David Allan, joint author. II. Title.

TG350.H43   624.2   78-9084
ISBN 0-471-04263-3

Printed in the United States of America

10 9 8 7 6 5 4 3 2 1

# Preface

The design and analysis of bridge structures have grown in the last decade with the changeover from conventional girder-slab bridges to complex interchanges requiring curved units. In addition, the engineer has come to deal with girders and supports (slant legged), steel decks (orthotropic), box girders, and straight cables and girders (cable stayed).

The design of such complex structures has often required imagination and ingenuity, since research may lag behind practical possibilities. This book partially remedies this difficulty by presenting up-to-date techniques for these complex systems and permitting a better understanding of the structural response.

Although the market has been flooded with literature on structural analysis, structural design books, especially those that show an integrated approach to the subject, have not been very numerous. We have two objectives in this book: (1) to make use of an integrated approach to the structural design of bridges and (2) to present design procedures for steel beam and girder bridges. The second objective entailed the presentation of the latest design concepts for steel bridges as well.

The design specifications used are those presently given in the AASHTO code. The AASHTO standard *Specifications for Highway Bridges* (11th ed., 1973) and the *Interim Specifications* (1974–1977) are followed in the design where applicable. The word Specifications in the text refers to these AASHTO Specifications. For problems for which the code does not specify a criterion, suitable design information has been adopted.

In addition to steel bridges the book also considers the design of the reinforced concrete slab and concrete piers and abutments.

Since bridge design and specifications in the United States still use the English system of measurements, this system has been retained in this book. A conversion table to metric equivalents is included to enable engineers in other countries to follow the design procedures.

We begin with the history and development of bridges. Then we go on to discuss the materials and loads used in bridge design. The chapters that follow present the analysis and design of various types of steel bridges; slab-steel beams, continuous beams, curved girders, and orthotropic steel box girders—straight, curved, and slant legged—are discussed. Analytical techniques, which do not require computers, are given in detail. Appropriate specifications are listed, followed by a complete design example. All of the bridge examples apply to moderate span structures.

Later we discuss the problems associated with bridge rating, giving an appropriate example. Then the requirement related to substructure design is presented.

The last chapter deals with the cable-stayed bridge. Details of the analysis and construction of this type of bridge are given.

The design examples present the usual procedure and form for the necessary design computations as developed in a bridge design office. The explanation of each step is not contained in the design example, but is presented in the text preceding the example. The student using this volume as a textbook will be trained in the development of design computations.

In general, the book is intended for senior-year college and first-year graduate-school students. The material, however, should also be useful to practicing engineers who work in modern bridge design.

C. P. Heins
D. A. Firmage

*College Park, Maryland*
*Provo, Utah*
*September* 1978

# Contents

**Conversion Factors**

| | English to SI | SI to English |
|---|---|---|
| Length | | |
| | 1 in. = 25.40 mm | 1 mm = 0.0394 in. |
| | 1 ft = 0.3048 m | 1 m = 3.281 ft |
| | 1 in. = 0.0254 m | 1 m = 39.37 in. |
| | 1 chain = 20.117 m | 1 m = 0.0497 chains |
| | 1 mile = 1.609 km | 1 km = 0.622 mile |
| Area | | |
| | $1 \text{ in.}^2 = 645.2 \text{ mm}^2$ | $1 \text{ mm}^2 = 0.00155 \text{ in.}^2$ |
| | $1 \text{ ft}^2 = 0.0929 \text{ m}^2$ | $1 \text{ m}^2 = 10.76 \text{ ft}^2$ |
| | $1 \text{ yd}^2 = 0.835 \text{ m}^2$ | $1 \text{ m}^2 = 1.196 \text{ yd}^2$ |
| Volume | | |
| | $1 \text{ in.}^3 = 16.387 \text{ mm}^3$ | $1 \text{ cm}^3 = 0.0610 \text{ in.}^3$ |
| | $1 \text{ ft}^3 = 0.0283 \text{ m}^3$ | $1 \text{ m}^3 = 35.31 \text{ ft}^3$ |
| | $1 \text{ yd}^3 = 0.765 \text{ m}^3$ | $1 \text{ m}^3 = 1.564 \text{ yd}^3$ |
| Moment of Inertia | | |
| | $1 \text{ in.}^4 = 41.62 \text{ cm}^4 = 416{,}200 \text{ mm}^4$ | $1 \text{ mm}^4 = 2.40 \times 10^{-6} \text{ in.}^4$ |
| | $= 0.4162 \times 10^{-6} \text{ m}^4$ | $1 \text{ m}^4 = 2.40 \times 10^6 \text{ in.}^4$ |
| Mass | | |
| | 1 lb = 0.454 kg | 1 kg = 2.205 lb |
| | 1 ton (2000 lb) = 907.2 kg | 1 Mg = 1.102 ton (2000 lb) |
| | 1 tonne (metric) = 1.102 ton (2000 lb) | |
| Force | | |
| | 1 lb = 4.448 N | 1 N = 0.2248 lb |
| | 1 kip (k) = 4.448 kN | |
| Stress | | |
| | $1 \text{ psi} = 6.895 \text{ kPa } (\text{kN/m}^2)$ | 1 MPa = 145.0 psi |
| | $1 \text{ ksi} = 6.895 \text{ MN/m}^2$ | $1 \text{ N/mm}^2 = 145.0 \text{ ksi}$ |
| | | $1 \text{ MN/m}^2 = 145.0 \text{ ksi}$ |

*Design of Modern Steel*
*Highway Bridges*

# 1

# Development and History of Bridges

## 1.1 BRIDGE DEVELOPMENT

People have always been interested in transporting themselves and their goods from one place to another. In early times waterways were used wherever possible. Today the cheapest means of moving goods is by boat; however, navigable waterways do not always go in the direction desired nor are they available in abundance. Thus it has been necessary to develop land transportation methods and means of crossing rivers and deep canyons. To move by land easily and quickly bridges have become a necessity, since they provide a means for moving produce throughout an area. Highway and railroad development have therefore become an absolute necessity for economic development.

Good highways and railways are also necessary in the waging of warfare. The Romans were aware of this, and a very important asset in their ability to extend their empire was the speed (for that day) with which they could move their legions from one place in the empire to another over their well-developed roads.

Hitler, in his plans for European conquest, developed the autobahn to move armies, manufactured goods, and raw materials quickly and in great quantities.

Europe developed economically earlier than most other areas of the world because of several factors, including the development of water and land transportation systems.

The rapid economic development of the United States could not take place until land transportation was developed. Even today, one factor that has caused many countries to lag behind in economic development is the lack of good land transportation systems.

## 1.2  HISTORY OF BRIDGES

Bridges have always been an important part of our environment. They have been major subjects of literature and art, both ancient and modern. Mythology has its mention of bridges; however, there is no reference to such in the Bible. Wars have been fought over bridges and in many cases the capture of a strategic structure has had a pronounced effect on the final outcome of the war. Bridges have been the center of village or city life. In ancient times, human sacrifices were performed at the start of a bridge-building project. These ceremonies were to appease the gods and thus ensure the safety of the structure. Appreciation for a bridge was no doubt greater in ages past than it is today when such objects of convenience are expected and demanded by the public. The knowledge and skill of the bridge builder is almost infinitely greater today than it was a few centuries ago, yet the place of the bridge builder in the community is nowhere near as great. Nevertheless, the personal satisfaction derived from the opening of a new bridge is just as gratifying to the creator.

The first bridges were simple beam spans of stone slabs or tree trunks, and for longer spans, single strands of bamboo or vine were stretched across the chasm and loops or baskets containing the traveler were pulled across the stretched rope. Gravity aided in the first half of the journey, but human strength was needed for the second half. From the single-strand crude suspension bridge, multiple-cable and floor systems evolved. Such crude bridges have been recorded in historical journals in Asia, South America, and Africa and such structures are still in use today in primitive areas.

Early cantilever bridges in which timber cantilever beams extended out from piers on both sides of the stream were built in China.

The earliest known arch has been found in excavations at Ur in the Middle East. These ruins have been dated at about 4000 B.C. Arch structures dating back to 3000 B.C. have been found in Egypt. These arches were all parts of buildings rather than bridges. The earliest surviving stone arch bridge is at Smyrna, Turkey. It dates from the ninth century B.C. Early Romans used the stone arch in bridges and buildings. The first known stone arch bridge in Rome is the Pons Solares built across the Teverone shortly after the seventh century B.C. The Romans built many timber bridges, but the remaining monuments to their bridge-building skill are the stone arch bridges scattered throughout those areas of Europe once ruled by Rome. Most famous of the remaining Roman bridges are the Pons Augustus at Rimini, Italy; the Puente Alcantara in Spain; and the Pont du Gard in southern France. They were built in the first or second century. The Pons Augustus has five semicircular arches, with spans varying from 23 to 28 ft. The Pont du Gard (Fig. 1.1) is part bridge, part aqueduct, with three tiers of arches rising to a height of 156 ft above the river. The largest arch has a clear span of approximately 80 ft. The topmost tier is 885 ft long and is the only tier in which mortar was used.

A most notable bridge was Trajan's bridge over the Danube. It was a colossal structure, with massive stone piers 150 ft high, 60 ft wide, and 50 ft thick. Timber arches spanned the distance between the piers. The total length was over 3000 ft. It took 17 centuries before the lower Danube was bridged again (at Turner Severin). Traces still remain of Trajan's bridge.

**Figure 1.1**  Pont du Gard Aqueduct, France.

No significant Roman bridges were built after the fourth century A.D. Creative engineering almost vanished from Europe for 400 years. Political units were very small and therefore could not support road and bridge building. Although bridge building in Europe hardly existed from the fourth to the eighth centuries, momumental structures were built in Persia, China, and India.

There is a written record of an iron-chain suspension bridge in India in the seventh century. Shortly after this time, iron-chain suspension bridges appeared in China. The writings of Marco Polo described many outstanding bridges in China and he was especially impressed by the quantity, as well as the quality, of Chinese bridges.

The first major bridge in Europe after the Dark Ages was the Pont d'Avignon. This bridge of legendary fame is attributed to Benoet (or Benezet), who was supposedly a member of the order of monks called Frères du Pont (Brothers of the Bridge). This brotherhood took on the responsibility of assuring the safety of travelers at bridges and ferries and later the actual construction of bridges. Only a portion of the Pont d'Avignon remains standing in the Rhone River at Avignon, France. The four beautiful stone arches still standing range in span from 101 to 110 ft, which is a remarkable length for 1187, the year of completion of the bridge. The total length was about 300 ft. Although the bridge saw many acts of war, it was the Rhone that was responsible for destroying the majority of the arches.

The most historic structure of the English speaking world was the Old London Bridge. This bridge, started in 1176 and completed 33 years later, stood for over 600 years. Peter of Colechurch was the builder of the bridge, but he died before its completion and was buried in the chapel on the bridge. The bridge, with its 19 heavy wide piers, spanned 936 ft. Almost half of this length was taken up by piers, resulting in a great increase in stream velocity between the piers. Subsequent scour and poor foundations resulted in constant repairs to the bridge. The children's nursery rhyme *London Bridge is Falling Down* has a basis of fact.

In addition to the chapel dedicated to Thomas Becket, each pier contained multistory buildings with shops and living quarters. The bridge was the center of London life and was the scene of many battles. The gate to the bridge, known as Traitor's Gate, was the favored spot for displaying the severed heads of those who lost the battles or were considered guilty of treason. The bridge was ugly and poorly constructed because very little overall planning went into its creation (actually it just grew). It had to be rebuilt constantly because of fires, floods, and wear. Nevertheless, there has never been a bridge that was the focus of life such as the

Old London Bridge was. The last century of its existence saw many alterations, but it was finally removed in 1831.

During the early life of the London Bridge, many other bridges were built in Europe that were to become historical edifices, including the Ponte Vecchio in Florence in 1345 (Fig. 1.2), the Ponte di Rialto in Venice in 1591, the Pont Neuf in Paris in 1607, and the Karlsbrücke bridge at Prague in 1503. These bridges are stone arches of moderate length except for the Karlsbrücke, which is nearly 2000 ft long. All four bridges are still standing today. The Vecchio and Rialto bridges were inspired by the Renaissance and are still admired by visitors to those cities. The Ponte Vecchio has been the center of Florentine life and the goldsmith shops on the bridge have been in existence through the centuries. Its segmental arches were very bold for the time in which it was built. It recently displayed its strength by withstanding the great flood of November 1966.

The Ponte de Rialto was built by Antonio da Ponte after winning the design competition. Its cost was $375,000, which was very expensive for a bridge with a span of only 88 ft. The high cost was due to the lavish furnishings and the difficult foundation construction. The foundation

**Figure 1.2**  Ponte Vecchio, Italy.

excavation had to be performed without danger to adjacent buildings. Da Ponte used 6000 wooden piles, each approximately 6 in. in diameter and 11 ft long in the base of each abutment. They were driven to refusal by a system that dropped a large stone weight on the pile. Da Ponte's beautiful work is the highlight of a trip down the Grand Canal in Venice.

As the building of new and monumental bridges was a mark of the Renaissance, other countries besides Italy were experiencing this creative activity. The Pont Neuf was the second stone arch bridge built over the Seine. This bridge of the early seventeenth century lasted for over 300 years. At the start of construction, there were numerous quarrels and delays over whether the bridge should have shops or not. Finally, only stalls were allowed. The bridge was plagued with foundation problems and the original builder, Du Cerceau, was removed from this position because of his religious affiliation. After completion, this bridge was a focal point of life in Paris and much of the history of this city for over 200 years transpired over or near the Pont Neuf. Napoleon Bonaparte was declared First Consul on this bridge. Robespierre took his last ride across this structure on his way to execution. As Paris grew and more bridges were built, the use of this structure as a passageway from the Right Bank to the Left Bank decreased.

Stone arch bridges have evolved from the heavy stone piers and semicircular arches of the Romans to the three-centered arch of Benezet (Pont d'Avignon), to the segmental arch of Gaddi (Ponte Vecchio), to the elliptical arch of Ammanati (Ponte Santa Trinita), to the thin arch and narrow piers of Perronet. Jean Perronet, the director of the first college of engineering, Ecole des Ponts et Chaussees, was probably the first man to understand the true significance of the arch. Thus, in the eighteenth century, the age of reason in structural engineering began. Perronet was responsible for the successful construction of many beautiful stone arch bridges, and his Pont de Neuilly has been called "the most graceful stone bridge ever built." This bridge had five boldly flattened arches, each with a span of 120 ft and piers only 13 ft thick. Orthodox engineers of the day claimed that the bridge would never stand; however it remained until 1956 when it was removed to make way for a modern highway bridge. However, Perronet's last bridge, the Pont de la Concorde, built 1791, remains a monument to arch bridges.

In the political and industrial revolutions of the eighteenth century. Great Britain was the home of three men who became very famous as bridge builders, John Rennie, Thomas Telford, and George Stephenson. At an early age, they all learned to work with their hands and became skilled mechanics, Rennie as a mill-wright, Telford a stone mason, and Stephenson an engine mechanic. Since their rise to fame from very humble beginnings is a stirring story beyond the scope of this book, only a brief account is given here.

Rennie received a college education at the University of Edinburgh, majoring in mathematics. His vacations were spent repairing mill machinery for tuition money. He was instrumental in helping Watt develop the steam engine and went on to the building of many bridges in Great Britain. His two most famous bridges were the stone arch structures, the Waterloo Bridge and the New London Bridge. The latter was built by Rennie's son after his death, but according to the father's design. The elder John Rennie refused knighthood, preferring not to have a title. His son John Rennie, Jr. became a well-known builder and accepted knighthood. Coming from a less humble beginning the son was more comfortable with a title.

The first iron bridge was erected at Coalbrookdale, England, near the site of the first iron smeltering. This bridge of a single span of 140 ft consisted of semicircular cast iron ribs supporting the roadway above the ribs. The bridge was designed by Thomas Pritchard and built by John Wilkinson and Abraham Darby. It is still standing today.

Thomas Telford was self-educated and worked on several stone bridges as a stone mason. He later became interested in iron bridges through acquaintance with Wilkinson and Darby. He built a beautiful steel trussed arch with a span of 150 ft and a 20-ft rise over the Spey in Scotland. This structure is considered the first modern iron arch bridge. Telford's major achievement was the suspension bridge over the Menai Straits in Wales. This was a wrought iron chain suspension bridge with a span of nearly 600 ft—an unheard-of length at that time. The bridge, consisting of 16 suspension chains, was under construction from 1818 to 1826. It, like some other suspension bridges in later days, had trouble with the wind and had to be repaired several times. In 1939 the wrought iron chains were replaced by steel cables so the bridge could support heavy truck traffic.

George Stephenson was the inventor of the steam locomotive. Stephenson, like Telford, was self-taught, mainly by studying his son Robert's lessons at night. Since bridges were needed for his railways, George began building them with his son Robert. Robert became more of a renowned bridge builder than his father. They both used the new materials, cast and wrought iron, in their bridges. Robert's most famous bridge was the Britannia Bridge at the Menai Straits in Wales. This bridge was composed of two continuous hollow tubes of wrought iron. The tubes, acting as beams through which the trains passed, had two spans of 230 ft and two of 460 ft. The bridge was finished in 1850, 4 years after the first foundation stone was laid. George Stephenson died in 1848 during the construction of the bridge. Robert lived only 11 more years. Robert was buried in Westminister Abbey alongside Thomas Telford. England gave fitting recognition to two men who were so important in the development of the country.

As the stone arch bridge was mothered by Europe, the truss, and in particular, the timber truss, was born in America. Palladio, an Italian architect, described four truss types in his *Treatise on Architecture*, in 1570. The timber truss was used occasionally in European buildings, but because of the lack of heavy timber in western Europe, the timber truss was rarely used for bridges. The covered timber bridge became very popular in early America for the carriage and wagon roads. The roof protected the load-carrying members from the elements and if the bridge was not subjected to the hazard of fire, it lasted for many decades. Early timber bridges were usually simple truss types, but later more sophisticated types were developed. One of the most spectacular timber bridges ever built was the Schuylkill River Bridge at Fairmount, Pennsylvania. This structure was built by a German immigrant, Lewis Wernwag, and had a trussed arch span of 340 ft. Iron rods were used for the diagonals of the truss. A counter system for the diagonals was used since the rods could not resist compression. This bridge, built in the early nineteenth century, was destroyed by fire in 1838. Wernwag and several other bridge builders constructed timber structures of all types in the eastern United States. Many new truss arrangements were developed, such as Theodore Burr's arch-truss, Ithiel Town's lattice truss, and Colonel Stephen Long's multiple king post. These designs were not developed from any theory of structural mechanics, but were the results of experience only, and such structures were built by "rule of thumb" principles. Shortly after the introduction of the Long multiple king post, William Howe modified it by using wrought iron rods for the verticals in the Long truss. Later the tension diagonal was omitted and the remaining diagonal resisted the shear by compression. Satisfactory connections at the ends of timber compression members were easier and more secure than those of the timber tension members.

A publication appeared in 1847 that advanced bridge design from the age of rule of thumb to the age of science. This publication was Squire Whipple's *A Work on Bridge Building*. In this writing, the author analyzed the stresses in a truss. He recognized that some members were in tension and some in compression. He suggested cheaper cast iron for the compression members and wrought iron for the tension members. Steel was known and recognized as having great tensile strength, but was considered too expensive. Shortly after Whipple's publication, treatises on structural mechanics appeared in England giving mathematical formulae for the determination of stresses in various types of trusses. In

1858 W. J. M. Rankine published his book entitled *Applied Mechanics.* This served as a bible to the bridge engineer for over 50 years.

The expansion of the railroads in America coincided with the introduction of the mathematical theory of truss analysis. The result was a great upsurge in the construction of truss railroad bridges, the majority of which were wrought iron. By experiments conducted about the middle of the nineteenth century, cast iron was found to have inadequate tensile strength. This great upsurge in bridge building brought forth many new types of trusses. Whipple introduced the truss form that carried his name, as did Wendell Bollman and Albert Fink. Later the Pratt and Warren trusses were developed.

As a result of the expansion of the railroads in the United States during the 1800s, large quantities of railroad bridges were built. Because many were of questionable design and materials, there were numerous failures. About 40 bridges a year failed in the 1870s and about half of these were timber spans. Many of the failures resulted in high loss of life. In 1887 a railroad bridge near Chatsworth, Illinois, collapsed while a double locomotive with 15 passenger cars was on the bridge. The lead locomotive made it to the other side, but the remaining portion of the train fell and burned. Eighty four people were killed.

The United States was not the only country plagued by bridge failures. On December 28, 1879, the famous Firth of Tay railroad bridge collapsed during a high wind while a train was on the bridge. All 75 people on the train were killed and many of the bodies were never recovered. Thirteen through truss spans of over 200 ft each fell into the Firth of Tay with the train. Sir Thomas Bouch, the designer of the bridge and an official of the railroad, was severely criticized for the failure. He died shortly after the disaster.

These railroad bridge failures brought about many changes in bridge design and construction. One result was the abandonment of wrought iron in favor of the new stronger metal—steel. A second result was the realization by the public and engineering profession that bridge design and construction required men of considerable training and dedication. Rules of thumb were no longer satisfactory and research in the mechanics of materials was needed. The science and art of structural engineering became important.

Through the years the suspension bridge has been used to span wide rivers or gorges. Its construction process has evolved from the use of the vine to the special cable process of today. Rope was used before the cable process was developed.

During the period 1839 to 1845, the Kettenbrueke was constructed in Budapest with a central span of 663 ft. This bridge was the work of an English engineer, W. Tierney Clark. A French engineer M. Vicat was the first to develop the method of constructing a wire cable in place instead of building it first on land. He used this method on a bridge across the Rhone in 1829. Now the process of evolution had gone from spinning the cable in place back to building the cable in the shop and erecting it in place already assembled. The Bethlehem Steel Company used this method for the first time on a major suspension bridge over the Narragansett Bay near Newport, Rhode Island. This bridge was completed in 1968.

Four memorable bridges were built in the latter half of the nineteenth century, three of them in the United States and the other in England. The Niagara Falls suspension bridge was designed and built by John Roebling from 1851 to 1855. It was the first successful railroad suspension bridge and had a span of 821 ft. The cables were composed of wrought iron ungalvanized wires. Many well-known engineers claimed the bridge was impossible when Roebling made his proposal. The bridge was replaced in 1896 because of increased railroad loadings.

The nineteenth century produced James B. Eads, a brilliant American engineer. He was born in 1820 at Lawrenceburg, Indiana. At the age of 13, he arrived in St. Louis, where he later gained fame for a bridge that had many "firsts." Eads' life was one of determination and courage. He coupled this with a desire to learn all he could about any subject in which he was involved. He had no formal schooling as an engineer, but went to any end to increase his knowledge. He first gained employment on a Mississippi River steamboat and later developed a ship for salvaging wrecks in the river. He became very successful and well-known in this work. He was very ingenious in developing mechanical equipment vital to his salvage operations.

Eads saw the coming of the Civil War and designed iron-clad boats for such an event. This was 2 years before the *Monitor* and the *Merrimac*. Although Eads' original ship designs had received little interest from the Navy, after the start of the Civil War, he entered a bid with the promise of a delivery in 65 days. He received the contract from Secretary Stanton of the War Department on August 7, 1861. In 2 weeks he had 4000 men working on the ships. He did not meet the 65-day deadline exactly, partly because of the lack of payments by the War Department; however, by December the ships were ready for action. Four ships were assigned to Grant's command to assist in the battle against Fort Henry. The shelling was so intense from Eads' gunboats (they were actually still his since payment had been delayed) that the Confederate forces abandoned the fight before Grant's army arrived. The first big Northern victory of the war was the capture of Fort Donelson by Grant's army with the able assistance of the gunboats. Eads received payment for his ironclads shortly thereafter.

By 1866 it became necessary to construct a bridge across the Mississippi at St. Louis. Experienced bridge engineers all hesitated, but James Eads, who had never built a bridge, stepped forward. He knew how to organize and handle men and how to mold and fabricate steel, and he also knew the river better than anyone else. The history of the construction of the St. Louis bridge is fascinating; however, the details are beyond the scope of this book. In summary, Eads selected the untried metal, steel, for his superstructure, which consisted of tubular steel arch ribs spanning the river in three large spans of 497, 515, and 497 ft. The two river piers were sunk by pneumatic caissons. This method had been used before in Europe, but not in America. Eads traveled to Europe to study the workings of a pneumatic caisson. He improved upon the European practice by inventing the sand pump and many other facilities. Despite some legal actions by those who did not want the bridge to be built or did not believe the bridge could be safely built, Eads moved ahead with his project. After resisting floods, ice, and tornadoes for 5 years, the bridge was opened to train traffic in July 1874 and is still in service today. The floor system was renovated in 1947 at which time lightweight concrete was used to decrease the dead load. In 1920 James B. Eads was elected to the American Hall of Fame—the first engineer thus honored.

What Eads Bridge was to the West, the Brooklyn Bridge was to become to the eastern United States, a symbol of man's achievement in the field of civil engineering. John Roebling foresaw the demand of wire cables for bridge building, so he established a wire mill in Trenton, New Jersey, in 1849. After building the suspension bridge at Niagara Falls, he was the logical person to design and construct the much-needed bridge across the East River in New York. Roebling's plan to erect a 1600-ft span was met with criticism since this was 50% greater than the longest span yet built—Roebling's suspension bridge at Cincinnati. Despite the prophets of doom, support for the structure was obtained.

During preliminary work, Roebling's foot was crushed and 3 weeks later he died from the injury. The work was turned over to his son, Washington A. Roebling, who had been trained by his father. He was a graduate civil engineer of Rensselaer Polytechnic Institute. Washington Roebling was a victim of the "bends," caused by working in pneumatic caissons used for sinking towers; he became partially paralyzed and had to direct the remainder of the construction from his bed using his wife as messenger. The Brooklyn Bridge was completed in the spring of 1883 at a total cost of $9 million. This structure remained the record span for several years. A postcard of the Brooklyn Bridge received by one of the present writers as a small boy first generated his interest in bridges.

The fourth major bridge of this period was a railroad bridge built in England. As has been mentioned, the Firth of Tay was an obstacle to travel from south to north along the east coast of Scotland. The Firth of Forth was an even greater obstacle than the Firth of Tay, since it was wider. The well-known engineers Benjamin Baker and John Fowler, both members of the latter's London engineering firm, developed the concept of a colossal structure across the Firth of Forth. Their design was to be a cantilever truss type of structure with two main spans of 1700 ft each (Fig. 1.3). The span was not the only item that was huge in this structure. The depth of the truss over the piers was 350 ft. The members of the truss were tubular sections of maximum diameter of 12 ft. The cost of the bridge was a tremendous $16 million and it had a total steel tonnage of 58,000. This is compared to Eads' St. Louis bridge of 11,000 tons. The Forth Bridge is truly a mammoth bridge, even by present-day standards. During the construction 47 men of a total labor force of 4500 lost their lives in accidents. Without the safety standards of today, this was not considered abnormal. Pneumatic caissons were used for sinking the piers and the newly invented electric arc lamp was of great aid in the work in the caissons. Construction was started in 1883 and the bridge was opened to traffic in 1889.

This impressive structure is still carrying trains today. Both Baker and Fowler were knighted for their engineering accomplishments. This bridge held the record for the longest cantilever span until it was exceeded by the Quebec Bridge in 1917. It was an objective for German bombers in World War II, but suffered only minor damage. Despite its size, it was only a small target from several thousand feet in the air.

The major bridge construction in the early twentieth century that attracted world-wide attention was that of the railroad bridge over the St.

**Figure 1.3**   Firth of Forth Bridge, Scotland.

Lawrence River near Quebec, Canada. The construction of this bridge resulted in tragedy. The design was a massive cantilever with a central span of 1800 ft. The consulting engineer was Theodore Cooper, who got his engineering start as James Eads' superintendent on the St. Louis bridge. Cooper then went on to become America's foremost bridge engineer and was the originator of the Cooper Railroad Bridge Loadings. The bids on Cooper's original design were more than the railroad company was willing to pay, so under great pressure, Cooper reduced the amount of steel in his bridge.

On the morning of August 29, 1907, the worst bridge construction disaster in history occurred. The nearly completed south cantilever arm suddenly collapsed and crashed into the river, carrying 86 workmen, only 11 of whom survived. In addition to the lives lost, 10,000 tons of fabricated steel and many thousands of man hours of labor were suddenly obliterated.

Theodore Cooper had received word in his New York office that the bridge had an unusual sag. He immediately telegraphed instructions to remove all workmen from the bridge, but his order was received too late—the structure had already fallen. The lengthy and detailed investigations that followed placed the cause of the failure on the buckling of a compression member caused by inadequate lacing bars. Like the Tay bridge failure, this tragedy also brought about better engineering by extended research into the strength of large compression members. Cooper, like others, was a victim of lack of sufficient knowledge by the engineering profession at that time. He retired to seclusion and died a few years later.

The story of the Quebec Bridge was not just one of tragedy, but also one of courage and perseverance. Several years after the failure, plans were prepared for a new bridge, requiring 2.5 times as much steel as the old bridge. The suspended span was to be raised from barges in the river to its final position at a height of 150 ft above the water. In 1916 the cantilever arms were successfully constructed and after the suspended span had been raised about 12 ft, it slipped out of its lifting device and fell into the river. Eleven men were killed in this second disaster. A year later, a new suspended span was successfully lifted into place. This bridge is still the longest-span cantilever-type bridge.

While the railroad required great activity in bridge building in the nineteenth century, the automobile was the reason for the construction of multitudes of bridges in the twentieth century. This period saw the construction of long-span bridges of every type. This was possible not only through more sophisticated methods of design, but better and stronger materials and more exacting procedures in fabrication and erection. Higher-strength steels were produced and quality control in steel manufacture and concrete mixing allowed for higher allowable design stresses. Engineering became more scientific. Bridge designers were given better and more education, and specifications and standards were adopted after careful consideration by qualified groups of engineers. The engineering societies became influential in promoting better engineering and in disseminating knowledge and experience.

It is impossible in the space available to mention all, or even a majority, of the outstanding bridges built in the twentieth century. New record span lengths were achieved in every category of bridge type. The record spans were naturally all highway-type bridges. In addition to the record span lengths, other developments, even more significant, took place. One such development was that of reinforced concrete. Some use of concrete had been made earlier, but not until the great surge in highway construction did concrete become very important in bridge construction.

Robert Maillart was responsible for several beautiful concrete arch bridges in Switzerland in the early 1900s. During this period, engineers and the public became aware that a bridge could not only be an object of utility but also a work of art. Many of the earlier bridges had certainly been beautiful, but in the nineteenth century, the pleasing proportions of bridges were more the result of happenstance than of an effort to achieve such an effect. In 1928 the American Institute of Steel Construction started making awards for the most beautiful steel bridges constructed each year in the United States. These awards are sought after by bridge engineers and have made the engineering profession aware that bridges can be beautiful. As the late D. B. Steinman, the well-known suspension-bridge designer, so aptly stated, "No one can remain unmoved at the sight of a beautiful bridge. A bridge is not only a stepping stone of civilization—it is also an expression of the aspirations of humanity."

Three bridges of significance completed in 1917, the same year as the Quebec Bridge, were the Sciotoville, Ohio Bridge, the Metropolis Bridge across the Ohio River, and the Hell Gate Arch across the East River, New York. The Sciotoville, Ohio Bridge was the first major continuous truss bridge built in the United States. This structure had a maximum span of 775 ft. It was not until about 40 years later that this span length was exceeded for a continuous truss. Because of the uncertainty of foundation analysis, the statically indeterminate continuous truss was avoided by bridge engineers. The statically determinate cantilever type was usually preferred until the development of the science of soil mechanics.

The Metropolis bridge was, until 1974, the record simple truss-span bridge with a single span length of 720 ft. In 1974 a suspended span of 822 ft of a cantilever truss was built in Chester, Pennsylvania. The single. simple span truss of this span length would not likely be considered an economical bridge under present-day costs of material and labor. With the current knowledge of soil mechanics and techniques of foundation construction, shorter continuous spans most often prove more economical.

The Hell Gate Arch is a railroad bridge carrying four tracks. The span of 977 ft was a record for arch bridges. The structure, designed under the direction of Gustav Lindenthal, was also the heaviest structure ever built up to that date. The dead load of the bridge is 52,000 lb/ft. This bridge is still an imposing sight as it sits astride the East River in New York City.

There were a number of fine bridges built after World War I when highway construction was greatly accelerated. However, it was not until 1929 that the record span of the Quebec Bridge was exceeded. The new record holder was the Ambassador suspension bridge in Detroit, Michigan, with a central span of 1850 ft. This record was not held for very long, as in 1931, the famous George Washington Bridge across the Hudson River at New York City was completed with a span of 3500 ft, nearly twice the span of the Ambassador Bridge. The chief design engineer on this bridge was O. H. Ammann. Ammann was born in Switzerland, but received his early experience in the United States working under Gustav Lindenthal.

The George Washington Bridge was originally designed to have two decks, automobiles on the eight-lane top deck and rapid transit on the lower deck. However, the lower deck was not built in the original construction. As a consequence, the usual stiffening truss was not added to the structure. This resulted in an unstiffened suspension bridge—the only major unstiffened suspension bridge in the world. The lower deck was added at a later date.

The world's two greatest steel arch bridges were completed in 1931 and 1932. The Kill-van-Kull Bridge at Bayonne, New Jersey, with a record span of 1652 ft was completed in 1931 and the Sydney Harbor Bridge was completed a few months later in 1932 with a span of 1650 ft. The Sydney Harbor Bridge is heavier than the Bayonne Bridge, since it carries a rail transit system, as well as automobile traffic. Both bridges are very imposing structures.

In these same years, other great bridges were underway in California, including the San Francisco–Oakland Bay Bridge, with its 6 miles of structure and the double suspension bridge, the Golden Gate Bridge, with its record clear span of 4200 ft. The former was completed in 1935 and the latter in 1937. The great center anchorage pier of the Bay Bridge is a remarkable engineering construction achievement. The pier was carried over 240 ft below the water surface. A caisson with dredging wells was used for this pier. The old pneumatic caisson had been ruled obsolete several years earlier.

Everything about the Golden Gate Bridge was a record at the time it was built. The bridge towers are 746 ft high and each main suspension cable is 3 ft-0.375 in. in diameter and is composed of 27,572 wires.

In the twentieth century, the suspension bridge became the single answer to problems requiring long clear spans. In 1940, however, the suspension bridge designers, as well as the entire world, received news of a most unusual happening. A suspension bridge of 2800 ft span near Tacoma, Washington, had vibrated in the wind with such large amplitudes that the suspender cables had failed and most of the roadway had fallen into the waters below. Smaller vibrations had been noted on this structure earlier, as well as on several suspension bridges that had large span-to-weight or span-to-width ratios. This bridge, dubbed "Galloping Gertie," had a length-to-width ratio of 72, over twice that of the next bridge. The bridge designer, Leon S. Moisseiff, had never imagined a structure this large vibrating with amplitudes of nearly 15 ft. Moisseiff had been the designer of many notable bridges and had first applied the deflection theory of suspension bridges in the design of the Manhattan Bridge in 1909. The combination of solid girder for the stiffening truss and solid roadway produced a profile to the wind that caused uplift forces great enough to induce the vertical and torsional vibrations. This happened at a wind velocity estimated at 40 mph.

As a result of this bridge failure, the aerodynamic stability of suspension bridges became the subject of much research. The wind tunnel became the research tool of the bridge engineer. The results of this research led to changes in suspension bridge design that were incorporated into bridges built after World War II.

Bridges have always been important to armies in their campaigns. In early history the armies usually constructed their military bridges out of timber or stone found in the vicinity of the river crossing. In the U.S. Civil War, the Union Army used prefabricated boats and timbers for some of their floating bridges. Speed in erection became very important. World War I saw little new development in military bridging since it was not a war of movement. However, World War II did lead to some new ideas in bridge construction. Hitler's generals knew that bridges that could be erected in a matter of minutes, or few hours at most, were very necessary for the type of blitzkrieg war they planned. The German Army introduced rubber inflatable boats for their floating bridge construction in the early stages of World War II. However, the two most remarkable bridges used in this war were the Bailey Bridge, developed by the British engineers, and the floating treadway bridge, developed by the U.S. army engineers. Both of these bridges were very instrumental in the successful campaigns of the Allied forces. The Bailey Bridge was an erector-set type of structure in which 10-ft long panels could be pinned together to form trusses of any length and different combinations of trusses could be used in a bridge depending on the span length and weight of the vehicle to be supported. The versatility of the bridge was one of its strong points, as was its speed of erection. A 60-ft bridge, capable of carrying a 40-ton tank could be erected on a prepared site within an hour. The Bailey Bridge was also used with special boats as a floating bridge for wide river crossings. However, its great worth was as a fixed bridge. This bridge has been extensively used since World War II for emergency bridging when permanent bridges have been washed away by floods. The writer has seen many Bailey bridges used in the less-developed countries of the world as more or less permanent bridges.

The most successful floating military bridge was the steel treadway bridge, which consisted of two main elements: the inflatable rubber boat with plywood and steel saddle and the two steel treadways that sat on the boats (pontons). The 15-ft-long treadway sections were carried in a special truck that had a rotating A-frame on the end that lifted the treadways onto the pontons. Each truck carried a specific number of pontons and treadways. The treadways could be rapidly pinned together by a double-pin arrangement to give a moment-resistant deck. Several hundred feet of bridge could be constructed in a few hours by a well-trained company of engineering troops.

During World War II a very remarkable bridge was constructed in a European country that was not involved in the war. This bridge, the largest concrete arch bridge ever built to that date, is the Sando Bridge in Sweden, with a 866-ft span. This bridge not only had a record span, but its clean lines and delicate arch made it an object of beauty. Its span was exceeded in 1963 by the Gladesville concrete arch bridge in Sydney Harbor, which has a span of 1000 ft.

With the end of World War II, the bridge engineers of the world were faced with a tremendous volume of bridge building. Central Europe had most of its bridges destroyed, the United States needed many new highways, and the undeveloped countries were looking for improved roads. Necessity again proved the mother of invention.

The decade from 1948 saw many new developments in bridge design and construction. One of the most significant of these developments was prestressed concrete. Although the concept of prestressed concrete was over 50 years old, it was not until Freyssinet of France and Magnel of

Belgium started using high-strength steel wires for prestressing tendons that the method become practical for highway bridges. A number of such bridges were built in Europe in the late 1940s, and the first such bridge in the United States was the Walnut Lane Bridge constructed in 1949 in Philadelphia. The European practice was generally to use the post-stressed method, while in the United States the pretension method generally proved more economical because of mass production in central casting yards. One of the most notable examples of the mass production technique was the construction of the Lake Ponchartrain Bridge near New Orleans completed in 1957. This bridge is a causeway-type structure, 24 miles long. The prestress bridge sections were produced in a central yard and floated by barge to the construction site. Each 56-ft single span of 33-ft width was cast and placed as a unit. The bridge was built in a total of 2 years. During the 1950s and the 1960s prestressed concrete bridges of many types were constructed all over the world. The Interstate Highway System in the United States required thousands of relatively short-span structures for which the prestressed concrete beam was utilized. In Europe longer spans of prestressed concrete were constructed.

In the late 1960s and in the 1970s a method of cantilever prestressed concrete construction was developed in which traveling forms were used. Remarkably long spans were achieved. Several countries built such record spans of around 800 ft. These bridges were all continuous girder bridges. Such continuous prestressing was also used for concrete cable-stayed bridges.

It appears that what can be achieved with prestressed concrete is limited only by the bridge engineers' ingenuity.

A second post-war development in Europe was the orthotropic steel deck plate bridge. In this type of construction a steel plate with a thin wearing surface of asphaltic material is the deck. The steel plate is stiffened longitudinally with ribs and transversely by closely spaced floor beams. The deck plate not only serves as the deck, but also as the top flange of the main longitudinal beams. The result has been a saving in steel, as well as a shallower longitudinal beam.

The first major bridge of this type was the Kurpfalz Bridge at Mannheim, Germany built in 1950. Other notable structures of this type are the Save River Bridge, Belgrade, with spans of 245, 856, and 246 ft and the Poplar Street Bridge across the Mississippi River in St. Louis. The San Mateo-Hayward Bridge over the south end of San Francisco Bay is the longest in the United States, with a central span of 750 ft. The longest span to date is the President de Silva Bridge in Rio de Janeiro, with central span of 984 ft.

All long-span steel bridges in western Europe built recently have the steel orthotropic deck. As fabrication and construction techniques improve, there will undoubtedly be more bridges of this type built. There is a considerable saving in steel, and cantilever construction is possible.

The bridge engineer has looked very fondly at aluminum. In the heat-treated alloys it has high strength, light weight, and corrosion resistance. In 1949 an all-aluminum highway bridge with a 290-ft arch span was built over the Sagenay River between Arvida and Shipshaw, Canada. However, because of the higher cost of aluminum, this metal has rarely been used for highway bridges.

The widespread use of high-strength steels in bridge construction is a recent development. The yield points of steels used in bridge construction vary from 36 (A36) to 100 ksi (A514). Many structures now contain several types of steels. The increased corrosion resistance of the higher-strength steels make them attractive to the bridge designer.

Rivets were the standard fastener for steel bridges for over 100 years. The development of the high-tensile-strength structural bolt and advanced techniques in welding in the 1950s began to change the method of fastening steel elements together. By 1965 most bridges were connected by either high-strength bolts or welding.

Two very interesting bridges are the Lake Maracaibo Bridge in Venezuela, completed in 1962, and the Fehmarnsund Bridge in West Germany, opened to traffic in 1963. The former is a 5.5 mile combination prestressed concrete and cable-stay structure. The latter is called a basket-handle arch because the pair of arches tilt inward and almost touch at the crown. The roadway is supported by an interesting arrangement of crossed cable hangers. The Fehmarnsund Bridge has a clear span of 815 ft.

The Chesapeake Bay Bridge and tunnel project completed in 1964 was one of the largest engineering projects of the world. This project, which involved several different types of bridges, is 17.5 miles long and was built at a cost of $140 million.

The suspension bridge reached several crowning achievements in the 1960s. The record for length of a single span held by the Golden Gate Bridge for nearly 30 years was exceeded by the Verrazano-Narrows Bridge across New York Harbor. This bridge, with a span of 4260 ft, was opened to traffic in 1965. Total cost of the structure was $325 million. The bridge was designed by O. H. Ammann's firm.

In 1957 the Mackinac Straits in Michigan was bridged by a large suspension bridge with a central span of 3800 ft. However, the total length was 8400 ft—a record total length. The side spans were also longer than usual.

Two very interesting suspension bridges were completed in Europe in 1966. The Severn Bridge in Great Britain has two unique features of design. The usual stiffening truss was abandoned in favor of a continuous-welded steel box only 10 ft deep and with a width equal to the four-lane roadway. This box provides sufficient torsional rigidity. Wind-tunnel tests of a model proved its aerodynamic stability when it was supported by diagonal suspenders instead of the usual vertical suspenders. The design proved economical for a bridge with a central span of 3240 ft and a total span of 5240 ft. The deck used was orthotropic plate with an asphalt wearing surface.

The record European span is a suspension bridge in Istanbul, Turkey, completed in 1973, with a central span of 3524 ft.

A suspension bridge with a world-record span is under construction near Hull, England. This bridge across the Humber River will have a central span of 4626 ft and a box girder of shallow depth.

The Taugus River Bridge in Lisbon, Portugal, has a main span of 3323 ft. One of the towers was founded on bed rock 260 ft below water level. This depth was a record for a bridge foundation. The total length of the stiffening truss is 7472 ft and the required expansion devices in the roadway at each end are capable of providing for a movement of $4\frac{1}{2}$ ft.

A dramatic major development in bridge design has taken place since the mid-1950s. The use of multiple straight cables from a tower to the deck grider, instead of the usual draped cable with vertical suspender

**Figure 1.4**  Theodor Heuss Bridge, W. Germany.

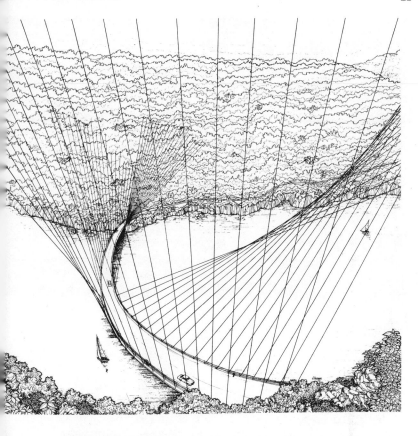

**Figure 1.5** Proposed Ruck-a-Chucky Bridge, California.

bles, was first used on a long-span bridge in Dusseldorf, West Germany. This bridge was completed in 1958 and had a central span of 853 ft ig. 1.4). The girder on this bridge was a steel box. Girders of both steel nd prestressed concrete have been used. Central spans of over 1000 ft ve been used for this type of bridge. More information on cable-stayed ridges is given in Chapter 12.

A bridge of most unusual characteristics is presently (1977) under esign. This structure is the Ruck-a-Chucky Bridge to be built over the iddle fork of the American River near Auburn, California. Because of e difficult site conditions this bridge will be curved at a 1500-ft radius nd will have a clear span of 1300 ft. Holding it in place will be many raight cables anchored to the rock of the steeply sloping mountains hich rise several hundred feet above the bridge. It will be the first bridge its kind (Fig. 1.5).

**Figure 1.6**   White Bird Canyon Bridge, Idaho.

Undoubtedly the public and even some engineers of every era belie
that the ultimate in bridge design and construction has been reached, th
we have arrived at the end of the road of new developments. For t
present, this, of course, remains to be seen; history would indica
otherwise.

Today the structural engineer has at his disposal the most powerf
analytical tool ever imagined—the digital computer. This instrument c
perform in a matter of minutes a volume of calculations that would ha
previously taken years. With this time-saving tool, many types of bridg
can be investigated and the optimum design ascertained. Also, mo
exacting analytical computations can be performed. Approxima
methods are superceded by more exact procedures.

New materials undoubtedly will be developed with higher strengt
thus making possible lighter, more beautiful bridges (Fig. 1.6).

The engineering work on the large bridge projects of today is
complex that many engineers are involved in the design and constructio
The credit cannot go primarily to one person as it did in the past. Lar
firms of specialists in many areas are involved. Nevertheless, one man h
to make the final decisions. These final decisions, of which there may

many, call for a wealth of technical knowledge and sound judgment based on many years of experience.

All challenging sites have not yet been bridged. Many more highways and railways are needed all over the world. The English Channel is under study. Roads through the Himalayas and the Andes will be required. The challenge is there, and only the qualified can succeed.

# CHAPTER 2

# Materials of Bridge Construction

## 2.1 INTRODUCTION

A modern highway bridge is usually constructed of many different materials. The selection of the best material for each item in the bridge is a major responsibility of the bridge engineer. In some cases the most suitable material is obvious, yet in other situations it may be very difficult to choose between two or more materials. The engineer must make the selection based on all the available information.

Steel and concrete are the two major materials used in bridge construction. Most all other materials employed in major bridge construction are used in specialties, such as railings, bearings, and expansion devices.

Wood has been used in a few major structures, but more in small mountain or farm bridges. The initial cost of timber bridges is relatively low; if properly designed and maintained, they can last a long time.

When concrete is used as a stress-carrying member it must be reinforced with steel rods. For bridge decks, concrete is predominant. However, for long-span bridges, there can be a saving in using the steel orthotropic deck with an asphalt wearing surface. Almost all bridges built recently have an asphalt wearing surface over the concrete deck. This surface acts as an added protection against deicing salt solutions penetrating to the steel reinforcing bars.

The bridge engineer must decide whether steel or concrete is most suitable for the main beams or girders of a bridge. For some bridges it may be quite obvious that one or the other material is preferable. The designers of the Golden Gate Bridge had little trouble deciding that the superstructure of this bridge would be steel. However, the preliminary designs of the 1300-ft Ruck-a-Chucky Bridge indicate that the cost of a steel box girder is very comparable to the concrete girder. In such cases it may be wise to design both types of bridges and let the bid quotations tell which is most economical. Economic conditions at the time of bidding may be the deciding factor in determining the relative costs between a steel and concrete bridge.

## 2.2  CONCRETE

Concrete is also the predominant material for curbs, sidewalks, parapets, and substructures, as well as for bridge decks. For many years concrete was used exclusively for bridge railings, but most present-day bridges use metal railings or a concrete parapet with an aluminum or steel railing above.

Concrete can be readily formed to almost any shape and can be made in a variety of compressive strengths by varying the amount of cement and water. The proper proportions in a concrete mix depend on the requirements of strength, durability, shrinkage, and creep. It would not be economical to produce a concrete far above the demands of the environment it will be subjected to in its lifetime.

A 28-day compressive strength of 3000 to 4000 psi is usually satisfactory for most requirements such as deck slabs, curbs, sidewalks, and substructure. Slightly higher strengths may be preferred if greater durability is desired. For the girders of medium-span cast-in-place or prestressed concrete bridges, strengths up to 6000 psi are common.

A recent consternation to highway officials in states in colder climates is the spalling of concrete decks. All concrete decks have minute cracks near the surface. Deicing salts in solution penetrate through these cracks to the steel reinforcing bars. This moisture forms iron oxide, which expands upon formation and the resulting expansive force causes pieces of concrete to fracture from the deck. Repair of such spalled surfaces can be very costly, as well as disruptive to traffic flow. A multitude of schemes have been proposed and tried for eliminating this problem. These schemes vary from increasing the cover of concrete over the reinforcing bars to using bars with zinc coating, to the use of a thin waterproof membrane plus asphalt surfacing. At the present time the last scheme appears to be the most effective and practical.

Concrete is an ideal material for substructures. Piers and abutments are subjected to moisture from flowing water or groundwater. Durable concrete will resist the action of moisture for many years if it is properly made. Concrete can be formed to the optimum shape. In some piers and abutments the quantity of concrete needed is based on stability more than stress. Portland cement concrete hardens under water, which is an advantage in some special cases of subsurface construction.

If concrete is not cured properly it will result in an inferior product. Bridge engineers should require that all concrete be properly proportioned, mixed, and cured. Great precision in developing and solving a mathematical model is of little value if the construction material does not match the design requirements.

Research is always underway to develop new and better types of concrete.

## 2.3    STEEL

### 2.3.1    History

The first use of iron in bridge building on a large scale came with the use of cast iron. Because of its low tensile strength and ductility it soon proved inadequate. Wrought iron replaced cast iron in bridges after about 1850. With the failure of many railroad bridges using cast and wrought iron, the need for a better bridge-construction material was recognized.

With the development of Bessemer steel in the first half of the nineteenth century, and the production of open-hearth steel at Birmingham, England in 1867, steel became available at reasonable cost and in sufficient quantity for bridge building.

Although steel was first used for the eye-bars on a suspension bridge over the Danube Canal at Vienna in 1828, the first all-steel bridge was a railroad bridge over the Missouri River at Glascow, South Dakota, in 1878. The use of steel for the Glascow Bridge is not considered a real first in bridge construction, since the Eads Bridge across the Mississippi, built in 1868 to 1974, made extensive use of steel. The main tubular arches were of steel furnished by Andrew Carnegie.

The Brooklyn Bridge, completed in 1883, was built with steel wires in the cables and in 1890 the monumental Firth of Forth railroad bridge in

Scotland was completed. This all-steel superstructure had a record span of 1700 ft. Tests were conducted on the steel in this bridge and the specifications stated that "the ultimate tensile strength was to be not less than 33 tons* per square inch."

Since the time of the Firth of Forth Bridge steel making and fabricating have undergone many changes. Today there is a wide variety of steels available for bridge fabricating. The bridge engineer needs to have knowledge of the physical properties of the various types of steel so that he can make the proper selection.

### 2.3.2 Grades of Steel

Steel is essentially composed of iron and a small percentage of carbon. In addition, there are a number of elements added in small quantities to produce the proper alloy. Other chemicals are present from the original ore, principally sulfur and phosphorus. These last two elements are generally deleterious to the quality of the steel, and specifications on structural steel limit the maximum amount of these elements.

There is not adequate space in this book to present the metallurgy of steel; however, it is readily available elsewhere (1).

The physical properties of steel are dependent on the kind and quantity of alloying elements, the amount of carbon, the cooling rate of the steel, and the mechanical working of the steel, such as rolling and stressing.

At the present time several grades of structural steel are available for use in bridge construction. Not all grades are available from all rolling mills. The designer should check the availability in his local area.

Structural steels can be categorized into three general classifications: (1) carbon steels, (2) high-strength low-alloy steels, and (3) heat-treated alloy steels. Figure 2.1 shows the stress–strain curves of a steel in each classification. The steels are identified by the numbers of the American Society for Testing Materials (ASTM) Specifications governing the different grades of steel. The American Association of State Highway and Transportation Officials (AASHTO) has designation numbers for the various grades that are controlled by the AASHTO Specifications. The various grades of steels are referred to herein by the ASTM designation.

1. *Carbon Steels.* The carbon steels are those with a copper content between 0.40 and 0.60% and maximum manganese content of 1.65%. No minimum content is specified for other alloying elements. Although there

---

* Undoubtedly British long tons.

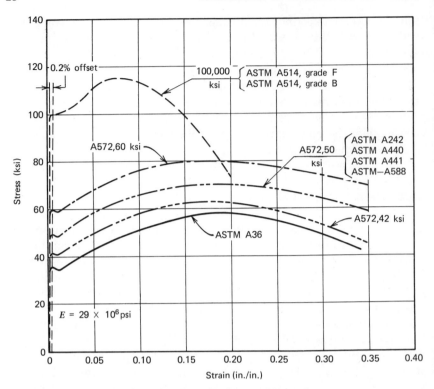

**Figure 2.1**  Stress strain curves for steel.

have been other carbon steels used in the past, today the carbon steel used in bridges is A36. This much-used steel has a minimum yield point of 36 ksi. The tensile strength can vary from 58 to 80 ksi. The minimum elongation is 20% in an 8-in. gauge length. This steel is readily welded (referred to as killed steel) and is usually the most economical grade for short-span bridges. It needs painting for protection from moisture.

2. *High-Strength Low-Alloy Steels.* The high-strength low-alloy steels are those that have a yield point greater than 40 ksi and are not heat treated to obtain the necessary stress levels. The older steels in this catagory are the ASTM A242, A440, and A441. They have yield strengths of 42 to 50 ksi. The higher value is for a material thickness of $\frac{3}{4}$ in. and there is a reduction in strength when a thicker material is used. For thicker material the effect of grain length due to rolling does not reach the inside steel. The A440 steel is not recommended for welding.

A572 and A588 steels are two new grades of steel that can be readily used for bridge construction. A572 steel has a specified minimum columbium and vanadium content. It can be obtained in yield strengths varying from 42 to 65 ksi. The higher strengths have a higher carbon content so that for welding it is recommended that grades 42, 45, and 50 (yield points) be used. There is an upper limit of thickness for each grade. AASHTO prescribes impact requirements for A572 of a minimum toughness of 15 ft-lb at 40°F as measured by longitudinal Charpy V-notch specimens. This steel has a greater resistance to atmospheric corrosion than A36 steel.

When high resistance to atmospheric corrosion is desired A588 steel is usually specified. This steel has a corrosion resistance four times that of A36 steel. It has been left unpainted (bare) for many bridges. This steel forms a thin iron oxide film on the surface when atmospheric moisture is present. Under appropriate atmospheric conditions this film is a durable and tightly adherent coating that resists any further penetration of moisture. The color of the oxide coating goes, with time, from a red-orange to a dark purple-brown. The coating forms faster and the color change progresses more rapidly in climates of high humidity. Weathering details can be obtained from the steel producers or the American Institute of Steel Construction (AISC). The iron oxide washes to some extent when water passes over the surface, thus staining surfaces of concrete substructures. If the appearance of such stains is objectionable provisions should be made to prevent such moisture from running down the concrete surfaces. These weathering steels have trade names such as Mayri R (Bethlehem) and Cor-Ten (U.S. Steel).

The A588 steels have a minimum yield point of 50 ksi up to and including a thickness of 4-in. This is an advantage over other low-alloy steels in thicknesses of 2 to 4 in..

A588 steel is readily weldable by the electric-resistance, submerged-arc, manual-arc, and gas metal-arc welding processes. Low-hydrogen electrodes are specified by the American Welding Society. If a weathered appearance of the finished weld is important, special electrodes should be used. High-strength bolts meeting ASTM A325 or A490 specifications and having a chemistry that will produce weather characteristics compatible with A588 steel are available.

3. *Heat-Treated Constructional Alloy Steel.* These steels contain alloying elements that exceed those contained in carbon steel and they are heat treated to obtain strength and notch toughness. The bridge steel in this category is A514. It has a yield strength of 100 ksi in thicknesses up to and including $2\frac{1}{2}$ in. and 90 ksi from $2\frac{1}{2}$ to 6 in. This steel has a high corrosion resistance.

Table 2.1 gives the minimum mechanical properties of structural steel plates and shapes.

# Table 2.1 Specified Minimum Mechanical Properties for Structural-Steel Plates and Shapes[a]

| AASHTO[b] Designation | ASTM Designation | Plate Thickness Range (in.) | Yield Stress (ksi) | Tensile Strength (ksi) | Elongation (%) In 2 in.[c] | Elongation (%) In 8 in. |
|---|---|---|---|---|---|---|
| | | *Carbon Steels* | | | | |
| M183 | A36 | To 8 inclusive | 36 | 58 | 23–21 | 20 |
| | | Over 8 | 32 | 58 | | |
| | | *High-Strength, Low-Alloy Steels* | | | | |
| M161 | A242 | To $\frac{3}{4}$. inclusive | 50 | 70 | | 18 |
| | | Over $\frac{3}{4}$ to $1\frac{1}{2}$, inclusive | 46 | 67 | | 19 |
| | | Over $1\frac{1}{2}$ to 4, inclusive | 42 | 63 | 24–21 | 16 |
| M222 | A588 Grade A | To 4, inclusive | 50 | 70 | 21–19 | 19 |
| | | Over 4 to 5, inclusive | 46 | 67 | 21–19 | |
| | | Over 5 to 8, inclusive | 42 | 63 | 21 | |
| M223 | A572 Grade 42 | To 4, inclusive | 42 | 60 | 24 | 20 |
| M223 | A572 Grade 50 | To 2, inclusive | 50 | 65 | 21 | 18 |
| M223 | A572 Grade 60 | To $1\frac{1}{4}$, inclusive | 60 | 75 | 18 | 16 |
| M188 | A441 | To $\frac{3}{4}$, inclusive | 50 | 70 | | 18 |
| | | Over $\frac{3}{4}$ to $1\frac{1}{2}$ inclusive | 46 | 67 | | 18 |
| | | Over $1\frac{1}{2}$ to 4 | 42 | 63 | 24–21 | 18 |
| | | Over 4 to 8 | 40 | 60 | 24 | |
| M187 | A440[d] | To $\frac{3}{4}$, inclusive | 50 | 70 | | 18 |
| | | Over $\frac{3}{4}$ to $1\frac{1}{2}$, inclusive | 46 | 67 | | 18 |
| | | Over $1\frac{1}{2}$ to 4, inclusive | 42 | 63 | 21 | 18 |
| | | *Heat-treated Constructional Alloy Steel* | | | | |
| M244 | A514 Grade F | To $2\frac{1}{2}$, inclusive | 100 | 115 | 18 | |
| | | Over $2\frac{1}{2}$ to 4, inclusive | 90 | 105 | 17 | |
| | A514 Grade B | $\frac{3}{16}$–$1\frac{1}{4}$, inclusive | 100 | 115 | 18 | |

### 2.3.3 Comparative Costs for Structural Steel

The bridge design must not only meet all strength requirements but also be as economical as possible. Therefore a knowledge of the relative costs of steels is needed. Costs vary from one location to another with variations in shipping costs and other factors. The bridge designer should consult with local fabricators to ascertain relative costs for the various steels. Table 2.2 shows relative costs and yield strengths costs of beams for various grades of steel. These values are approximate and are based on A36 steel as unity. Although A588 shows a relative cost greater than 1, the saving in cost of painting may offset this to a degree.

It may be very feasible to fabricate the structure with several grades of steel. This would prove economical for long-span trusses and also likely for the longer-span plate and box girders. In the latter types of bridges the webs could be of one grade of steel and the flanges of a higher-strength steel.

**Table 2.2  Relative Costs and Yield Strengths**

| Steel | Relative Cost[a] | Relative Yield Point[b] | Cost/Yield Point |
|---|---|---|---|
| A36 | 1.0 | 1.0 | 1.0 |
| A572-50 | 1.15 | 1.39 | 0.83 |
| A441 | 1.18 | 1.28 | 0.92 |
| A588 | 1.33 | 1.39 | 0.96 |
| A514 | 1.73 | 2.78 | 0.62 |

[a] Based on 1-in.-thick plates August 1977.
[b] Varies with plate thickness, values are for 1 in. thickness.

---

[a] The following are approximate values for all the structural steels: Modulus of elasticity, $29 \times 10^3$ ksi; shear modulus, $11 \times 10^3$ ksi; Poisson's ratio, 0.30; yield stress in shear, 0.57 times yield stress in tension; coefficient of thermal expansion, $6.5 \times 10^{-6}$ in./in. °F for the temperature range −50 to +150°F; density, 490 lb/ft³.
[b] The AASHTO Material Specification has a notch toughness requirement not called for in ASTM.
[c] The minimum elongation values are modified for some thicknesses in accordance with the Specification. Where two values are shown for the elongation in 2 in., the first is for plates and the second for shapes.
[d] Not intended for welded applications; this is not low-alloy steel but is included under this heading for convenience.

## 2.3.7  Brittle Fracture

In recent years there have been failures of some bridges as a result of brittle fracture of the steel. The most disastrous of such was the Point Pleasant Bridge (1). The following factors increase the possibility of brittle fracture increases:

1. Use of higher-strength steels.
2. Increasing thickness of steel material.
3. Lower service temperatures.
4. Reduced factors of safety.
5. More complex arrangement of parts of structures with increase in possibility of high stress concentration.
6. Increased use of welding.

Although brittle fracture can take place in bolted or riveted structures, welding can result, if not skillfully done, in initial imperfections that can initiate brittle fracture.

Brittle fracture can be defined as a failure of a steel element under a condition of brittle behavior as normally present in structural steels. A complete explanation of the phenomena of brittle fracture is available in other sources (2). The material toughness of the steel should be sufficient to ensure against brittle fracture under the service conditions. Material toughness is defined as the ability to deform plastically in the presence of a notch. The notch toughness depends on service temperature, loading rate, plate thickness, steel composition, and heat treatment.

Brittle fractures start from flaws of various kinds. Cracks are present in all large and complex structures. Good fabrication practice and inspection can limit the original size of flaws, but minute cracks from rolling, cutting, drilling, and welding can be present. Therefore brittle fracture can initiate from a crack if the conditions are such that the size of the crack will propagate with time and stress. The design engineer, the inspection personnel, and the fabricator should exercise care so that the undetected cracks are extremely small and conditions are not such as to amplify the crack size. The following factors are important in design and fabrication to prevent brittle fracture:

1. Flaws should be restricted in the finished steel. Good fabrication and inspection are necessary.

2. Any steel should have a high toughness strength and should be subjected to qualifying tests before becoming part of the finished structure.
3. The greater the stress intensity at a flaw, the greater the possibility of crack propagation. Therefore stress concentrations at possible flaws increase the stress level, as do residual stresses. Designing in higher-strength steels increases the probability of brittle fracture. Therefore high-strength steels should have high notch-toughness values.
4. Fatigue stresses can increase the size of a flaw. Hence the designer should make proper provision for lower stress levels if the number of stress cycles will be high.
5. The crack toughness of structural steels decreases with decreasing temperature. Therefore structures in regions of very low temperature require greater care to prevent brittle fracture.
6. In thick plates a triaxial state of stress that reduces the ductility of the steel occurs and the notch toughness is reduced. This is so even though the metallurgical characteristics of the steel are not changed.
7. The crack toughness of steels decreases with increasing loading rate. Structural elements subject to high-impact stresses are more susceptible to brittle fracture.

Fracture mechanics concepts (3) can be used in the determination of design parameters to prevent the occurrence of brittle fracture in steel bridges. No doubt better methods wll be developed in the future to reduce the probability of structural failure by brittle fracture of the steel.

In 1973 the AASHTO Committee on Bridges and Structures adopted modifications to the AASHTO Material Specifications. These modifications are based on a 15 ft-lb Charpy V-notch (CVN) impact value (see Table 2.3). It is noted that three service groups are given based on minimum service temperatures. For lower minimum service loads the CVN value is required at a lower temperature. The requirements are more severe for A588 steels at thicknesses over 2 in. when welded and for all thicknesses of A514 steels. It is noted that the level of required toughness is a CVN impact value of 15 ft-lb, at a test temperature that is 70°F above the minimum service temperature. This increase in test temperature of 70°F accounts for the fact that the CVN test is an impact test, while the service loading is an intermediate loading rate. The impact test is much easier to conduct and analyze than an intermediate loading rate test would be. The rationale behind this loading-rate shift has been proven by research (4).

**Table 2.3   AASHTO Impact Specifications for Bridge Steels**

| ASTM Designation | Thickness (in.) | Temperature Zone 1[a] | Temperature Zone 2[a] | Temperature Zone 3[a] |
|---|---|---|---|---|
| A36 | | 15 at 70°F | 15 at 40°F | 15 at 10°F |
| A572 | Up to 4, mechanically fastened | 15 at 70°F | 15 at 40°F | 15 at 10°F |
| | Up to 2, welded | 15 at 70°F | 15 at 40°F | 15 at 10°F |
| A440 | | 15 at 70°F | 15 at 40°F | 15 at 10°F |
| A441 | | 15 at 70°F | 15 at 40°F | 15 at 10°F |
| A242 | | 15 at 70°F | 15 at 40°F | 15 at 10°F |
| A588[b] | Up to 4, mechanically fastened | 15 at 70°F | 15 at 40°F | 15 at 10°F |
| | Up to 2, welded | 15 at 70°F | 15 at 40°F | 15 at 10°F |
| | Over 2, welded | 20 at 70°F | 20 at 40°F | 20 at 10°F |
| A514 | Up to 4, mechanically fastened | 25 at 30°F | 25 at 0°F | 25 at −30°F |
| | Up to $2\frac{1}{2}$, welded | 25 at 30°F | 25 at 0°F | 25 at −30°F |
| | Over $2\frac{1}{2}$ to 4, welded | 35 at 30°F | 35 at 0°F | 35 at −30°F |

[a] Zone 1: Minimum service temperature 0°F and above (−18°C and above). Zone 2: Minimum service temperature from −1 to −30°F (−19 to −34°C). Zone 3: Minimum service temperature from −31 to −60°F (−35 to −51°C).

[b] If the yield point of the material exceeds 65 ksi, the temperature for the CVN value for acceptability is reduced by 15°F for each increment of 10 ksi above 65 ksi.

## 2.4   FASTENERS

Steel bridges may be connected by riveting, bolting, or welding. Riveting is only used for some shop connections. It has been almost replaced by the use of high-strength bolts. Bolting is the fastest method of field connection of structures. Arc-welding of structural parts has increased greatly and has changed the basic design of several types of bridges, such as steel girder bridges.

### 2.4.1   Rivets

This type of fastener must be power driven by pneumatically or electrically driven hammers. Two grades of rivets are permitted. The structural steel rivets should conform to ASTM A502 grade 1 (AASHTO M288 grade 1) or ASTM A502 grade 2 (AASHTO M288 grade 2). The grade 2 rivet has a higher allowable stress in shear for bearing-type connections.

A control factor in the use of hot-driven rivets is the temperature of the rivets at the time of driving. The Specifications require them to be heated to a "cherry red." Care should be taken not to overheat or burn. Since all riveting on bridges is done in the shop control of the temperature at driving is easily achieved.

### 2.4.2 Bolts

This fastener comes in two general categories and three grades. This first category is low-carbon steel bolts (low strength) that are turned or ribbed and meet ASTM A307 specification. They are not generally used in joints of main members. They should not be used in joints subjected to fatigue.

The second category is high-strength bolts. The lower-strength grade is controlled by ASTM A325 and the higher-strength grade by ASTM A490. The A 325 type 1 bolt is produced from medium-carbon steel, the type 2 bolt is from low-carbon martensite steel, and the type 3 bolt is produced from weathering steel similar to A588 and A242 steels. Under present AASHTO Specifications all types carry the same allowable stresses.

The A490 bolt is made from alloy steel. It can carry approximately one third greater allowable stress than the A325 bolt.

All high-strength bolts carry markings on the head to indicate the type of bolt, that is, A325 and A490. Type 1 bolts may be marked with three radial lines 120° apart, type 2 bolts can be marked with three radial lines 60° apart, and type 3 bolts have the A325 underlined and the manufacturer may add other identifying marks such as the letters WR. A490 bolts are made of weathering steel.

The usual bolt diameters used in bridges are $\frac{3}{4}$, $\frac{7}{8}$, 1, and $1\frac{1}{8}$ in. Bolts are obtainable in $\frac{1}{8}$-in. increments up to $1\frac{1}{2}$ in., except A325 type 2 is obtainable up to 1 in. maximum. Bolt lengths are in increments of $\frac{1}{4}$ in.

### 2.4.3 Welding

In the past decade welding has become the predominant method for the connection of parts of steel bridges. This is especially so with respect to shop fabrication. For field erection high-strength bolting is preeminent in the United States, but in western Europe the use of welding for field connecting is quite common. The development of semiautomatic and fully automatic welding machines has been a big factor in the development of welding in bridge structures. Machine welds can be faster and more economical and have a higher degree of quality control. All of the structural steels discussed earlier in this chapter are weldable, but welding is not recommended for A440 and some types of A514. The level of carbon and manganese in A440 steel makes it undesirable for welding. A514 steel is a heat-treated steel. Certain types of A514 steel can be welded. A572 is a high-strength steel that is a good selection when weldable steel is desired. Table 2.2 shows its favorable cost/strength ratio: The proper electrode should be selected for welding the various grades of steel. The steel producers can provide the recommended electrode.

Field fabrication by welding requires special precautions with regard to temperature of material, sequence of welding, prevention of distortion, and so forth. The designer of steel bridges should be knowledgeable about the welding process, as well as about types of welds and design of welds.

## 2.5 MISCELLANEOUS MATERIALS

Many other materials are required to construct a bridge in addition to steel and concrete. Such material might be cast iron, cast steel, bronze, copper, aluminum, neoprene, or asphalt. These materials are used in specialties such as bridge railings, drainage scuppers, expansion devices, bearings. A new bridge over the Seine River in Paris has a handrail of plate glass panels framed in copper alloy posts and rails. The author has designed bridges that have equestrian paths. To provide a nonslip surface for horse shoes the plastic Tartan was used. The bridge engineer is required to be knowledgeable about the development of other new materials.

AASHTO Specifications make reference to required ASTM or AASHTO Specifications for other materials. When specialty materials are required it is well to confer with the manufacturers of the specialty items for evaluations of their products. Requirements should not be more severe than service demands, yet service life should be sufficient without excessive maintenance.

## REFERENCES

1   C. V. Scheffey, "Pt. Pleasant Bridge Collapse," *Civil Eng.* July 1971, p. 41.

2   A. Hanson, and J. G. Parr, *The Engineer's Guide to Steel*, Addison-Wesley, 1965, Ch. 4.

3   S. T. Rolfe, Designing to Prevent Brittle Fractures in Bridges, Specialty Conference on Safety and Reliability of Metal Structures, American Society of Civil Engineers, 1972, pp. 175–216.

4   S. T. Rolfe, The New AASHTO Material Toughness Requirements, Specialty Conference on Metal Bridges, American Society of Civil Engineers, 1974, pp. 156–172.

5   S. M. Barsom, *The Development of AASHTO Fracture-Toughness Requirements for Bridge Steels*, American Iron and Steel Institute, 1975.

3

# Loads on Bridges

## 3.1 DEAD LOADS

The dead load of a highway bridge consists of the weight of the structure plus any equipment that is attached to the structure. Some bridges have to carry water or utility lines that may have appreciable weight. A paradox of structural design is that the true dead load of the structure cannot be determined until the bridge is designed and a final design cannot be accomplished unless the true dead load is known. It is therefore necessary to make a preliminary estimate of the dead load and then perform the design based on the estimated value. The weight of the structure can then be calculated and compared with the previously estimated weight. The two weights most likely will not agree. It is then necessary to perform a second cycle of design based on the new dead load. If there has been any change in sizes of members, at the completion of this second cycle, the dead load is again calculated. This process of design refinement is repeated until the designer is sure that the final design calculations of the structure utilize the "as-built" weight of the bridge. An experienced design engineer usually arrives at a convergence in one or two cycles. If a computer program is used in the design process, several cycles can be performed in a short time.

It is possible to arrive at a final value of dead load for one part of the structure before proceeding with the design of a supporting part. For example, the floor system can be designed before the main girders or trusses, and the final weight of the entire superstructure can be determined before it is necessary to begin the design of the substructure. It is therefore obvious that it is not difficult to select a preliminary dead load for any part of the structure. Some data have been published giving empirical equations for determining the estimated weight of bridge structures. However, most equations that have been published are very limited in application or are obsolete as a result of changes in design specifications. A study of similar bridges is a good means of obtaining preliminary dead-load estimates.

The following unit weights of materials are commonly used in highway bridges.

| | |
|---|---|
| Concrete | $150 \, lb/ft^3$ |
| Steel | 490 |
| Cast iron | 450 |
| Asphalt paving | 140 |
| Timber | 35–60 |
| Stone masonry | 150–175 |
| Aluminum | 170 |

The distribution of the dead load is a significant factor in the design of bridges. In most instances the dead load is assumed to be uniformly distributed along the length of a structural element, such as a slab, beam, or truss. Continuous bridges may have main load-carrying elements of varying depth. Considering the weight as uniformly distributed for such nonprismatic bridges may lead to some error, but it is usually negligible. Any error would generally be on the side of safety. Other cases of dead-load distributions are discussed in the appropriate sections of this book.

Adequate allowance should be made in the design for future additions to the structure, such as wearing surface and utility lines.

## 3.2  LIVE LOADS

Highway bridges should be designed to safely support all vehicles that might pass over them during the life of the structure. It is not possible for the designer to know what vehicles will use the structure or what the required life of the bridge will be. To ensure the safety of the structure, some form of control must be maintained so that the designer has to provide sufficient strength in the structure to carry present and future predicted loads. The regulation of vehicles using the bridge has to be such that excessive weight vehicles are prohibited from crossing the structure. Design control is provided in the United States by AASHTO, which specifies the design live load, and traffic regulation is provided by state laws regulating the weights of motor vehicles.

The present design vehicles contained in the AASHTO Specifications (1) were adopted in 1944. The loadings consist of five weight classes, namely: H10, H15, H20, HS15, and HS20. The design vehicles for each of the five classes are shown in Fig. 3.1. These vehicles were not selected to resemble any particular vehicle in existence, but are hypothetical. Any actual vehicle that would be permitted to cross a bridge should not produce stresses greater than those caused by the hypothetical vehicle.

The lighter loads, H10 and H15, are used for the design of lightly traveled state roads while the H20 and HS20 are used for national

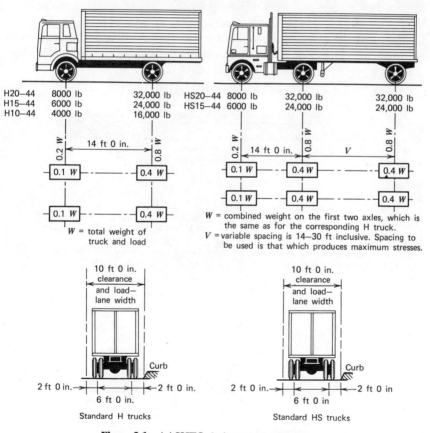

**Figure 3.1** AASHTO design vehicle loadings.

highways. The HS20 is used for the design of bridges on the Interstate Highway System. An additional alternate loading was instituted for this system. This loading consists of two axles spaced at 4 ft and weighing 24 k each (Fig. 3.1).

The HS truck loadings show a variable spacing of the two rear axles from 14 to 30 ft. The correct spacing is the length that produces the maximum effect. For stresses in simple span bridges, this spacing is the minimum value of 14 ft. However, for continuous spans a spacing greater than 14 feet may produce the maximum effect. The influence diagram indicates the proper spacing of the axles for maximum stresses.

In addition to the truck loadings, the Specifications contains equivalent loadings (Fig. 3.2) to be used in place of the truck loadings when they produce a greater stress than the truck. Prior to the 1944 Specifications, the design live load consisted of the basic *H* trucks preceded and followed by a train of trucks weighing three-quarters as much as the basic truck. In 1944 the HS truck was developed and the equivalent lane loading took the place of the train of trucks. Presently only *one* truck is to be used per lane per span. For longer spans the equivalent loading

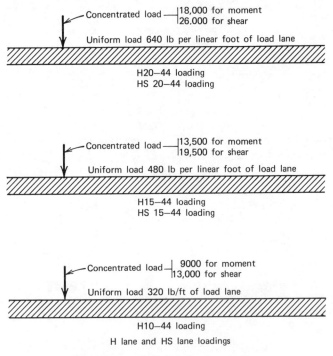

Figure 3.2   AASHTO design lane loadings.

produces greater stresses than the single truck. For instance, the H20 truck produces the greater bending moment in simple beam spans for spans up to 56 ft. For spans longer than 56 ft the equivalent lane loading produces the greater bending moment. The span length at which the loading changes for shear calculations is different than that for moment calculations. The equivalent lane loading approximates the shear and moment produced on long spans by a train of trucks as used prior to 1944.

The concentrated load used in the equivalent lane loadings is different for moment than for shear. Only one concentrated load is used in a simple span or for a positive moment in continuous spans. Two concentrated loads are used for a negative moment. The equivalent lane load is placed so as to produce the maximum stresses. The uniformly distributed load can be divided into segments when applied to continuous spans. Both the concentrated load and the uniform load are distributed over a 10-ft lane width on a line normal to the centerline of the lane.

Although the AASHTO Bridge Design Specifications was prepared for use in the United States, all or parts of the Specification have been adopted by other countries. However, those countries that have had a well-developed highway system for a long period of time have their own specification. The design load requirements of several countries are shown in Fig. 3.3.

Each country may have its own characteristic vehicle types and military loadings. Whether super-heavy loads are carried over the highways or the railroads has an effect on the selection of the maximum design vehicle. The inability of the railroads in England to carry very heavy loads such as large turbines has required the adoption of an abnormal heavy loading specification in that country.

Design vehicles for any country should be selected very carefully. Bridges should not be damaged as a result of normal traffic or an occasional single overload, nor should they become obsolete in a few years because of heavier vehicles on the highways. In a like manner, the design vehicle should not be more severe than the heaviest vehicles that will use the structure during its lifetime. Such a situation is wasteful of the resources of a country. Mature engineering judgment is needed in the selection of proper design loadings.

Studies have been conducted in the United States to determine proper highway loadings, and additional studies are presently in progress. If the span is long, the actual live loading consists of a train of vehicles of various lengths and weights. The presence of a single heavy vehicle is of less importance than the loading spectrum over the entire span. For design purposes a train of vehicles can be equated to an equivalent

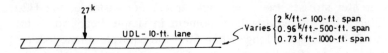

(a)  British  HA  loading

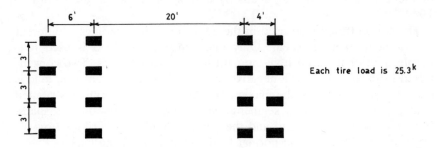

Each tire load is 25.3$^k$

(b)  British  HB - Abnormal  loading

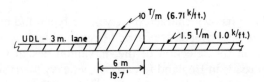

(c)  German  loading   DIN  1078

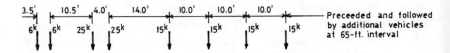

(d)  Indian - Class A train of vehicles

Note :
A meaningful comparison would have to include dynamic allowance, multiple lane loading, distribution to structural elements, etc.

**Figure 3.3**   Design live loads of various countries.

uniform lane load. The proper equivalent uniform lane load can only be determined after extensive studies have been conducted on weight and spacing of actual vehicles using the highways or those that may use the bridge during its lifetime.

For a comparison of design loads as adopted by various countries to have meaning, the impact factor, the method of distributing the live load, as well as allowable stresses must be included. Such comparisons are complex and would show different results for different spans. Generally speaking, AASHTO loadings are somewhat lighter than European loadings for short-span bridges, but are comparable for long spans.

The AASHTO Specifications permits a reduction to 0.90 of total load when a bridge is designed for three lanes and a reduction to 0.75 of total load when four or more lanes are loaded.


## 3.3  DYNAMIC EFFECT OF VEHICLES

It is well-known that a vehicle moving across a bridge at a normal rate of speed produces greater stresses than if the vehicle were in a static position on the structure. This increment in stress can be called the *dynamic effect*. The terminology for the dynamic effect among bridge designers and bridge design specifications is *impact*. This latter term is not scientifically correct, since it denotes one body striking another, which only takes place in a bridge when a wheel falls into a "chuck hole." The total dynamic effect is a result not only of vehicle wheels striking deck imperfections, but also of the live load being applied to the structure in a very short period of time. It can be proven by simple calculations of the theory of dynamics that a load instantly applied to a beam causes stresses of twice the magnitude as when the same load is static on the beam. In actual load applications to highway bridges, the total live loading is never instantaneous, but is applied to the structure in a finite time period. The dynamic effect due to the sudden loading is variable for all structural elements of a bridge.

In addition to the true impact effect and the sudden loading effect, there is also a third effect, which is caused by the vehicle vibrating on its springs. Uneven roadway surfaces contribute to this effect. The vibration of the vehicle on its springs induces vibrations in the structure. The magnitude of stresses is dependent on the relative masses of the vehicle and the bridge, the natural frequency of the structure, and the damping characteristics of the bridge.

An attempt to make an analytical study of the dynamic effect of a particular vehicle crossing a specific bridge leads to many complications because of quantities that can be evaluated only very approximately. Since it is difficult to make a confident analysis, the past procedure has been to assign an approximate quantity for the dynamic effect. This quantity is determined by a very simple equation. The AASHTO Specification gives the equation

$$I = \frac{50}{125 + L} \tag{3.1}$$

for determining the "impact factor." The stresses due to impact are then calculated by multiplying the live load stress by the value of $I$ obtained from equation 3.1. The impact stresses are then added directly to the live-load stresses to obtain the total stresses due to vehicle loads. The quantity $L$ is defined as the loaded length of the structure. This value is more specifically the length of structure that would be loaded to produce maximum stress if the live load consisted of a uniformly distributed load. Further explanations of $L$ for continuous structures are given in the appropriate chapters.

The maximum value of the impact factor suggested by AASHTO is 0.30. This corresponds to a simple span length of bridge of 41.7 ft. All spans less than this length use the value 0.30 for impact.

Most bridge designers would prefer to use a more scientific approach than equation 3.1; however, there has been a very limited amount of research on this subject, and, until more studies are forthcoming, equation 3.1 will have to suffice. The authors believe that except in unusual cases the value obtained from equation 3.1 is conservative. The percent of stress due to the dynamic effect is relatively small for long-span bridges.

## 3.4  LONGITUDINAL FORCES

When vehicles brake or accelerate while on a bridge, longitudinal forces are transmitted from the wheels of the vehicle to the deck. The magnitude of the longitudinal force depends on the amount of acceleration or deceleration. The maximum longitudinal force results from a sudden braking of a vehicle. The magnitude of this force is dependent on the weight of the vehicle, the velocity of the vehicle at the instant of braking, and the time interval to come to a complete stop. The force is given by

$$F = \frac{W}{g}\left(\frac{\Delta V}{\Delta t}\right) \tag{3.2}$$

where $W$ = weight of vehicle (k)

$g$ = acceleration of gravity = 32.2 ft/sec²

$\Delta V$ = change in velocity (ft/sec) in the time interval $\Delta t$ (sec).

If a truck of weight $W$ was traveling at 60 mph (88 ft/sec) and the brakes were suddenly applied, the longitudinal force would be dependent on the time it took the truck to change from a velocity of 60 mph to a velocity of zero. Assuming the vehicle came to a complete stop in 6 sec and the deceleration was uniform over this time interval, the longitudinal force would be

$$F = \frac{W}{32.2} \times \frac{88}{6} = 0.46\,W$$

Since the coefficient of friction of the rubber tires on a dry roadway is approximately 0.75, the above force can be transmitted to the roadway.

The AASHTO Specification provides for a longitudinal force of 5% of the live load in all lanes carrying traffic headed in the same direction. The live load as specified is the equivalent uniform lane load plus the concentrated load for moment. For spans of less than 84 ft the total equivalent lane load weight is less than the weight of the HS20 truck. It appears that the AASHTO provision is low for short-span two-lane structures. For long-span multiple-lane structures, the AASHTO provision is more realistic. For instance, for a 400-ft-span bridge of two lanes on the Interstate System (traffic in both lanes in the same direction), the longitudinal force from AASHTO would be

$$F = 0.05(0.64 \times 2 \times 400 \div 2 \times 18) = 27.4 \text{ k}$$

The comparative force from one HS20 truck suddenly braking on the structure (deceleration of 14.7 ft/sec²) would be

$$F = 0.46 \times 72 = 33 \text{ k}$$

There is the possibility of more vehicles than one truck simultaneously braking on a bridge with a span of 400 ft.

The Specification provides for the center of gravity of the longitudinal force to be applied 6 ft above the roadway surfaces. The longitudinal force adds very little stress to any members of the superstructure but is important in the design of the bearings and substructure (see Chapter 10). An additional longitudinal force due to the friction on expansion bearings should be considered in the substructure design. This frictional force is selected after consideration of type of bearings and possible service maintenance. The use of Teflon sliding bearings, which have a low coefficient of friction, minimizes the longitudinal force due to bearing friction.

### 3.5   CENTRIFUGAL FORCE

When a body travels on a curvilinear path, it produces a force perpendicular to the tangent of the path. This force is given by

$$F = \frac{W}{g}\frac{v^2}{r} \tag{3.3}$$

where  $W$ = weight of the body
       $g$ = acceleration of gravity = 32.2 ft/sec$^2$
       $v$ = velocity of the body
       $r$ = radius of the path of the body

If the velocity $v$ is expressed in miles per hour, then equation 3.3 can be written as

$$F = 0.0668\frac{Wv^2}{r} \tag{3.4}$$

Equation 3.4 is given in the AASHTO Specification in slightly different form. This centrifugal force is to be applied 6 ft above the roadway surface. When the reinforced concrete slab or steel grid deck is keyed or attached to its supporting members, the deck can be assumed to resist this centrifugal force within its plane. This force can then be transmitted to the substructure by bracing at the end of the span. The application of the centrifugal force at 6 ft above the deck can produce vertical forces in the main supporting girders or trusses.

### 3.6   WIND LOADS

Wind loads have been the concern of bridge design engineers for many years, but the determination of the effect on a bridge is very complex. The design wind loads as contained in specifications are approximate for any particular structure. The problem of wind loads for a particular structure is very complex because of the many variables that affect the wind force, such as size and shape of the bridge, probable angles of attack of the wind, shielding effects of terrain, and the velocity–time relationship of the wind.

Studies on wind loads on bridges (2, 3) have yielded some information on this subject, and summaries of the latest studies are contained in references 4 and 5.

The wind load is a dynamic force. A peak wind velocity may be reached in a short period of time and may remain for a time interval or decay rapidly (gust). If the time interval to reach peak pressure is equal to or greater than the natural frequency of the structure, the wind load can be treated for all practical purposes as a static load equal to the peak pressure. This is usually the condition for most bridges.

The wind pressure on a solid object can be expressed by the equation

$$p = C_D \rho \frac{v^2}{2}$$

where $C_D$ = drag coefficient
$\rho$ = density of the air
$v$ = wind velocity

The total force on a bridge is the pressure multiplied by the effective area. The effective area can be taken as the area as seen in elevation multiplied by $C_A$, the area coefficient. The value of $C_A$ depends on the angle of attack of the wind, the type of bridge and floor structure, and the distance between beams or girders. Another factor, expressed as coefficient $C_\psi$, that affects the total horizontal force is the yaw angle (angle of wind direction in the horizontal plane measured from a perpendicular to the plane of the superstructure). The total horizontal force can be expressed as

$$H = C_D C_A C_\psi A_T \rho \frac{v^2}{2}$$

For girder bridges reference 3 gives the horizontal force acting transversely on the bridge as

$$H = C_H A_H \rho \frac{v^2}{2} \tag{3.5}$$

where $A_H$ is the total projected area as seen in elevation and $C_H$ varies from 1.36 to 1.87 for the bridges tested. This corresponds to pressures from 34.8 (two-girder bridge) to 47.8 psf (three- and four-girder bridge).

The values given above are for a maximum angle of attack of $\pm 15°$, which is considered maximum for most bridges.

Reference 3 also discusses uplift forces and resulting overturning moments. These factors are usually neglected in most design specifications.

The AASHTO Specification gives basic wind forces for a velocity of 100 mph as

girders and beams        50 psf

A minimum force of 300/lb/ft should be used on girder and beam spans.

The unit pressure is to be applied to the total area of the structure as seen in elevation at an angle of 90° to the longitudinal axis of the structure.

An addition wind force is also contained in the Specifications. This is a wind force on the live load equal to 100 lb/ft applied 6 ft above the roadway. When this additional load is applied, only 30% of the wind load given above is applied to the structure itself.

The above loadings are only the loadings applied to the superstructure. The above values of wind force are considered in the substructure design for occasions when the wind is perpendicular to the bridge but winds at other angles are considered also for skew (yaw) angles of up to 60°. When the wind is at a skew, longitudinal forces on the substructure should also be considered. The reader is referred to the AASHTO Specification for the details on the perpendicular and longitudinal forces for various skew angles and also the wind on the substructure itself.

The AASHTO Specification also provides for consideration of a vertical overturning force of 20 psf on the deck and sidewalk area at the windward quarterpoints of these areas.

The wind forces discussed above are treated as static forces. This treatment is usually satisfactory when the structure is very rigid. However, for flexible structures, this procedure is not sufficient. Suspended structures require a more investigative procedure. A steady-state wind may set up an effect called "vortex shedding" if the structure has a long period of natural frequency and a minimum of natural damping. The vortex shedding refers to the production of vortices as the wind passes around the structure producing pressures and suctions alternating on each side of the obstruction to the wind. This alternating direction of forces at a right angle to the wind causes the structure to vibrate in a direction perpendicular to the wind. If the period of shedding of vortices is the same as the natural period of the structure, large amplitudes of vibration, that are sufficiently large to damage the structure take place. These vibrations occurred in the case of the Tacoma Narrows suspension bridge in 1940.

For stiff structures, the wind velocity would have to be greater than normal for resonance to occur. However, for flexible structures, resonance can occur at low wind velocities. Such an occurrence for a tower-type structure at a velocity of 10 mph is known to one of the authors.

The bridge designer should be aware of the vortex-shedding phenomenon, even though it is unlikely to occur in most bridges. Several members of truss bridges have vibrated in the wind as a result of vortex-shedding effects and repairs have been necessary.

## 3.7 EARTHQUAKE FORCES

Until recently the effects of earthquakes on highway bridges were largely ignored or only given consideration with regard to the design of substructures. Superstructures most likely have adequate strength to resist any inertia force in the vertical or horizontal direction. However, in recent earthquakes in Alaska, southern California, and Guatamala, several bridges were destroyed. The failure was not caused by the collapse of any element of the superstructure, but rather by (1) the superstructure shaking off the bearings and falling to the ground; (2) the structural failure of piers; (3) the tipping of piers as a result of large earth movements; or (4) the loss of strength of the soil under the substructure as a result of the vibrations induced in the ground.

In times of earthquakes, it is important that bridges remain in place so that critical disaster relief vehicles such as ambulances and fire trucks can function.

The effect of an earthquake on a structure depends on the elastic characteristics of the structure and the distribution of the weight. A rigorous analysis is complex and involves the application of structural dynamics. In addition, it is necessary to know the expected ground motion under the substructure. The usual procedure is to greatly simplify the problem by considering that the earthquake produces lateral forces acting in any direction at the center of gravity of the structure and having a magnitude equal to a percentage of the weight of the structure or any part of the structure under consideration. These lateral loads are then treated as static loads.

In 1975 AASHTO published an Interim Specification that listed a new earthquake loading procedure. This new procedure is much more rigorous than the former. The earthquake force is still treated as an equivalent static force if the supporting members are approximately equal. This equivalent static force is treated as a horizontal force acting at the center of gravity of the structure. This new earthquake criterion in AASHTO is too long to repeat here, but a brief explanation of the procedure is given so that it can be more readily followed.

The equivalent static force $EQ$ is given by the formula

$$EQ = CFW$$

where $W$ = total dead weight of structure
$\quad$ $F$ = framing factor (value given in AASHTO) depending on the type of pier
$\quad$ $C$ = combined response coefficient = $ARS/Z$

The definition of each of the factors contributing to the value of $C$ is given in the AASHTO Specifications. However, values of $C$ can be determined from graphs that have as variables the depth of alluvium to rock-like material, maximum peak rock acceleration, and period of frequency of the structure. This frequency period is given by the basic equation for a system with single degree of freedom:

$$T = 0.32 \frac{W}{P}$$

where $W$ is again the total dead weight of the structure and $P$ is the force required to cause a 1-in. horizontal deflection of the whole structure.

Minimum values of $C$ occur when the period $T$ is very small—usually less than 0.2 sec or above 1.0 sec.

For the design of restraints to ensure that the superstructure will maintain its correct position with relation to the substructure

$$EQ = 0.25(DL)$$

where $DL$ is the dead load of the superstructure. All restraints that act simultaneously resist the lateral force $EQ$.

A very important consideration in the design of bridges subjected to seismic action is the effect of water on piers during an earthquake. The forces due to towers submerged in water were studied at the University of California, Berkeley (7). This study showed that for the analysis of earthquake forces, an added mass, in addition to the weight of the pier, must be included. This added mass depends on the ratio of column radius to water height. Its value can be considerable—as much as the mass of water displaced by the piers. A careful dynamic analysis should be made for bridge piers in active seismic areas, especially when the piers are submerged.

### 3.8   STREAM FLOW PRESSURE

Substructures constructed in a region of flowing water should be designed to withstand water pressure. Such pressure could cause the pier to slide or overturn. Of considerable concern is the scour around the bottom of the piers. The original London Bridge (demolished in 1831) had such wide piers that the increased stream velocity caused a severe scouring condition around them. Remedial work continued throughout the life of the structure.

The design for every bridge over a stream should involve the careful study of possible stream velocities. Almost all major streams in developed countries have flow records from which velocities can be obtained. If such information is lacking, the engineer must make the best estimate possible using whatever data are available.

The pressure of flowing water against a body can be predicted by the equation

$$P = KV^2 \tag{3.6}$$

where  $P$ = the unit pressure in lb/ft$^2$
  $V$ = the maximum possible velocity of the water in ft/sec
  $K$ = constant depending on the weight of the water and the shape of the pier.

The AASHTO Specification gives the following values for $K$

  $K = 1\frac{3}{8}$  piers with square ends

  $K = \frac{1}{2}$  piers with ends angled at 30° and less

  $K = \frac{2}{3}$  piers with circular ends.

Although it is known that the velocity of the water is variable with depth, it is satisfactory to use a constant pressure for the full depth of the water.

## 3.9  FLOATING ICE PRESSURE

In cold climates floating ice can cause very high forces against piers. Bridges have been completely demolished from the pressure of floating ice.

The AASHTO Specification suggests a pressure of 400/lb/in$^2$ of contact area of ice and substructure. The thickness of ice and its point of application on the piers should be determined by the engineer after as complete an investigation as possible.

The value of 400 psi is based on an estimated crushing pressure of ice of this value. Higher values for the crushing strength of ice have been determined for low temperatures. A crushing strength of over 800 lb/in$^2$ has been recorded for ice at 2°F. A study of ice pressures against dams (6) due to a rise in temperature—the condition producing the greatest pressure—has led to the conclusion that the 400 psi is an overly severe loading for dams. However, the condition for an expanding ice sheet behind a dam is quite different from floating ice being forced against a bridge pier by flow of water and possible wind drag on the ice. Very little is known with regard to these factors and until more information is forthcoming, the engineer should use the generally accepted value of 400 psi.

## 3.10  MISCELLANEOUS

There are additional forces that may be applied to some types of structures under particular conditions. Such forces are caused by temperature changes, shrinkage, elastic shortening, and earth pressures. For quantitative values of these items, the engineer should consult the Specifications or make a judgment based on a study of the particular conditions.

## 3.11  LOADING COMBINATIONS

The whole or parts of a bridge may be subjected to several of the previously mentioned loads simultaneously. The engineer has to decide what combination of loads and magnitudes will most likely be applied simultaneously.

The AASHTO Specification has set up combinations of loads that can be considered as acting on a structure simultaneously. For some combinations, the normal allowable stresses can be increased, which is an indirect way of saying that only a percentage of the sum of the maximum effects is considered, since the possibility of the peak value of these events occurring simultaneously is remote. For instance, when dead load, earth pressure, effects of buoyancy, stream flow, and wind loads are applied simultaneously, the allowable stresses can be 125% of the normal allowable stresses. This is the same as saying that only 80% of the peak values of these loads would be considered as acting together.

In the eleventh (1973) edition of the AASHTO Bridge Design Specifications, a concept of ultimate strength design was introduced into bridge design in the United States. This new procedure does not replace the working design method, but is a supplement to it. The 1975 Supplement to the Specifications introduced the following general equation for the combination of loads:

$$\text{group } N = \gamma \ [\beta_D \cdot D + \beta_L (L + I) + \beta_C \cdot CF + \beta_E \cdot E + \beta_B \cdot B + \beta_S \cdot SF + \\ \beta_W \cdot W + \beta_{WL} \cdot WL + \beta_L \cdot LF + \beta_F \cdot F + \beta_R (R + S + T) \\ + \beta_{EQ} \cdot EQ + \beta_{ICE} \cdot ICE] \quad (3.7)$$

where  $D$ = dead load
$\quad\quad\ L$ = live load
$\quad\quad\ I$ = impact due to live load
$\quad\quad\ E$ = earth pressure
$\quad\quad\ B$ = buoyancy
$\quad\quad\ W$ = wind force on structure

$WL$ = wind load on live load (100 lb/ft)
$LF$ = longitudinal force from live load
$CF$ = centrifugal force
$F$ = longitudinal force due to bearing friction or shear
$R$ = rib shortening (arches or frames)
$S$ = shrinkage
$T$ = force due to temperature change
$EQ$ = earthquake
$SF$ = stream flow pressure
$ICE$ = ice pressure
$N$ = loading group number
$\gamma$ = load factor
$\beta$ = coefficient

The values of load factor ($\gamma$) and coefficients ($\beta$) (Table 3.1) depend on the group loading as well as on whether working stress (allowable stress) design or load factor design is used.

The Specifications designate nine different loading groups to be used in bridge design. The groups I, II, and III apply to both bridge superstructures and substructures. Groups IV, V, and VI are primarily applicable to

**Table 3.1  Table of Coefficients $\gamma$ and $\beta$**

| Group | $\gamma$ | $D$ | $L+I$ | $CF$ | $E$ | $B$ | $SF$ | $W$ | $WL$ | $LF$ | $F$ | $R+S+T$ | $EQ$ | $ICE$ | % |
|---|---|---|---|---|---|---|---|---|---|---|---|---|---|---|---|
| | | | | | | | | $\beta$ Factors | | | | | | | |
| | | | | | | | Working Stress Design | | | | | | | | |
| | 1.0 | 1 | 1 | 1 | $\beta_E$ | 1 | 1 | 0 | 0 | 0 | 0 | 0 | 0 | 0 | 100 |
| | 1.0 | 1 | 0 | 0 | 1 | 1 | 1 | 1 | 0 | 0 | 0 | 0 | 0 | 0 | 125 |
| I | 1.0 | 1 | 1 | 1 | $\beta_E$ | 1 | 1 | 0.3 | 1 | 1 | 1 | 0 | 0 | 0 | 125 |
| V | 1.0 | 1 | 1 | 1 | $\beta_E$ | 1 | 1 | 0 | 0 | 0 | 0 | 1 | 0 | 0 | 125 |
| | 1.0 | 1 | 0 | 0 | 1 | 1 | 1 | 1 | 0 | 0 | 0 | 1 | 0 | 0 | 140 |
| I | 1.0 | 1 | 1 | 1 | $\beta_E$ | 1 | 1 | 0.3 | 1 | 1 | 1 | 1 | 0 | 0 | 140 |
| II | 1.0 | 1 | 0 | 0 | 1 | 1 | 1 | 0 | 0 | 0 | 0 | 0 | 1 | 0 | 133 |
| III | 1.0 | 1 | 1 | 1 | 1 | 1 | 1 | 0 | 0 | 0 | 0 | 0 | 0 | 1 | 140 |
| X | 1.0 | 1 | 0 | 0 | 1 | 1 | 1 | 1 | 0 | 0 | 0 | 0 | 0 | 1 | 150 |
| | | | | | | | Load Factor Design | | | | | | | | |
| | 1.3 | $\beta_D$ | 1.67 | 1 | $\beta_E$ | 1 | 1 | 0 | 0 | 0 | 0 | 0 | 0 | 0 | n.a. |
| A | 1.3 | $\beta_D$ | 2.20 | 0 | 0 | 0 | 0 | 0 | 0 | 0 | 0 | 0 | 0 | 0 | n.a. |
| | 1.3 | $\beta_D$ | 0 | 0 | $\beta_E$ | 1 | 1 | 1 | 0 | 0 | 0 | 0 | 0 | 0 | n.a. |
| I | 1.3 | $\beta_D$ | 1 | 1 | $\beta_E$ | 1 | 1 | 0.3 | 1 | 1 | 1 | 0 | 0 | 0 | n.a. |
| V | 1.3 | $\beta_D$ | 1 | 1 | $\beta_E$ | 1 | 1 | 0 | 0 | 0 | 0 | 1 | 0 | 0 | n.a. |
| | 1.25 | $\beta_D$ | 0 | 0 | $\beta_E$ | 1 | 1 | 1 | 0 | 0 | 0 | 1 | 0 | 0 | n.a. |
| I | 1.25 | $\beta_D$ | 1 | 1 | $\beta_E$ | 1 | 1 | 0.3 | 1 | 1 | 1 | 1 | 0 | 0 | n.a. |
| II | 1.3 | $\beta_D$ | 0 | 0 | $\beta_E$ | 1 | 1 | 0 | 0 | 0 | 0 | 0 | 1 | 0 | n.a. |
| III | 1.3 | $\beta_D$ | 1 | 1 | $\beta_E$ | 1 | 1 | 0 | 0 | 0 | 0 | 0 | 0 | 1 | n.a. |
| X | 1.20 | $\beta_D$ | 0 | 0 | $\beta_E$ | 1 | 1 | 1 | 0 | 0 | 0 | 0 | 0 | 1 | n.a. |

arches and frames. The last three group loadings are for the design of substructures.

The concept of the group loadings is based on which loads could probably act simultaneously. In working stress design, an increase in allowable stress is permitted for some group loadings. The rationale for this increase is that when the peak values occur simultaneously (very few times in the life of a structure), the factor of safety can be reduced.

In the application of load factor design, the limit of stress is generally the yield point stress for steel and the limit for strength is usually a percentage of the 28-day compressive strength for concrete. The bridge designer should consult the appropriate sections of the Specifications for the proper limit of stress. Later chapters of this book show the use of both working stress and load-factor design.

## REFERENCES

1 American Association of State Highway and Transportation Officials, *Standard Specifications for Highway Bridges*, 1973.
2 J. M. Biggs, "Wind Loads on Truss Bridges," *Proc. Am. Soc. Civil. Eng.* Separate No. 201, July 1953.
3 J. M. Biggs, S. Namyet, and J. Adachi, "Wind Loads on Girder Bridges," *Proc. Am. Sec. Civil. Eng.* Separate No. 587, January 1955.
4 W. W. Pagon, "Wind Forces on Structures, Plate Girders and Trusses," *J. Struc. Div. Am. Soc, Civil Eng.* Vol. 84, No. ST4, July 1958.
5 "Wind Forces on Structures," final report task committee on Wind Forces, *Trans. Am. Soc. Civil Eng.* 1961, pp. 1124–1198.
6 E. Rose, "Thrust Exerted by Expanding Ice Sheet," *Trans. Am. Soc. Civil Eng.*, 1947.
7 C-Y. Liaw and A. K. Chopra, "Earthquake Response of Axisymmetric Tower Structures Surrounded by Water," University of California, Berkeley, Report No. EERC 73–25, October 1973.

# 4

# Design of Slab-Steel Beam Bridges

## 4.1 INTRODUCTION

The majority of all bridges are of the slab and beam type. This construction consists of several beams that span in the direction of the roadway and are topped with a reinforced concrete deck (Fig. 4.1). The longitudinal beams can be made of several different materials, but are usually of concrete or steel. The design of steel beams with reinforced concrete decks is presented here.

To design this type of bridge the number and spacing of beams are first selected. The slab is then designed. Once the slab has been designed and the curbs, parapet, and railing have been selected the dead load on the beams can be determined. The distribution of live load to the beams is then calculated from the Specifications.

A decision must be made in this type of bridge whether to use a mechanical shear-resistant device to connect the slab to the beams. Such a device ensures that the slab will act compositely with the steel beams and thus will assist the beams in carrying the longitudinal bending moments. This results in smaller steel beams. The saving in beam steel must be compared with the cost of the shear connectors. On beam spans

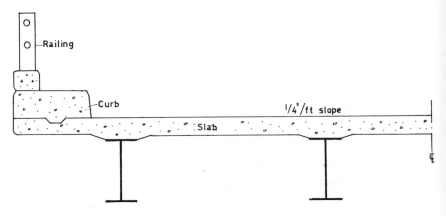

**Figure 4.1**   Cross section of steel beam and concrete slab bridge.

of less than 40 ft, the shear connectors may cost more than the saving in beam steel; on longer spans, however, the composite slab and beam bridge is definitely more economical.

## 4.2   DESIGN OF REINFORCED CONCRETE SLABS

The procedure for the design of stringers, longitudinal beams, floor beams, and slabs is given in Section 3—"Distribution of Loads" of the AASHTO Bridge Specifications. For design purposes, the slab is considered as a 1-ft-wide beam continuous over several supporting beams or stringers. The AASHTO Specification gives the live-load bending moment in a simple span for a 1-ft width of slab as

$$M = \pm \frac{(S+2)}{32} P \qquad \text{(spans 2–24 ft inclusive)} \qquad (4.1)$$

where $S$ is the effective span length in feet and $P$ is 12,000 pounds for the H15 and HS15 loading and 16,000 lb for the H20 and HS20 loading.

In slabs continuous over three or more supports, the moment, as determined by the equation above, is multiplied by 0.8 for both positive and negative values. For continuous spans over more than two supports the effective span $S$ is the clear span for slabs monolithic with concrete beams. For slabs resting on steel stringers, the effective span $S$ is the center-to-center distance of beams minus one-half of the stringer flange width.

In addition to the live-load moment just discussed, there is a moment caused by the dynamic effect. The impact percentage is given by $I = [50/(L+125)](100)$ (maximum is 30%). The impact allowance for slabs will always be 30% since $L$ is relatively small for slabs.

The AASHTO Specification does not cover the dead-load moments in slabs. For continuous slabs an amount of moment equal to $wS^2/10$ is no doubt sufficient. The dead-load moment is usually so small that an attempt at a more exact value is not warranted.

It is necessary to provide reinforcing steel in the longitudinal direction as well as the area of reinforcing steel required by the transverse moment. The AASHTO Specification does not give a method for determining longitudinal moments but requires an amount of steel in the longitudinal direction equal to $220/\sqrt{S}$ times the amount of transverse steel. The maximum value required is 0.67 times the transverse steel area and is the percentage used for $S \leq 10.8$ ft. Since the longitudinal bending moment results in tension in the bottom of the slab, this amount of longitudinal steel should be placed in the bottom of the slab.

The amount of reinforcing steel calculated above is to be used in the middle half of the span between the beams, and in the outer quarters of the slab not less than 50% of the amount used in the middle is required. In addition to this distribution steel in the bottom, reinforcing bars are required in the top of the slab, parallel to traffic for temperature, shrinkage, and tie steel. The minimum amount should be $\frac{1}{8}$ in.$^2$ of reinforcement per foot of the slab with an 18-in. maximum spacing.

Design Example 4.1 gives the procedure for reinforced concrete slab design in accordance with the AASHTO Specification. After the slab bending moment is determined, the area of steel is calculated from the usual formula for rectangular concrete beams. Although there is some reinforcing steel in the top of the slab midway between longitudinal beams and in the bottom over the beams, it is usually not necessary to relate the design to a concrete beam with compression steel. This compression steel is so near the neutral axis in the usual thickness of slab that it offers very little bending resistance.

Most states require a minimum structural thickness of slab, usually from $6\frac{1}{2}$ to $7\frac{1}{2}$ in. In 1974 the Specifications introduced a minimum thickness criterion [1.5.40(B)] stating that slabs with main reinforcement parallel or perpendicular to traffic should have a minimum thickness of $(S + 10)/30$ where $S$ is the span length of slab in feet so defined in the Specifications. For simple spans, the thickness should be about 10% greater. In addition, an allowance for wearing may be taken into consideration. Some state design specifications also make an allowance for extra dead weight for future asphalt paving.

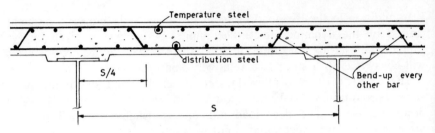

**Figure 4.2**   Reinforcing steel in bridge slab.

Spalling of concrete decks has become a serious maintenance problem, particularly in the northern states that use salt and other chemicals for ice removal. Cover should be sufficient to help prevent moisture from penetrating to the reinforcing steel. In most northern states, an asphalt cover of 1 to 2 in. over a waterproof membrane is used on bridge decks.

Since the bending moment in the slab caused by concentrated wheel loads is positive in the central region between beams and negative above and adjacent to the beams, the majority of the steel reinforcing has to shift from the bottom of the slab to the top. Although this may hold in theory, in reality it is not practical. Because the wheel load may be anywhere on the slab, the point of inflection has a variable position. Therefore some reinforcing steel, both top and bottom, should be straight for the full width of the slab. The usual practice is to bend one of each of two adjacent bars and leave the other bar straight, thus achieving some extra economy (see Fig. 4.2). When every other bar is bent, the steel area in the bottom of the slab, midway between beams, is twice that in the top of the slab, and over the beams, the reverse of this is true.

## Design Example 4.1

SLAB DESIGN

*Given:*

1. Span, center-to-center of bearings; 30 ft.
2. Load: two lanes of HS20.
3. Standard weight aggregate is used with maximum size—1 in.
4. Materials: concrete, $f'_c = 3000$ psi; reinforcing steel, $f_s = 20,000$ psi (grade 40).
5. Bridge cross section: as shown in Fig. 4.3.

First the span of slab must be determined. Then the flange width of the beam is estimated to be 12 in. with beams at 6 ft 9 in. center-to-center.

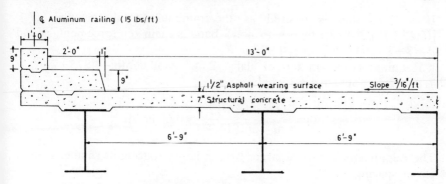

**Figure 4.3**  Steel beam bridge.

Then

$$\text{slab span} = 6.75 - \frac{12}{2 \times 12} = 6.25 \text{ ft}$$

It is estimated that a 7-in. slab (total concrete) will be required with a $1\frac{1}{2}$-in. asphalt wearing and waterproof surface. Then

$$\text{dead weight of slab} = \frac{7}{12}(0.15) + \frac{1.5}{12}(0.125) = 0.10 \text{ k/ft}$$

$$\text{dead-load bending moment} = \pm\frac{0.1(6.25)^2}{10} = \pm 0.39 \text{ k-ft/ft}$$

$$\text{live-load bending moment} = 0.8\left[\frac{S+2}{32}\right]P_{20} = 0.8\left(\frac{8.25}{32}\right)(16)$$

$$= \pm 3.30 \text{ k-ft/ft}$$

$$\text{impact bending moment} = 0.3(3.30) = \pm 0.99 \text{ k-ft/ft}$$

$$\text{total bending moment} = \pm 4.68 \text{ k-ft/ft}$$

and, using the standard flexure equation,
For allowable stress design of rectangular beams

$$d = \sqrt{\frac{2M}{f_c bJK}}$$

Since bending moment $M$ is not precise, $J$ and $K$ can be taken as $\frac{7}{8}$ and $\frac{3}{8}$, respectively:

$$f_c = 0.4f'_c = 0.4(3000) = 1200 \text{ psi}$$

$$d = \sqrt{\frac{2(4.68)(12)}{1.2(12)(\frac{7}{8})(\frac{3}{8})}} = 4.875 \text{ in.}$$

If a 7-in. slab is desired as the minimum total thickness $(7 >$ $[(6.25 + 10)/30](12)$, then with no. 5 bars as main reinforcement, actual $d = 7 - 2 - \frac{5}{8}(\frac{1}{2}) = 4.69$ in. $< 4.875$.

Required cover on top of slabs: 2 in.; total depth: $4.875 + 2 + \frac{5}{16} =$ $7.19 = 7\frac{1}{4}$ in.; actual $d = 4.94$ in.

$$A_s = \frac{M}{f_s J d} = \frac{4.68(12)}{20(\frac{7}{8})(4.94)} = 0.65 \text{ in.}^2/\text{ft}$$

The reinforcement can be larger bars at larger spacing or smaller bars at closer spacing:

$$\text{no. 5 bars at } 5\frac{1}{2}\text{-in. spacing} = 0.68 \text{ in.}^2/\text{ft}$$

$$\text{no. 4 bars at } 3\frac{1}{2}\text{-in. spacing} = 0.69 \text{ in.}^2/\text{ft}$$

$$\text{no. 6 bars at 8-in. spacing} = 0.66 \text{ in.}^2/\text{ft}$$

The no. 4 bars produce too close a spacing and are more expensive. Either the no. 5 or no. 6 are satisfactory. If no. 6 is used, $d = 4.875$ in. and $As = 0.66$ in.$^2$

Bond and shear do not need checking in bridge slabs when thickness and steel area have been calculated by way of AASHTO procedures. The main reinforcing (transverse) is so arranged that every other bar is bent, with the bent bar at the top of the slab (2-in. cover) over the longitudinal beams and at the bottom of the slab (1-in. cover) between beams. The bending point is approximately $6.75/4 = 1.7$ (say, 1 ft 9 in.) from the longitudinal beam.

The longitudinal or distribution reinforcing steel is now calculated. This amount is a percentage of the main reinforcement:

$$\frac{220}{\sqrt{s}} = \frac{220}{\sqrt{6.25}} = 88\% \qquad (\text{maximum} = 67\%)$$

Longitudinal $A_s = 0.67(0.67) = 0.45$ in.$^2$ Use no. 5 with $8\frac{1}{4}$ in. spacing.

The only other item to be determined is the temperature of steel at the top of the slab. The requirement is $\frac{1}{8}$ in.$^2/$ft of slab width, with maximum spacing at 18 in. Number 4 bars at 18-in. spacing meet this requirement. This steel is placed at the top of slab just under the main transverse steel. It serves as tie steel for the main reinforcing at the top of the slab. The slab and its reinforcing steel are shown in Fig. 4.2. The reinforcing steel in the curb and parapet is usually a state standard. Such typical arrangements are as shown in Fig. 4.4.

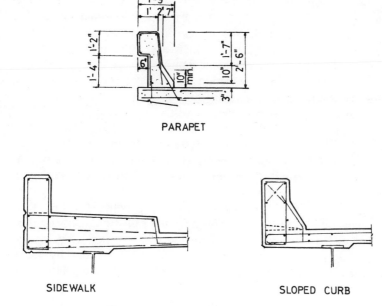

PARAPET

SIDEWALK                                      SLOPED CURB

**Figure 4.4**   Curb and parapet details.

## 4.3   DESIGN OF STEEL BEAMS (NONCOMPOSITE)

The simple beam span bridge of less than 30 ft is the most common bridge in all countries.

The designer must choose between several beams at close spacing or fewer beams at a larger spacing. The latter selection usually results in a saving of total amount of steel, provided the spacing is not so large that a concrete slab thicker than the minimum is required. Under AASHTO HS20 or H20 loading, the maximum spacing of steel beams for a 7-in. thick slab is approximately 8 ft, considering a 28-day concrete strength of 3000 psi.

In the short-span range, rolled steel beams are usually more economical than beams fabricated by welding three plates together. For long spans, the use of shear connectors between the steel beam and concrete slab is economical. This type of construction, termed "composite," is discussed later in this chapter.

The ratio of depth of the steel beam to the span length, as required by AASHTO, should not be less than 1:25. If depths less than these are used, the sections are increased so that the maximum deflection is not greater than if this ratio had not been exceeded. The deflection caused by live load plus impact is limited to less than $\frac{1}{800}$ of the span, center-to-center of bearings. The Specification states: "When bridges have cross bracing or diaphragms sufficient in depth and strength to insure lateral distribution of loads, the deflection may be computed for the standard loading, considering all beams or girders as acting together and having equal deflection." Theoretically, it would take diaphragms of infinite moment of inertia to achieve this equal deflection of all beams. However, since the deck slab aids considerably in distributing live loads and since the $\frac{1}{800}$ limitation is somewhat arbitrary, an equal distribution of live load for deflection computations is not unreasonable provided the diaphragms are more than just shallow spacers.

Design Example 4.2 shows the basic procedure for the design of short-span steel beam bridges. To utilize the full strength of the compression flange, it is necessary to embed the compression flange in the slab concrete to give it full lateral support. This is easily done by the proper forming of the deck slab. The general dimensions of the bridge are the same as that for Design Example 4.1.

In Design Example 4.2, a method of distributing the dead load to the exterior girder different from that of AASHTO is used. In this method, the dead load on the exterior beam is the reaction on the exterior beam of the slab, curb, and railing, assuming the slab acts as a simple beam between the exterior and the adjacent beam. Moments are then taken about the center line of the adjacent beam to determine the reaction on the exterior beam. This method is common practice among bridge engineers. AASHTO Specifications state that if the diaphragms are sufficiently stiff, all dead load placed after the slab has hardened can be equally distributed to all longitudinal beams. Since diaphragms in short-span steel beams usually consist of shallow channels, this equal distribution of dead load is not often used.

The wheel-load distribution for moment, for the exterior steel beams, is different than for concrete beams. For four or more steel beams supporting a concrete floor slab, the wheel load to an outside beam is $S/5.5$ (for $S = 6$ ft or less) or $S/4.0 + 0.25S$ (for $6$ ft $< S < 14$ ft) where $S$ is the beam spacing from the outside beam to the adjacent beam.

For the end-shear computations of the interior beam, no lateral distribution factor of the wheel loads adjacent to the end of the beam is considered. The slab is considered as a simple beam between supporting steel beams for determination of the end-beam reactions. However, this procedure is not followed in the case of the exterior beam since it gives considerably less live load to the exterior beam than the procedure used in computing the bending moment. Therefore distribution factor for moment is used for the shear determination for the exterior beam.

After the bending moments for the exterior beam are calculated, the same quantitites are determined for the interior beams. The dead load carried by an interior beam is the width of slab (and any wearing surface on this amount of slab) equal to the center-to-center spacing of longitudinal beams. The estimated beam weight is also included.

The proportion of each wheel load (both front and rear) to be applied to any interior beam is given in Section 1.3.1 of the Specifications. For a concrete slab on steel beams, the fraction is $S/5.5$, where $S$ is the center-to-center spacing of beams. This distribution factor is the same for any interior beam.

After the bending moments are calculated, the required section modulus is determined. The Specifications state that "the size of the exterior beam shall not be less than the interior beam." In Design Example 4.2, the exterior beam is 8 lb/ft heavier because the dead load on the exterior beam is almost twice the dead load of an interior beam. If the dead load of curb, railing, and so forth were equally distributed to all beams, then the exterior and interior beams would probably be of the same size. If the bridge consisted only of one span, then beams of equal size most likely would be used.

The transverse slope of the deck can be achieved by building a variable-depth haunch in concrete over the beams and/or variable-depth steps on the piers and abutments to give a different elevation to each beam.

After the beams are selected, they are checked for deflection. In the short-span beam, the heavy rear axle is placed at midspan to produce the deflection. The conjugate beam method is used to calculate the deflection. The deflection caused by the front wheel is very small. If it is neglected, the deflection can simply be calculated using the equation

$$LL\ \Delta = \frac{PL^3}{48EI} = \frac{19.7(30)^3(12)^3}{48(30)(10)^3(2096)} = 0.305 \text{ in.}$$

The live load plus impact deflection must be less than $\frac{1}{800}$ of the span.

The end shear is calculated next. Only very rarely do shear stresses dictate the size of the steel beams. The design example shows that the web shear stress is very low. This value is an average web stress determined by dividing the end shear by the depth of beam multiplied by web thickness. It is compared against an allowable shear stress of 12 ksi for A36 steel.

For beams with thin webs and heavy wheel concentrations, it may be necessary to add web stiffeners at the bearings. The design of the bridge bearings is not covered in this chapter, but is treated in Chapter 10.

Spacing of diaphragms should not exceed 25 ft and one should be located at each end of the span. These end diaphragms should support the edge of the slab or the slab should be thickened at the end of the span to support the concentrated wheel load.

### Design Example 4.2

STEEL BEAM BRIDGE—NONCOMPOSITE

*Given:*

1. Concrete slab: as shown in Fig. 4.5.
2. Span: 30-ft center-to-center of bearings.
3. Load: two lanes of H20 traffic.
4. Diaphragms: C15x 33.9 ends and midspan.
5. Cross section: with sidewalk, in Fig. 4.5 as shown.
6. Beams: A36 steel

EXTERIOR BEAM

Distributed dead loads

$$\text{Parapet: } 0.75 \times 0.15 \times \frac{8.83}{6.75} \qquad\qquad 0.15 \text{ k/ft}$$

$$\text{Railing: } 0.015 \times \frac{8.83}{6.75} \qquad\qquad 0.02$$

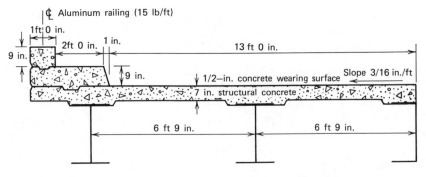

**Figure 4.5** Noncomposite steel beam bridge.

Curb and sidewalk: $(3.04 \times 0.75)0.15 \times \dfrac{7.8}{6.75}$      0.40

Slab: $(0.625 \times 9.33)0.15 \times \dfrac{4.66}{6.75}$      0.60

Estimated weight of beam      $\underline{0.10}$

     1.27 k/ft

Sidewalk live load: $2 \times 0.06 \times \dfrac{7.33}{6.75}$      0.13 k/ft

Concentrated dead load (diaphragm): $0.034 \times \dfrac{6.75}{2}$      0.12 k at

     midspan

Dead-load moment: $\dfrac{1.40(30)^2}{8} + \dfrac{0.12(30)}{4}$      158.4 ft-k

Wheel loads/beam: $\dfrac{4.25}{6.75} = 0.63$ or $\dfrac{6.75}{4.0 + 0.25(6.75)}$      1.19 (use)

Live-load moment: $\dfrac{20(13.6)^2}{30} \times 1.19$      146.7 ft-k

Impact moment: $146.7 \times 0.3$      44.0 ft-k

     Total moment: 349.1 ft-k

INTERIOR BEAM

Distributed dead loads

     Slab: $(0.625 \times 6.75)0.15$      0.63 k/ft
     Weight of beam      $\underline{0.08}$

     0.71 k/ft

Concentrated dead load (diaphragm): $0.034 \times 6.75$      0.23 k

Dead-load moment: $\dfrac{0.71(30)^2}{8} + \dfrac{0.23(30)}{4}$      81.6 ft-k

Wheel loads per beam: $\dfrac{S}{5.5} = \dfrac{6.75}{5.5} = 1.23$

Live-load moment: $\dfrac{20(13.6)^2}{30} \times 1.23$      151.7 ft-k

     Impact moment: $151.7 \times 0.3$      45.5 ft-k

     Total moment: 278.8 ft-k

DESIGN OF BEAMS

Allowable bending stress is 20 ksi since compression flange is fully restrained. *The section moduli are:*

Exterior beam required section modulus: $(349.1 \times 12/20) = 209.5$

$W27 \times 84$ section modulus: 211.7 in.[3]

Interior beam required section modulus: $(278.8 \times 12/20) = 167.5$ in.[3]

$W24 \times 76$ section modulus: 175.4 in.[3]

A check of deflection can be made for the $W24 \times 76$ interior beam by placing the rear axle at midspan (Fig. 4.6). Then

$$R_L = \frac{150 \times 7.5 \times 20 + 150 \times 7 \times 10.33 + 14.6(0.5 \times 0.67 + 7 \times 5.67)}{30}$$

$$= 1131 \text{ k-ft}^2$$

$$EI\Delta = 1131 \times 15 - 150 \times 7.5 \times 5 = 11{,}315 \text{ ft}^3\text{-k}$$

$$LL\Delta = \frac{11{,}315 \times 12^3}{30 \times 10^3 \times 2096} = 0.31 \text{ in.}$$

$$\text{total } LL + I \text{ deflection} = 1.3 \times 0.31 = 0.40 \text{ in.}$$

$$\text{allowable } \Delta = \frac{30 \times 12}{800} = 0.45 \text{ in.} > 0.40$$

In accordance with the AASHTO Specification, the live load should be equally distributed to all beams, however the diaphragms $(15C \times 33.9)$ are too shallow to achieve a nearly equal distribution. Nevertheless, the $W24 \times 76$ beam would meet deflection limitation.

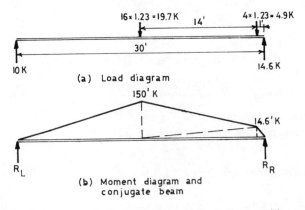

**Figure 4.6** Live loading for deflection evaluation. (a) load diagram; (b) moment diagram and conjugate beam.

END SHEAR

Exterior beam

Dead loads: $1.27 \times 15$                                    19.0 k

Live load: $\left(16 + 4 \times \dfrac{16}{30}\right)1.19$              21.6

Impact: $21.6 \times 0.3$                                    $\dfrac{6.5}{47.1 \, k}$

Interior beam

Dead load: $0.71 \times 15 + 0.1$                   10.8 k

Live load: $16 \times 1.52 + 4 \times 1.23 \times \dfrac{16}{30}$       26.9

Impact: $26.9 \times 0.3$                                    $\dfrac{8.1}{45.8 \, k}$

Average web shear

Exterior beam: $\dfrac{47.1}{26.7 \times 0.463} = 3.8 \text{ ksi} < 12$

Interior beam: $\dfrac{45.8}{23.9 \times 0.44} = 4.4 \text{ ksi} < 12$

Because of the low web shear stress, no stiffeners are required. The length of bearing should be such that web crippling does not occur.

## 4.4  DESIGN OF STEEL BEAM BRIDGES BY LOAD FACTOR METHOD

The load factor method, as an alternate design approach was first introduced into the AASHTO Specifications in the eleventh edition (1973) and a few minor changes were made in the later Interim Specifications. It is a method for designing structural members for multiples of the design loads.

The rationale behind this new procedure is simply that the factor of safety for dead loads need not be as large as the factor of safety for live load and impact. The dead-load stresses can be predicted much more accurately than the maximum live load plus impact stresses. On short-span bridges the live load plus impact stresses are much greater than the dead-load stresses, while for long-span bridges the opposite is true. Economy can be achieved in long-span bridges by this new approach, while the likelihood of a large overstress in short-span bridges is very much lessened.

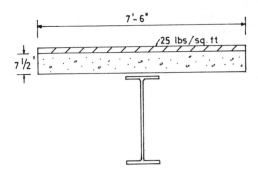

**Figure 4.7** Typical noncomposite cross section.

In this method the analysis is based on the elastic behavior of the structure. It is not a plastic design approach, using mechanism action through the development of plastic hinges. The design strength for steel is set equal to the yield strength ($F_y$) of the steel as given in the Specifications. The value of $Mu$ is set equal to $F_y$ times $Z_x$, the plastic section modulus.

An illustration of this method is given in Design Example 4.3. This design is for a short-span bridge and does not show the magnitude of economy achieved in the Load Factor Method; however, it does show the procedure to be used for a bridge of any length.

The slab and beam bridge have the longitudinal beams designed by the Group I loading. This loading says that the maximum moments, shear, and forces are determined by the formula

$$1.3[D + \tfrac{5}{3}(L + I)]$$

which indicates that on the dead load there is a 1.3 factor of safety and on the live load plus impact there is a factor of safety of 2.17. These values are in contrast to a factor of safety of 1.8 on both dead and live plus impact loads in the allowable stress design method.

A 60-ft simple beam span is chosen as the bridge for Design Example 4.3. A noncomposite design is used because Design Example 4.4 covers the composite design. In actual practice it is quite likely that a 60 ft beam would be more economical as a composite beam. Only the design of an interior beam is described. A typical slab is used (Fig. 4.7).

**Design Example 4.3**

*Given:*
1. Span: 60 ft center-to-center of bearings.
2. Load: HS20.

3. Material: A36 Steel.
4. Slab: $7\frac{1}{2}$ in. total
5. Wearing surface: 25 lb/ft$^2$.
6. Beam spacing: 7 ft 6 in.

Dead load

| | |
|---|---:|
| Wearing surface: 7.5(0.025) | 0.19 k/ft |
| Slab: 7.5/12(7.5)(0.150) | 0.70 |
| Beam (estimated) | 0.16 |
| | 1.05 k/ft |

Dead-load bending moment: $\dfrac{wl^2}{8} = \dfrac{1.05(60)^2}{8}$     473.0 ft-k

Wheel-load distribution: $S = \dfrac{7.5}{5.5} = 1.36$ wheel loads

Live load bending moment (Fig. 4.8):

$$1.36\left[\frac{36(27.67)^2}{60} - 4(14)\right] = \qquad 548.0 \text{ ft-k}$$

Impact: $\dfrac{50}{125+60} = 0.27$

Live load + impact bending moment: 1.27(548)     696.0 ft-k

Total load factor bending moment: $1.3[473 + \frac{5}{3}(696)]$     2122 ft-k

Required plastic section modulus:

$$Z_x = \frac{M_u}{F_y} = \frac{2122(12)}{36} = 708 \text{ in.}^3$$

$$W36 \times 182, \qquad Z = 718 \text{ in.}^3 > 708 \text{ in.}^3$$

The values of plastic section modulus $Z_x$ are tabulated in the *American Institute of Steel Construction Manual of Steel Construction*. Comparing

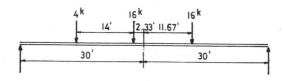

**Figure 4.8** Live loading for maximum moment.

with allowable stress method:

$$\text{total bending moment} = 473 + 696 = 1169 \text{ ft-k}$$

$$\text{required section modulus} = \frac{1169(12)}{20} = 702 \text{ in.}^3$$

The $S$ of 702 requires a larger section than that required by the load factor method. A value of W33 × 220 ($S\hat{x} = 742$) would be necessary. For this structure the resulting beam size is only slightly different by the two methods, because the dead load and live load plus impact are nearly equal. A greater difference between the two methods results when the span is longer.

In a complete design the shear is checked by the load factor method and the live load plus impact deflection under the working loads is calculated and compared to the $L/800$ limit.

## 4.5  DESIGN OF COMPOSITE STEEL BEAM BRIDGES

A considerable saving of steel can be realized in a stringer-type beam bridge if the concrete slab is used to aid the steel beam in resisting the longitudinal bending moment in the beam. This joint action of slab and beam takes place if the horizontal shear between the beam and slab can be resisted by the composite concrete–steel beam. The friction between the two materials resists some of the shear but before the ultimate strength of the steel beam is reached slippage would occur. Mechanical devices of sufficient shear capacity to develop the ultimate strength of concrete and steel beam acting as an integral unit are needed.

Three different style shear connectors have been used in the past. A small section of steel channel with its longitudinal axis transverse to the axis of the beam and welded to the top flange has been used. A reinforcing steel rod bent in the form of a circular spiral and laid along the top flange of the beam and welded to it at the contact points has also been used.

The most generally used shear connector is a specially made shear stud resembling a bolt without threads and with a round head. It is attached to the compression flange by an automatic welding process. Figure 4.9 shows beams with attached shear studs. This type of shear connector is more quickly welded than the other types mentioned. An automatic welding gun in which the studs are inserted and held against the flange of a steel beam is used. A pull of the trigger closes the circuit, thus supplying the necessary voltage to melt the weld material in the end of the stem. It takes approximately 1 sec to weld the stud.

**Figure 4.9** Attaching shear studs to steel beam.

A shear connector must be of such a design that the slab is held down on the beam flange so that no separation of the two materials can occur.

The design of the concrete deck slab in a composite beam bridge is no different from that for a noncomposite steel beam bridge. In Design Example 4.4 the deck slab is not designed, since the procedure would be the same as that in Design Example 4.1.

Composite beam bridges can be constructed by two methods. One method would be to place shoring under the steel beams before the concrete is placed in the deck forms. The shoring must be of sufficient strength so that there is insignificant deflection of the steel beams when the wet deck concrete is placed. The second method of construction does not use shoring but lets the steel beams deflect and support the wet concrete by bending in the beams. Which method is used dictates the design procedure.

When shoring is used, the composite beam resists all the dead-load moments, as well as live-load and impact moments. When no shoring is used, the steel beam section alone (no composite action) is required to carry the dead load consisting of beam weight and weight of the wet concrete slab. Additional dead loads, such as curb, parapet, railing, and asphalt wearing surface, if placed after the deck has hardened, are carried by composite beam action.

The required amount of steel in the beams is less when the bridge is constructed with beam shoring than when it is built without shoring. However, the saving in steel, in the majority of bridges, is less than the cost of supplying and erecting the shoring. Both types of construction should be investigated for the least cost.

Design Example 4.4 is based on construction without shoring. The procedure for the design of the beams is to first determine the dead-load bending moments for those loads that the beam will carry without composite action. The dead-load bending moment with composite action is determined next followed by the live-load and impact bending moments.

The effective width of the slab that is to be part of the composite section is determined. This effective width is the least value of

1. One-fourth the span.
2. Center-to-center distance of beams.
3. Twelve times the least thickness of slab.

The stresses in the composite section are determined from the standard $Mc/I$ equation in which the value of $I$ is determined on the basis of the transformed moment of inertia. The transformed moment of inertia is calculated by transforming the effective concrete slab area into an equivalent steel area. This transformation is accomplished by dividing the effective width of slab by the value of the ratio of the modulus of elasticity of the beam steel to the modulus of elasticity of the slab concrete. This ratio $E_s/E_c$ is given the symbol $n$. The modulus of elasticity of the steel has a specific value (29,000,000 psi), but this is not so for the concrete. If a compressive test were conducted on a concrete cylinder and a strain measurement were taken, the value of strain for a given stress would be time dependent. This increase in strain with time is called creep and is characteristic of all concretes. The creep results in a lower value of modulus of elasticity for concrete subject to long-time loading. Dead loads placed on the bridge after the deck slab has hardened (in the case of nonshoring), or all dead loads other than steel beam weight, in the case of shored construction are considered to be long-time loads. Therefore the value of $n$ for these long-time dead loads is greater than the $n$ for short-time loading such as live load and impact. The AASHTO Specification gives the value to be used for $n$ when the loading is of short duration.

This value is a function of the 28-day compressive strength of the concrete. For the long-time dead loads the value of $n$ to be used is three times the specified value for short-time loading.

If only a rolled steel beam with the nominal slab thickness of 7+ in. is used as composite, the centroidal axis of the composite section is just below the steel compression flange. The moment of inertia is increased but the distance to the extreme fiber in tension is also increased and thus the value of the section modulus ($S$) is increased by only a small amount (nowhere near the percentage increase of $I$). To have a more efficient steel section the tension flange must be larger than the steel compression flange, thus keeping the centroidal axis of the composite section near the mid-depth of the steel beam. In welded girders of three plates the bottom plate can be made wider and thicker than the compression flange plate. In using a rolled steel beam, a cover plate should be welded to the tension flange for best economy of steel.

Economic studies have shown that for best economy the lightest and shallowest beam with the largest cover plate possible gives the best design. The Specifications limit the thickness of the cover plate to two times the flange thickness. The cover plate can be wider or narrower than the flange plate. Some designers prefer a plate narrower than the flange. The ending of a narrower plate is simpler; transverse welds at the ends of the plate are not required. This design is preferable from a standpoint of fatigue strength of the steel rolled section at this location. Some cover-plated beams have actually failed in fracture of the tension flange at the end of the cover plate.

If the cover plate is narrower than the flange there should be sufficient room (equal to size of fillet weld plus $\frac{1}{8}$ in.) on each side of the plate to lay a fillet weld. If the plate is wider, it should be at least 2 in. wider than the flange.

The Specifications state that "any partial length cover-plate shall extend beyond the theoretical end by the terminal distance, or it shall extend to a section where the stress in the beam flange is equal to the allowable fatigue stress for 'Base Metal adjacent to or connected by fillet welds,' whichever is greater." The terminal distance is two times the cover-plate width for plates not welded across their ends, and $1\frac{1}{2}$ times for plates welded across their ends. The designer should observe other phases of the Specifications relating to cover-plated beams.

After the beam section is determined, the next step is to design the shear studs. The number and spacing of the studs is dependent on the magnitude of the horizontal shear on the plane between the concrete slab and the steel beam.

The shear connectors are designed for two loading criteria; fatigue and ultimate strength. First, the connectors are designed on the basis of fatigue strength. Since the stress cycle would be caused by live load and impact only, the shear in the beam due to live load plus impact is calculated for approximately five locations along the beam between the support and midspan. The horizontal shear ($S$) per linear inch of beam is determined from

$$S_r = \frac{V_r Q}{I}$$

where     $V_r$ = range of shear due to live loads plus impact. Range of shear is the sum of the positive and negative shear due to live load plus impact

$Q$ = statical moment, about the centroidal axis, of the composite section of the transformed area of the slab, or the area of steel reinforcement embedded in the concrete for negative moment

$I$ = moment of inertia of the transformed area of the beam or girder

Since the linear shear per inch times the spacing of the shear connectors at any position along the beam must be resisted by the shear connectors in each row, the following equation can be written:

$$Z_r(n) = S_r(s)$$

where     $Z_r$ = fatigue shear strength of one shear connector

$n$ = number of shear connectors (in the case of studs) in a transverse row on the beams

$s$ = maximum spacing of shear connectors at any point along the beam corresponding to the respective value of $S_r$

The value of $Z_r$ for channel and stud shear connectors is given in the AASHTO Bridge Design Specification. The value of $Z_r$ depends on the number of cycles of maximum stress the connector is expected to receive during its lifetime. Values are given for $1 \times 10^5$, $5 \times 10^5$, and $2 \times 10^6$ cycles. In a bridge with a 50-year life this corresponds to 5.5, 27.5, and 110 maximum stress cycles per day. The designer must estimate the probable number of stress cycles per day.

The maximum spacing of the connectors is then

$$s = \frac{Z_r n}{S_r} = \frac{Z_r n I}{V_r Q}$$

The spacing at several positions along the length of the beam is calculated and then the details of spacing can be determined using four to five different spacings along the beam span. Stud connectors with diameters of $\frac{3}{4}$, $\frac{7}{8}$, or 1 in. are commonly used. Studs are available in increments of diameter of $\frac{1}{8}$ in. and lengths of $\frac{1}{2}$ in. The larger the stud the greater the spacing. Spacing of studs along the beam should not exceed 24 in. and minimum spacing should not be less than about 5 in. for good design.

The clear depth of concrete over the tops of the shear connectors should not be less than 2 in. and the connector must extend at least 2 in. into the slab. The minimum spacing of stud connectors in a transverse row is $1\frac{7}{8}$ inches. The height-to-diameter ratio of a stud connector should not be less than 4.

Design Examples 4.4 and 4.5 show the detailed design computations for a composite stringer-type bridge. Design Example 4.4 is for a bridge in which no shoring is used and 4.5 is the same design when the steel beam is supported by shoring during construction. Only an interior beam is designed. The design of an exterior beam would be similar. Where there is not sufficient width of slab to meet the effective slab width criterion, only the available slab width is used in determining the section properties of the exterior composite beam.

Two methods have been used in the selection of the spacing (pitch) of shear connectors. In method 1 the average range of shear over the entire length of beam is used, as well as the average $I/Q$. The spacing then determined is used throughout the length of the beam.

In method 2 the spacing is determined on the basis of the shear requirement at several locations along the beam. This, of course, gives closer spacing at the end, where the range of shear is greatest, and greater spacing near the middle of the beam. From a plot of required spacing, (Fig. 4.15) the detail spacing can be determined along the length of the beam so that at no location is the actual spacing greater than the minimum spacing shown by the curve.

The Specifications require that the ultimate shear strength of the total shear connectors be equal to or greater than the ultimate compressive strength of the concrete slab or the ultimate tensile strength of the steel beam, whichever is smaller. At ultimate load it is considered that the slab has an average compressive stress of $0.85f'_c$ and the steel beam has a uniform stress equal to $F_y$.

Many bridge design organizations have computer programs that select various configurations of rolled beams and cover plates that carry the computed bending moments without overstress. Such programs can save considerable time. It is expected that a bridge designer be capable of doing hand calculations.

The required length of cover plate is determined by first obtaining the bending moments at several locations inward from the bearing. Stresses are then calculated. In Design Example 4.4, bending moments and stresses are calculated for distances 5, 10, 15, and 20 ft from the bearing. It is seen that somewhere between 10 and 15 ft from the support the maximum bending stress exceeds the limit of 20 ksi. By straight-line interpolation, the point having a stress of 20 ksi without a cover plate is determined. This is the theoretical point of cutoff of the cover plate. To this length must be added the development length at each end. For cover plates not welded across the ends the minimum development length is twice the width of cover plate.

## Design Example 4.4

STEEL BEAM BRIDGE—COMPOSITE

In this example the design of an interior beam, rather than the complete structure, is considered.

*Given:*

1. Span 57 ft center-to-center of bearings.
2. Structural slab: $7\frac{1}{4}$ in.
3. Beam spacing: 7 ft (five beams).
4. Loading: HS20 no sidewalk live load.
5. Beams: A36 steel.
6. Bridge constructed without shoring.
7. An allowance for future paving of 25 psf.
8. Haunch: 1 in. over the beams.
9. All other data as in Design Example 4.2.

Dead load A (no composite action)

Slab: $\dfrac{7.25}{12} \times 7.0 \times 0.15$ 　　　　　　　　0.634 k/ft

Haunch: $\dfrac{1}{12} \times 2.0 \times 0.15$ 　　　　　　　　0.025

Steel: beam, diaphragm (assumed) 　　　$\underline{0.150}$
　　　　　　　　　　　　　　　　　　　　　0.809 k/ft

Dead load B (composite action; equally distributed to all other beams)

Curb, parapet, and railing 　　　　　　　0.23 k/ft

Asphalt paving: $\dfrac{0.025 \times 26}{5}$      $\dfrac{0.13}{0.36 \text{ k/ft}}$

Dead-load moments A: $\dfrac{0.809(57)^2}{8}$      329 ft-k

Dead-load moments B: $\dfrac{0.36(57)^2}{8}$      146 ft-k

Wheel loads per beam: $\dfrac{7.0}{5.5} = 1.27$

Maximum live load moment (Fig. 4.10):

$$\left[\frac{36}{57}(26.17)^2 - 56\right] 1.27 = 480 \text{ ft-k}$$

Impact: $\dfrac{50}{125 + 57} = 0.275$

Impact moment: $0.275 \times 480 = 132$ ft-k

Total live load + impact moment: 612 ft-k.

Effective flange width for composite action; $\frac{1}{4} \times 57 = 14.2$ ft

Distance center-to-center of beams: 7.0 ft (minimum)

$$12 \times \frac{7.25}{12} = 7.25 \text{ ft}$$

Transformed width of flange: $\dfrac{7.0 \times 12}{10} = 8.4$ in. $(n = 10)$

$$\frac{7.0 \times 12}{30} = 2.8 \text{ in. } (n = 30)$$

A beam of $W33 \times 118$ with a $9\frac{1}{2}$ in. $\times \frac{3}{4}$ in. plate is used (Fig. 4.11) with an effective flange width of 8.4 in. for $n = 10$ and 2.8 in for $n = 30$.

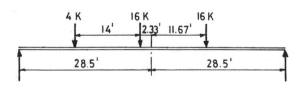

**Figure 4.10** Live loading for maximum moment.

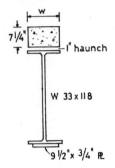

9 1/2"x 3/4" PL    **Figure 4.11**   Typical cross section.

| | $A$ | $Y$ | $AY$ | $AY^2$ | $I_0$ |
|---|---|---|---|---|---|
| $WF =$ | 34.71 | −16.43 | −570 | 9370 | 5886 |
| $PL =$ | 7.125 | −33.23 | −237 | 7875 | — |

$\sum = 41.83$ in.$^2$          −807   17,245   5886

$$\bar{y} = -19.29$$
$$I = 23,130 - 15,570 = 7560 \text{ in.}^4$$

With slab at $n = 30$:

| | $A$ | $Y$ | $AY$ | $AY^2$ | $I_0$ |
|---|---|---|---|---|---|
| slab = | 20.30 | +4.625 | +94 | 434 | 89 |

$\sum = 62.13$          −713   17,679   5975

$$\bar{y} = -11.48$$
$$I = 23,655 - 8185 = 15,470 \text{ in.}^4$$

With slab at $n = 10$:

| | $A$ | $Y$ | $AY$ | $AY^2$ | $I_0$ |
|---|---|---|---|---|---|
| slab = | 60.90 | +4.625 | +282 | 1303 | 267 |

$\sum = 102.73$          −525   18,548   6153

$$\bar{y} = -5.11 \text{ in.}$$
$$I = 24,700 - 2680 = 22,020 \text{ in}^4.$$

The weight of steel with the plate is 142 lb/ft, so the estimate of 150 lb/ft is adequate. The stress values are calculated next.

Steel stress (maximum in bottom fiber) (tension)

$$f_t = \frac{329 \times 12 \times 14.32}{7,560} = 7.48 \text{ ksi (dead load A)}$$

$$= \frac{146 \times 12 \times 22.13}{15,470} = 2.51 \text{ ksi (dead load B)}$$

$$= \frac{612 \times 12 \times 28.50}{22,020} = \underline{9.51 \text{ ksi}} \text{ (live load + impact)}$$

$$19.50 < 20 \text{ ksi (allowable)}$$

Steel stress (compression)

$$f_c = \frac{7.48 \times 19.29}{14.32} = 10.08 \text{ ksi (dead load A)}$$

$$= \frac{2.51 \times 11.48}{22.13} = 1.30 \text{ ksi (dead load B)}$$

$$= \frac{9.51 \times 5.11}{28.50} = \underline{1.71 \text{ ksi}} \text{ (live load + impact)}$$

$$13.09 \text{ ksi} < 20 \text{ ksi (restrained)}$$

Concrete stress (compression)

$$f_c = \frac{146 \times 12 \times 19.73}{15,470 \times 30} = 0.074 \text{ ksi (dead load B)}$$

$$= \frac{612 \times 12 \times 13.36}{22,020 \times 10} = \underline{0.446 \text{ ksi}} \text{ (live load + impact)}$$

$$0.520 < 1.2 \text{ ksi}$$

SHEAR

Shear in beam
  Dead load A: 0.81 k/ft (no composite action)
  Dead load B: 0.36 k/ft (composite action)
  Live-load wheel distribution:

$$1 + \frac{1.0}{7.0} + \frac{3.0}{7.0} = 1.57 \text{ (rear axle)} = \frac{7}{5.5} = 1.27 \text{ (other axles)}$$

Shear at end of beam

Dead load: $V_{(A)} = 0.81 \times 28.5 = 23.1 \text{ k}$ (for total reaction only)
Live load: $V_{(B)} = 0.36 \times 28.5 = 10.3 \text{ k}$

Live load: $V = 1.57 \times 16 + 1.27\left(16 \times \dfrac{43}{57} + 4 \times \dfrac{29}{57}\right) = 43.1\,\text{k}$

Impact: $V = 0.275 \times 43.1 = 11.9\,\text{k}$

Range of shear (live load + impact): $43.1 + 11.9 = 55.0\,\text{k}$

Range of shear at 6 ft from end of beam (Figs. 4.12 and 4.13)

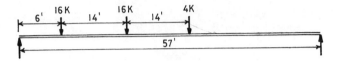

**Figure 4.12**  Live loading for shear at 6 ft—positive.

Positive live load $V = 1.27\left(16 \times \dfrac{51}{57} + 16 \times \dfrac{37}{57} + 4 \times \dfrac{23}{57}\right)$     33.5 k

Impact $V = 0.284 \times 33.5$     $\dfrac{9.5}{43\,\text{k}}$

Negative live load $V = 1.27\left(4 \times \dfrac{51}{57} - 4\right)$     −0.5 k

Impact $V = 0.3 \times (-0.5)$     $\dfrac{-0.2}{-0.7\,\text{k}}$

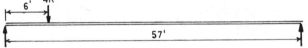

**Figure 4.13**  Live loading for shear at 6 ft—negative.

Range of $V = 43 - (-0.7) = 43.7\,\text{k}$

Range of shear at 12 ft from end of beam (Fig. 4.14)

Positive live load $V = 1.27\left(16 \times \dfrac{45}{57} + 16 \times \dfrac{31}{57} + 4 \times \dfrac{17}{57}\right)$     28.7 k

Impact $V = 0.294 \times 28.7$     $\dfrac{8.4}{37.1\,\text{k}}$

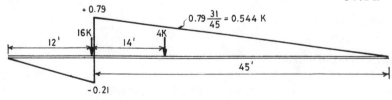

**Figure 4.14**  Live loading for shear at 12 ft.

Negative live load $V = 1.27[16(-0.21)+4(0.544)]$       $-1.5$ k

Impact $V = 0.3(-1.5)$       $\dfrac{-0.5}{-2.0 \text{ k}}$

$$\text{Range of } V = 37.1 - (-2.0) = 39.1 \text{ k}$$

Range of shear at midspan of beam

Positive live load: $V = 1.27\left(16 \times \dfrac{1}{2} + 16 \times \dfrac{14.5}{57} + 4 \times \dfrac{0.5}{57}\right)$      $15.4$ k

Impact: $V = 0.3 \times 15.4$       $\dfrac{4.6}{20.0 \text{ k}}$

Negative live load: $V = 1.27\left(16 \times \tfrac{1}{2} + 16 \times \dfrac{42.5}{57} + 4 \times \dfrac{14.5}{57} - 32\right) - 14.0$ k

Impact: $V = 0.3(-14.0)$       $\dfrac{-4.2}{-18.2 \text{ k}}$

Range of $V = 20.0 - (-18.2) = 38.2$ k

*Design of Stud Shear Connectors—Fatigue.* Use 5 in. $\times \tfrac{7}{8}$ in. shear studs, $h/d > 4.0$. Then

$$Z_r = \alpha d^2; \qquad \alpha = 10{,}600 \quad \text{for} \quad 500{,}000 \text{ cycles}$$
$$Z_r = 10{,}600(\tfrac{7}{8})^2 = 8200 \text{ lb}$$

At end of beam, where shear is maximum, there is no cover plate. The section properties at this location are:

$$Y = -3.0 \text{ in. (from top of wide flange section)}$$
$$I = 15{,}960 \text{ in.}^4 \qquad Q = 60.9(4.6 + 3.0) = 465 \text{ in.}^3$$

Where cover plate is present:

$$Q = 60.9(4.6 + 5.1) = 592 \text{ in.}^3$$

Shear per inch is given by

$$\frac{V_r Q}{I} = S_r$$

$$n Z_r = S_r \times (\text{spacing})$$

Using $n = 2$ studs per row,

$$\text{spacing} = \frac{2 Z_r}{S_r} = \frac{2 Z_r I}{V_r Q} = \frac{16.4 I}{V_r Q} = \frac{16.4}{S_r}$$

| Distance from End (ft) | $V_r$ (k) | $I$ (in.⁴) | $Q$ (in.³) | $S_r$ (k/in.) | Spacing (in.) |
|---|---|---|---|---|---|
| 0 | 55 | 15,960 | 465 | 1.60 | 10.2 |
| 6 | 43.7 | 15,960 | 465 | 1.27 | 12.9 |
| 12 | 39.1 | 22,020 | 592 | 1.05 | 15.6 |
| ⊄ | 38.2 | 22,020 | 592 | 1.03 | 15.9 |

Method 1 involves equal spacing of shear connectors throughout length of beam.

$$\text{Average } V_r = (55 + 38.2)/2 = 47 \text{ k}$$

$$\text{Average } I/Q = (15{,}960/465 + 22{,}020/592)^{\frac{1}{2}} = 35.7 \text{ in.}$$

$$\text{Spacing} = 16.4(35.7)/47 = 12.5 \text{ in.}$$

Using $12\frac{1}{2}$ in. spacing

$$\text{rows} = (57 \times 12)/12.5 + 1 = 56$$

$$\text{number of studs} = 56 \times 2 = 112$$

Method 2 involves variable spacing using above table (see Fig. 4.15).

*Ultimate Strength—Shear Connectors.* The total minimum number of connectors $(N)$ is given by

$$N = \frac{P}{\phi S_u}$$

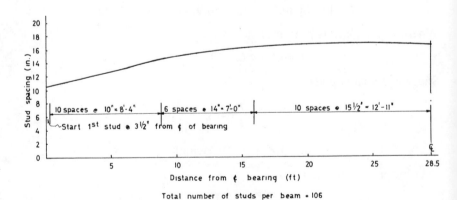

Total number of studs per beam = 106

**Figure 4.15**  Shear stud locations.

where $\phi$ = reduction factor = 0.85

$\quad P = P_1$ or $P_2$ (whichever is smaller)

$\quad P_1 = A_s F_y$

$\qquad = 41.83 \times 36 = 1506\ k$

$\quad P_2 = 0.85 f'_c bc$

$\qquad = 0.85 \times 3 \times 84 \times 7.25 = 1553\ k$

$\quad P = 1506\ k$

$\quad S_u = 930 d^2 (f'_c)^{1/2}$

$\qquad = 930(\tfrac{7}{8})^2 \sqrt{3000} = 39{,}000\ lb$

$\quad N = 1506/0.85 \times 39 = 45.2$ (say 46 studs)

The toal number of studs between support and center line as required for fatigue is $106 > 2 \times 46 = 92$. Use studs as required for fatigue.

LENGTH OF COVER PLATE

Stresses at various locations $(x)$ along the beam are now determined. For a simple beam of 57 ft span:

$$DL \text{ moment} = 28.5 wx - \frac{wx^2}{2}$$

The influence diagram for bending moment is given in Fig. 4.16. At $x = 10$ ft, $y = 8.25$. Then

$$LL \text{ moment} = 16\left(8.25 + 8.25\frac{33}{47}\right) + 4\left(8.25\frac{19}{47}\right)$$

$$= 238 \text{ ft-k}$$

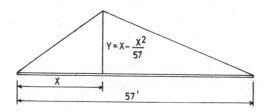

**Figure 4.16**  Bending moment influence diagram.

At $x = 15$ ft, $y = 11.05$. Then

$$LL \text{ moment} = 16\left(11.05 + 11.05\frac{28}{42}\right) + 4\left(11.05\frac{14}{42}\right)$$

$$= 310 \text{ ft-k}$$

At $x = 20$ ft, $y = 12.98$. Then

$$LL \text{ moment} = 16\left(12.98 + 12.98\frac{23}{37}\right) + 4\left(12.98\frac{9}{37}\right)$$

$$= 350 \text{ ft-k}$$

The distribution factor is 1.27.

| $x$ | DL Moment A (ft-k) | DL Moment B (ft-k) | LL Moment (ft-k) | Impact Moment (ft-k) | LL + Impact Moment (ft-k) |
|-----|-----|-----|-----|-----|-----|
| 10 | 190 | 85 | 302 | 83 | 385 |
| 15 | 225 | 113 | 394 | 108 | 502 |
| 20 | 299 | 133 | 445 | 122 | 567 |

The section modulus of the beam without cover plate is given for three cases:

Beam only ($W33 \times 118$): 358 in.$^3$

Beam with slab ($n = 30$): 482 in.$^3$

Beam with slab ($n = 10$): 535 in.$^3$

The bending stresses (maximum tension) without cover plate are given by the table below.

| Distance from Bridge (ft) | DL A (ksi) | DL B (ksi) | LL + Impact (ksi) | Total (ksi) |
|-----|-----|-----|-----|-----|
| 10 | 6.4 | 2.1 | 8.6 | 17.1 |
| 15 | 8.5 | 2.8 | 11.3 | 22.6 |
| 20 | 10.0 | 3.3 | 12.7 | 26.0 |

The theoretical cutoff of cover plate occurs between 10 and 15 ft from the bridge. Straight-line interpolation is sufficiently accurate. Thus

$$d = 10 + \frac{2.9}{5.5}(5) = 12.6 \text{ ft}$$

Therefore

$$\text{length of cover plate} = 57 - 2(12.6) = 31.8$$
$$\text{terminal distance} = 2 \times (\text{cover-plate width})$$
$$= 2(9\tfrac{1}{2}) = 19 \text{ in.}$$
$$\text{total cover-plate length} = 31.8 + 2(1.58) = 34.94, \qquad \text{say 35 ft}$$

It is also necessary to check for fatigue strength at the end of the cover plate. The 1974 Interim Specifications introduced new procedures for this phase of beam design. In this procedure the stress category is determined from the figure in Table (2) of the Specifications. For base metal at the end of partial length welded cover plates the stress category is E. The Specifications show that the allowable range of stress for category E and 500,000 cycles is 12.5 ksi. Since the beam is a simple span, the bending moment is always positive and the stress range is just the live load plus impact stress. The live load plus impact stress at $(57 - 35)/2 = 11$ ft from the support is seen from the above table to be less than 12.5 ksi (only 11.3 at 15 ft). The 35-ft length of cover plate is satisfactory.

WELDING OF COVER PLATE TO ROLLED SECTION

Shear per unit length in fillet weld at the end of the cover plate is

$$f_v = VQ/I$$

The theoretical cutoff point of the cover plate can be taken as 12 ft from the end of the beam. The fillet welds at this location are designed for both total stress and fatigue stress. For total stress, the incremental values of $f_v$ for dead loads A and B and live load plus impact are calculated at 12 ft from the support:

Dead load A: $V = (28.5 - 12)0.81 = 11.7 \text{ k}$

$\qquad Q = 7.125(33.23 - 19.29) = 99.3 \text{ in.}^3 \qquad$ (steel section only)

$\qquad I = 7560 \text{ in.}^4$

$$f_v = \frac{11.7(99.3)}{7560} = 0.15 \text{ k/in.}$$

Dead load B: $V = 16.5(0.36) = 5.2 \text{ k.}$

$$Q = 7.125(33.23 - 11.48) = 155 \text{ in.}^3$$
$$I = 15,470 \text{ in.}^4$$
$$f_v = \frac{5.9(155)}{15,470} = 0.06 \text{ k/in.}$$

Live load plus impact: $V = 37.1$ k (positive)

$$Q = 7.125(33.23 - 5.11) = 200.4 \text{ in.}^3$$

$$I = 22,020 \text{ in.}^4$$

$$f_v = \frac{37.1(200)}{22,020} = 0.34 \text{ k/in.}$$

Total $f_v = 0.55$ k/in.

The minimum size of fillet weld for maximum thickness of $\frac{3}{4}$ in. material is $\frac{1}{4}$ in.

From the 1977 Interim Specifications, the allowable basic shear stress is given by $F_v = 0.27F_u$, where $F_u$ is the tensile strength of the electrode material or the connected material, whichever is lower. For A36 steel, the $F_u$ of selected electrode material should be at least equal to 58 ksi. Then

$$F_v = 0.27(58) = 15.7 \text{ ksi}$$

With a shear on the throat of two $\frac{1}{4}$-in. welds of $0.15 + 0.06 + 0.34 = 0.55$ k/in., the unit stress is

$$f_v = \frac{0.55}{2(\frac{1}{4})(0.707)} = 1.56 \text{ ksi} < 15.7$$

The welds are now checked for fatigue stress. The range of shear at 12 ft. from the support is 39.1 k. Then

$$f_v = \frac{39.1(200)}{22,020} = 0.36 \text{ k/in.}$$

The unit stress is then

$$f_v = \frac{0.36}{2(\frac{1}{4})(0.707)} = 1.0 \text{ ksi}$$

The allowable fatigue stress in a longitudinal fillet weld is 27.5 ksi (category B—500,000 cycles). The $\frac{1}{4}$-in. fillet weld is more than adequate to carry the stress. The development length of cover plate based on weld strength is calculated below.

$$\text{Strength of cover plate} = 9\frac{1}{2}(\frac{3}{4})(20)$$

$$= 142.5 \text{ k}$$

The strength of two $\frac{1}{4}$-in. fillet welds is

$$2(\frac{1}{4})(0.707)(15.7) = 5.55 \text{ k/in.}$$

The length of weld before total stress in cover plate reaches 20 ksi is

$$L = \frac{142.5}{5.5} = 26 \text{ in.}$$

Since one-half of the theoretical length of cover plate is 12.6 ft, the required development length of 26 in. is adequately met by the 12.6 ft length.

## Design Example 4.5

STEEL BEAM BRIDGE—(COMPOSITE WITH SHORING) (see Fig. 4.17)

The conditions are the same as for Design Example 4.4 except shoring is used to support dead load of slab until the concrete deck has obtained strength (usually 7 days).

Dead load A (no composite action) steel beam, diaphragms = 0.140 k/ft.

Dead load B (composite action)

| | |
|---|---|
| Slab | 0.634 k/ft |
| Haunch | 0.025 |
| Curb, parapet, railing | 0.230 |
| Asphalt paving | 0.130 |
| | 1.02 k-ft |

Dead load moments A: $\dfrac{0.14(57)^2}{8} = 56.9 \text{ k ft-k}$

Dead load moments B: $\dfrac{1.02(57)^2}{8} = 414.2 \text{ k ft-k}$

Live load plus impact moments: 612 ft-k (see p. 77)
Effective flange width: 7.0 ft
Transformed width is 3.4 in. for $n = 10$; and 2.8 in for $n = 30$

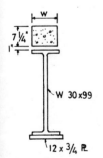

$7\frac{1}{4}$"

1"

W 30×99

12 × 3/4 ℞   **Figure 4.17**   Typical cross section.

|      | $A$  | $Y$    | $AY$  | $AY^2$ | $I_0$ |
|------|------|--------|-------|--------|-------|
| WF   | 29.1 | −14.82 | −431  | 6391   | 4000  |
| PL   | 9.0  | −30.2  | −270  | 8111   | —     |
| $\sum =$ | 38.1 |    | −701  | 14,502 | 4000  |

$$\bar{Y} = -18.4$$

$$I = 18,502 - 12,898 = 5604 \text{ in.}^4$$

With slab at $n = 30$:

|          | $A$  | $Y$   | $AY$ | $AY^2$ | $I_0$ |
|----------|------|-------|------|--------|-------|
| slab =   | 20.3 | +4.62 | 94   | 434    | 89    |
| $\sum =$ | 58.4 |       | −607 | 14,936 | 4089  |

$$\bar{Y} = -10.39$$

$$I = 19,025 - 6310 = 12,715 \text{ in.}^4$$

With slab at $n = 10$:

|          | $A$  | $Y$   | $AY$ | $AY^2$ | $I_0$ |
|----------|------|-------|------|--------|-------|
| slab =   | 60.9 | +4.62 | 282  | 1303   | 267   |
| $\sum =$ | 99.0 |       | −419 | 15,805 | 4267  |

$$\bar{Y} = -4.23$$

$$I = 20,072 - 1772 = 18,300 \text{ in.}^4$$

Steel stress (maximum in bottom fiber—tension): is given by:

$$f_t = \frac{56.9 \times 12 \times 11.99}{5,604} = 1.46 \text{ ksi (dead load A)}$$

$$f_t = \frac{414.2 \times 12 \times 20.0}{12,715} = 7.82 \text{ ksi (dead load B)}$$

$$f_t = \frac{612 \times 12 \times 26.16}{18,300} = \underline{10.50 \text{ ksi}} \text{ (live-load impact)}$$

$$19.78 \text{ ksi} < 20.0$$

The saving in weight of steel when shoring is used is

$$(118 + 24.2) - (99 + 30.6) = 12.6 \text{ lb/ft}$$

**REFERENCE**

1 American Institute of Steel Construction, *Bridge Fatigue Guide*, 1977.

# 5

# Continuous Beam and Plate Girder Bridges

## 5.1 INTRODUCTION

When the chasm to be bridged is sufficiently long to require multiple spans, the bridge designer can either select a series of simple spans or he can design a bridge that is continuous over the piers. As expected, there are advantages and disadvantages to both schemes.

The advantages of simple beam spans as opposed to continuous spans are:

1. Less engineering (statically determinate) is needed.
2. Generally no field splices are required, thus faster field erection is possible.
3. Support settlements do not increase stresses in beams.
4. Expansion devices must provide for single span only.

The advantages of continuous spans as opposed to simple spans are:

1. A reduction of material in the superstructure, or longer spans for the same material and thus possibly fewer piers.
2. Less deflection and vibration are present.

3. Longitudinal forces on the superstructure can be transmitted to the abutments instead of partially carried by the piers.
4. Fewer expansion devices are needed.
5. Fewer sets of bearings are used, thus there is the possibility of narrower piers.
6. A more pleasing appearance can be achieved because of possible variation in span length and depth of girders.

At the start of a design, the engineers have to make a selection on the basis of the advantages listed above. As each site is different, the weighing of each advantage is different for each bridge. Alternate designs can quite accurately predict the cost difference between simple and continuous spans. If foundation conditions are good, and other site characteristics indicate medium or long spans, the continuous structure shows the least cost. For short spans, there is little cost difference and the speed and simplicity of erection may favor the simpler spans. Where precast, prestress concrete beams are used, the simple span is most often favored. Cast-in-place concrete beams can be easily formed as continuous beams and the saving in weight and more pleasing appearance are definite advantages.

The continuous beam or girder bridge, whether of reinforced concrete or steel, is usually designed with a variable moment of inertia. For short spans, the continuous steel beam is generally a wide flange rolled beam with welded cover plates in the region of maximum moments. The longer spans require welded girders of varying depth web plates. Regardless of the span or material, the continuous beams are nonprismatic.

A continuous bridge can have two or more spans; however, five continuous spans is usually the maximum. Expansion and longitudinal forces on the substructure become difficult problems for long bridges. Two-span continuous bridges have only slight economy over simple spans. The usual continuous bridge structure has three spans with the center span one-fifth to one-third longer than the end spans.

## 5.2  ANALYSIS

Before the selection of the cross section of the beam at several positions along the span can be made, it is necessary to determine the total maximum moments and shears along the entire length of the bridge. To obtain these values, it is necessary to develop influence lines for moment and shear at several positions along the bridge. If the bridge has midpoint symmetry, influence lines for moment and shear at the support points and about six intermediate points per span are sufficient to plot the maximum moment and shear diagrams.

The development of the ordinates to the influence diagrams involves a considerable amount of numerical work. A computer program and the use of a digital computer could save a considerable amount of time. However, because the beams are nonprismatic, such a program is not simple. The detail hand calculations are developed in this chapter for the benefit of the novice bridge designer who needs to understand the calculations even if he has a computer and program available. The systematic procedure as outlined here should help in keeping the detail hand calculations to a minimum.

Design Example 5.1 illustrates the steps in the design of a three-span continuous steel plate girder bridge. Because of foundation conditions, aesthetics, and economy, the selection of the span lengths, center-to-center of bearings, are 99, 132, and 99 ft. Such a bridge can have two different design arrangements. One arrangement is a series of girders (probably four for a two-lane bridge) with a deck slab resting directly on the girder flanges. The other arrangement has two girders and a floor system consisting of floor beams and stringers with the slab supported by the stringers and the girders. This latter arrangement requires more detailed fabrication but can result in less steel since the two single girders can be made deeper. This arrangement is used in Design Example 5.1 so that the design of a floor system can be demonstrated. Such a floor system design can be similar whether used in a girder, truss, or arch bridge.

Floor beams are spaced at 16 ft 6 in. The stringers are continuous over one floor beam and are not composite with the concrete slab. The total thickness of the concrete slab is $7\frac{1}{2}$ in. The cross section of the bridge is shown in Design Example 5.1 (Fig. 5.8).

The moment-distribution method is used for the hand calculation in this design example, since it is probably the most direct method. If the girder were a constant section the procedure would be quite direct. However, for economy of material the welded girder has changes in cross section to meet the bending moment requirements. Therefore the frame constants used in the moment-distribution method are not the standard values for prismatic members. A study of girder bridges has shown that the carry-over factor for the middle span of a three-span girder varies from about 0.54 to 0.60. In the design example 0.56 is used. The differences in total bending moments and shears is only slightly affected by the carry over between the above values.

The relative stiffness of each span is slightly different from the value of $I/L$, since $I$ varies along the spans. The distribution factors between the end span and middle span can be closely estimated by increasing the $I$ of the middle span by 10% and reducing the end span by 10%. Experience has shown that these factors are very close to the actual values for standard plate girder bridge designs. A change of a few percentage points results in very little change in bending moments and shears and rarely any change in girder cross section.

The amplification factor for fixed-end moments (FEM) has the most effect on the girder design. Here again the values of fixed-end moments can be those for a prismatic beam increased by a given amount. Previous designs have shown that the fixed-end moment in the end span is approximately 20% more for the usual girder design than that for a prismatic beam and about 10% more for the middle span. A study has shown that increasing these factors to 30 and 20%, respectively, results in an increase in total negative bending moment of 5% and a decrease in total positive bending moment of a similar percentage.

Since the moment of inertia of the girder varies along the lengths, the relative stiffnesses, carry-over factors, and fixed-end moments are adjusted from those values used for prismatic members. Therefore the assumptions made relative to these values are:

1. Carry-over factor, middle span = 0.56.
2. Relative stiffness at interior support: same as constant-section beams with an increase of 10% for middle span and a decrease of 10% for the end span.
3. Fixed-end moments: values are computed as for a constant section and are adjusted by the following amounts:
   a. End span: large end is increased by 20%.
   b. Middle span: both ends are increased by 10%.

The assumptions above should be checked at the completion of the girder design.


## 5.3  DESIGN TECHNIQUE

First the sizes of stringers and floor beams are determined. Then the girder itself is designed. The dead load supported by a stringer is the weight of the slab plus the stringer weight. (Refer to Design Example 5.1.)

The stringers are designed for the maximum positive or negative moment. Since, in Design Example 5.1, the total positive and negative moments are nearly equal, the stringers are designed as prismatic sections. Cover plates are used only if there is a considerable difference in magnitude in the positive and negative moments.

The floor beams are considered simply supported with a span length equal to the distance center-to-center of the girders. Since the girders have a limited torsional resistance, the simple beam assumption for the floor beams is valid regardless of the type of floor beam–girder connection.

The loads applied to the floor beam are the reactions from the stringers. To determine these reactions, the slab is considered as a three-span continuous beam over the stringers. The reactions of the stringers are then applied to the floor beams. The equation used for determining the reactions is taken from reference (1). The stringers are considered to be unyielding supports. However, this is a slightly conservative assumption. The floor beams are designed on the basis of the limiting compression stress. Since the stringers provide resistance to lateral buckling of the floor beams, the unsupported length of the compression flange of the floor beam is quite small. The selected wide flange section is checked for shear and deflection.

The girder is a three-span nonprismatic section. The moment distribution method is used to develop the ordinates to the influence diagrams. As the girder is symmetrical and simply supported at the ends, only two distributions are needed for the three-span structure. An arbitrary fixed-end moment at the $B$ end of span $BC$ is assumed at $+100$. At the end of four cycles of distribution, the moments over the supports for the arbitrary fixed-end moments are obtained. The moments over the supports for loads anywhere on the structure can be determined from the following equations (reference pp. 116 and 117):

$$M_{ba} = 0.51\,M_{F_{ba}} - 0.49\,M_{F_{bc}} + 0.15\,M_{F_{cb}} + 0.15\,M_{F_{cd}}$$
$$M_{bc} = -0.51\,M_{F_{ba}} + 0.49\,M_{F_{bc}} - 0.15\,M_{F_{cb}} - 0.15\,M_{F_{cd}}$$
$$M_{cb} = -0.15\,M_{F_{ba}} - 0.15\,M_{F_{bc}} + 0.49\,M_{F_{cb}} - 0.51\,M_{F_{cd}}$$
$$M_{cd} = +0.15\,M_{F_{ba}} + 0.15\,M_{F_{bc}} - 0.49\,M_{F_{cb}} + 0.51\,M_{F_{cd}}$$

where $M_{ba}$, $M_{bc}$, $M_{cb}$, and $M_{cd}$ = internal support moments

$M_{F_{ba}}$, $M_{F_{bc}}$, $M_{F_{cd}}$, and $M_{F_{cd}}$ = fixed-end moments
due to a unit load.

To determine the moments over the supports, it is only necessary to compute the fixed-end moments for any specific loading. The fixed-end moments are computed for a unit load placed successively at each position of floor beam, until the midpoint of the structure has been reached. As mentioned in the previous section, the fixed-end moments are increased 20 and 10% for the end and middle spans, respectively.

**Table 5.1  Ordinates for Influence Diagram—Bending Moment**

| Moment at Point | Unit Load | | | | | | | | |
|---|---|---|---|---|---|---|---|---|---|
| | 1 | 2 | 3 | 4 | 5 | 6 | 7 | 8 | 9 |
| 1 | +12.72 | +9.50 | +6.35 | +3.63 | +1.47 | 0 | −1.18 | −1.98 | −2.07 |
| 2 | +9.44 | +19.0 | +12.7 | +7.27 | +2.93 | 0 | −2.37 | −3.97 | −4.13 |
| 3 | +5.90 | +12.0 | +19.05 | +10.9 | +4.40 | 0 | −3.55 | −5.95 | −6.20 |
| 4 | +2.37 | +5.0 | +8.90 | +14.5 | +5.87 | 0 | −4.73 | −7.93 | −8.27 |
| 5 | −1.17 | −2.0 | −1.25 | +1.70 | +7.08 | 0 | −5.92 | −9.92 | −10.3 |
| 6 | −4.70 | −9.0 | −11.4 | −11.2 | −7.70 | 0 | −7.1 | −11.9 | −12.4 |
| 7 | −3.90 | −7.55 | −9.56 | −9.39 | −6.45 | 0 | +7.84 | +1.17 | −1.73 |
| 8 | −3.10 | −6.10 | −7.73 | −7.58 | −5.20 | 0 | +6.28 | +14.2 | +8.95 |
| 9 | −2.30 | −4.65 | −5.89 | −5.76 | −3.95 | 0 | +4.71 | +10.8 | +19.6 |
| 10 | −1.65 | −3.20 | −4.05 | −3.95 | −2.70 | 0 | +3.15 | +7.35 | +13.8 |

**Table 5.2  Ordinates for Influence Diagram—Shear**

| Shear Points | Unit Load | | | | | | | | |
|---|---|---|---|---|---|---|---|---|---|
| | 1 | 2 | 3 | 4 | 5 | 6 | 7 | 8 | 9 |
| 0–1 | +0.786 | +0.576 | +0.385 | +0.220 | +0.089 | 0 | −0.072 | −0.111 | −0.125 |
| 1–2 | −0.214 | +0.576 | +0.385 | +0.220 | +0.089 | 0 | −0.072 | −0.111 | −0.125 |
| 2–3 | −0.214 | −0.424 | +0.385 | +0.220 | +0.089 | 0 | −0.072 | −0.111 | −0.125 |
| 3–4 | −0.214 | −0.424 | −0.615 | +0.220 | +0.089 | 0 | −0.072 | −0.111 | −0.125 |
| 4–5 | −0.214 | −0.424 | −0.615 | −0.780 | +0.089 | 0 | −0.072 | −0.111 | −0.125 |
| 5–6L | −0.214 | −0.424 | −0.615 | −0.780 | −0.911 | 0 | −0.072 | −0.111 | −0.125 |
| 6R–7 | +0.046 | +0.088 | +0.111 | +0.108 | +0.076 | 0 | +0.905 | +0.785 | +0.647 |
| 7–8 | +0.046 | +0.088 | +0.111 | +0.108 | +0.076 | 0 | −0.095 | +0.785 | +0.647 |
| 8–9 | +0.046 | +0.088 | +0.111 | +0.108 | +0.076 | 0 | −0.095 | −0.215 | +0.647 |
| 9–10 | +0.046 | +0.088 | +0.111 | +0.108 | +0.076 | 0 | −0.095 | −0.215 | −0.352 |

at Point

| 10 | 11 | 12 | 13 | 14 | 15 | 16 | 17 | 18 | 19 |
|---|---|---|---|---|---|---|---|---|---|
| −1.93 | −1.58 | −1.07 | −0.52 | 0 | +0.38 | +0.55 | +0.55 | +0.43 | +0.23 |
| −3.87 | −3.17 | −2.13 | −1.03 | 0 | +0.77 | +1.10 | +1.10 | +0.87 | +0.47 |
| −5.80 | −4.75 | −3.20 | −1.55 | 0 | +1.15 | +1.65 | +1.65 | +1.30 | +0.70 |
| −7.73 | −6.33 | −4.27 | −2.07 | 0 | +1.53 | +2.20 | +2.20 | +1.73 | +0.93 |
| −9.67 | −7.92 | −5.33 | −2.58 | 0 | +1.92 | +2.75 | +2.75 | +2.17 | +1.17 |
| −11.6 | −9.5 | −6.4 | −3.1 | 0 | +2.3 | +3.30 | +3.30 | +2.60 | +1.40 |
| −3.35 | −3.67 | −2.96 | −1.54 | 0 | +1.05 | +1.49 | +1.46 | +1.15 | +0.64 |
| +4.90 | +2.15 | +0.47 | +0.03 | 0 | −0.20 | −0.33 | −0.38 | −0.30 | −0.13 |
| +13.2 | +6.97 | +3.92 | +1.59 | 0 | −1.45 | −2.14 | −2.21 | −1.75 | −0.89 |
| +21.4 | +13.8 | +7.35 | +3.15 | 0 | −2.70 | −3.95 | −4.05 | −3.20 | −1.65 |

at Point

| 10 | 11 | 12 | 13 | 14 | 15 | 16 | 17 | 18 | 19 |
|---|---|---|---|---|---|---|---|---|---|
| −0.117 | −0.096 | −0.065 | −0.031 | 0 | +0.023 | +0.033 | +0.033 | +0.026 | +0.014 |
| −0.117 | −0.096 | −0.065 | −0.031 | 0 | +0.023 | +0.033 | +0.033 | +0.026 | +0.014 |
| −0.177 | −0.096 | −0.065 | −0.031 | 0 | +0.023 | +0.033 | +0.033 | +0.026 | +0.014 |
| −0.117 | −0.096 | −0.065 | −0.031 | 0 | +0.023 | +0.033 | +0.033 | +0.026 | +0.014 |
| −0.117 | −0.096 | −0.065 | −0.031 | 0 | +0.023 | +0.033 | +0.033 | +0.026 | +0.014 |
| −0.117 | −0.096 | −0.065 | −0.031 | 0 | +0.023 | +0.033 | +0.033 | +0.026 | +0.014 |
| +0.500 | +0.353 | +0.215 | +0.095 | 0 | −0.076 | −0.108 | −0.111 | −0.088 | −0.046 |
| +0.500 | +0.353 | +0.215 | +0.095 | 0 | −0.076 | −0.108 | −0.111 | −0.088 | −0.046 |
| +0.500 | +0.353 | +0.215 | +0.095 | 0 | −0.076 | −0.108 | −0.111 | −0.088 | −0.046 |
| +0.500 | +0.353 | +0.215 | +0.095 | 0 | −0.076 | −0.108 | −0.111 | −0.088 | −0.046 |

When the moments over the supports have been determined, the moments at all points along the span can be computed by combining the proportioned value of the negative moment diagram with the simple-beam moment diagrams.

The values of girder moment and shear are tabulated next. These moments and shears are for a unit load placed successively from point 1 to point 19, inclusively (Fig. 5.14). These values are given (columns) in Tables 5.1 and 5.2. The tabulated values (rows) are the ordinates to the influence diagrams for moment and shear, respectively. The values as tabulated are at the locations of the floor beams.

Plots of several of the influence diagrams are shown in Figs. 5.1 and 5.2. These plots illustrate the shape of the diagrams but they can be used to check the accuracy of the work if a larger scale is used. The diagrams can be used directly to determine the maximum moment and shear at any specific point on the structure. The area of the diagram is used for uniformly distributed loads and the values of the ordinates are used for concentrated loads. Determining the area under the influence line is quite time consuming, so a different approach is preferred. The total maximum moments and shears are found by considering the dead load and the live load as concentrated loads at the panel points (floor beam locations). The total dead-load moment or shear can be determined by taking the algebraic sum of the product of the influence line ordinates and the panel-point concentrated dead loads.

The panel-point concentrated live loads are calculated either from the truck loading or the equivalent lane loading. The AASHTO Specification

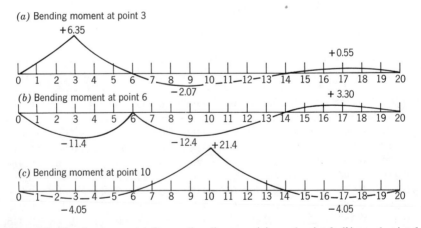

**Figure 5.1**  Bending moment influence line diagrams: (a) panel point 3; (b) panel point 6; (c) panel point 10.

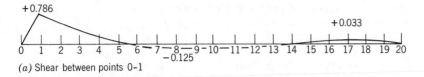

(a) Shear between points 0-1

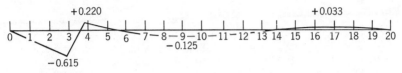

(b) Shear between points 3-4

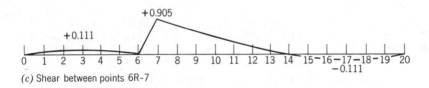

(c) Shear between points 6R-7

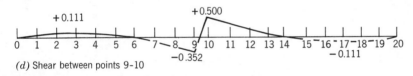

(d) Shear between points 9-10

**Figure 5.2** Shear influence line diagrams: (a) panel points 0–1; (b) panel points 3–4; (c) panel points 6R–7; (d) panel points 9–10.

states that either the truck loading or the equivalent lane loading is to be used, whichever produces the maximum stress. In case of doubt, the moments and shears for both loadings should be computed. In either case, the distribution of the live load to the girders has to be considered.

Table 5.3 gives the summation of the positive and negative ordinates of the influence diagrams, as well as the maximum positive and negative ordinates. This tabulation is needed for determining the dead- and live-load moments and shears when the equivalent lane loading is used. The AASHTO Specification states that when equivalent lane loading for continuous spans is used, only one concentrated load is used for a positive moment, but for a negative moment a concentrated load is placed in each of two spans. Therefore, for points 5 to 10 inclusive along the girder, the negative maximum ordinate column is the sum of the maximum negative

**Table 5.3   Summation of Influence Line Ordinates**

| | Bending Moment | | | | | Shear | | | | |
|---|---|---|---|---|---|---|---|---|---|---|
| | Summation of Ordinates | | | Maximum Ordinate | | Summation of Ordinates | | | Maximum Ordinate | |
| Point | + | − | $\Sigma$ | + | − | + | − | $\Sigma$ | + | − |
| 1 | 35.81 | 10.33 | +25.48 | 12.72 | 1.98 | 2.185 | 0.617 | +1.568 | 0.786 | 0.125 |
| 2 | 55.65 | 20.67 | +34.98 | 19.00 | 4.13 | 1.399 | 0.891 | +0.508 | 0.576 | 0.214 |
| 3 | 58.70 | 31.00 | +27.70 | 19.05 | 6.20 | 0.823 | 1.255 | −0.432 | 0.385 | 0.424 |
| 4 | 45.23 | 41.33 | +3.90 | 14.50 | 8.27 | 0.438 | 1.870 | −1.432 | 0.220 | 0.615 |
| 5 | 19.54 | 56.06 | −36.52 | 7.08 | 11.55[a] | 0.218 | 2.650 | −2.432 | 0.089 | 0.780 |
| 6 | 12.90 | 106.00 | −93.10 | 3.30 | 23.80[a] | 0.129 | 3.561 | +3.432 | 0.033 | 0.911 |
| | | | | | | 3.929 | 0.429 | +3.500 | 0.905 | 0.111 |
| 7 | 16.14 | 50.10 | −33.96 | 7.84 | 13.23[a] | 3.024 | 0.524 | +2.500 | 0.785 | 0.111 |
| 8 | 36.98 | 31.05 | +5.93 | 14.20 | 8.11[a] | 2.239 | 0.739 | +1.500 | 0.647 | 0.215 |
| 9 | 60.79 | 30.99 | +29.80 | 19.60 | 8.10[a] | 1.592 | 1.091 | +0.500 | 0.500 | 0.352 |
| 10 | 70.00 | 31.10 | +38.90 | 21.40 | 8.10[a] | 1.352[b] | 1.352[b] | 0 | 0.500 | 0.500 |
| | Reaction at end | | | | | 2.865 | 0.617 | 2.068 | 1.000 | 0.215 |

[a] Sum of maximum ordinates in two spans.
[b] Includes only one-half of ordinate at point 10.

ordinates in two spans. No such provision is stated for shear, so only the value of the single maximum ordinate is tabulated.

The influence lines for shear are labeled as the shear between two consecutive panel points, because the shear remains constant between floor beams, since loads can only be applied to the girders at the floor beams. The influence line changes from a value of one sign at one point to another value of the opposite sign at the adjacent point. In a multiple-girder system with slab bearing directly on the girder instead of stringers and floor beams, the influence diagram for shear changes suddenly from minus to plus at the location in question.

Tables 5.4A and 5.4B give the dead load, live load, and impact moments. The live-load values for the equivalent lane loading are given in Table 5.4A and the truck loading in Table 5.4B. The dead-load moment is obtained by multiplying the panel-point concentrated load of 37.68 k by the algebraic sum of the ordinates of the influence diagram. The moment, due to the uniformly distributed live load, is calculated by multiplying the sum of the positive or negative ordinates by the 13.2-k panel point load. The columns in Table 5.4A, giving values for live-load

**Table 5.4A  Girder Bending Moments—Equivalent Loading**

| Point | Dead-Load Bending Moment (ft-k) | Uniform Dist. Load + | Uniform Dist. Load − | Concen. Load + | Concen. Load − | Total + | Total − | Impact Percentage + | Impact Percentage − | Impact Bending Moment (ft-k) + | Impact Bending Moment (ft-k) − | Total Bending Moment (ft-k) + | Total Bending Moment (ft-k) − |
|---|---|---|---|---|---|---|---|---|---|---|---|---|---|
| 1 | +960 | 473 | 136 | 286 | 45 | 759 | 181 | 22.3 | | 169 | | 1888 | |
| 2 | +1318 | 734 | 273 | 428 | 93 | 1162 | 366 | 22.3 | | 259 | | 2739 | |
| 3 | +1044 | 775 | 409 | 429 | 140 | 1204 | 549 | 22.3 | | 268 | | 2516 | |
| 4 | +147 | 597 | 546 | 326 | 186 | 923 | 732 | 22.3 | 20.8 | 206 | 152 | 1276 | 737 |
| 5 | −1376 | 258 | 740 | 159 | 260 | 417 | 1000 | 22.3 | 20.8 | 93 | 208 | | 2584 |
| 6 | −3508 | 170 | 1399 | 74 | 536 | 244 | 1935 | | 20.8 | | 402 | | 5845 |
| 7 | −1280 | 213 | 661 | 176 | 298 | 389 | 959 | | 20.8 | | 199 | | 2438 |
| 8 | +223 | 488 | 410 | 320 | 182 | 808 | 592 | 19.5 | 20.8 | 158 | 123 | 1189 | 492 |
| 9 | +1123 | 802 | 409 | 441 | 182 | 1243 | 591 | 19.5 | | 242 | | 2608 | |
| 10 | +1466 | 924 | 411 | 482 | 182 | 1406 | 593 | 19.5 | | 274 | | 3146 | |

moment due to a concentrated load, are determined by multiplying the value of maximum ordinate in Table 5.3 by the concentrated load of 22.5 k.

The live-load moments in Table 5.4B are determined by superimposing the truck loading on the influence diagram. The middle 32-k axle load is placed at the location of the peak ordinate of the influence diagram. The

**Table 5.4B    Girder Bending Moments—HS20 Truck Loading**

| Point | Dead Load Bending Moment | 32-k Axles + | 32-k Axles − | 8-k Axle + | 8-k Axle − | Live-Load Bending Moment (ft-k) + | Live-Load Bending Moment (ft-k) − | Impact Bending Moment (ft-k) + | Impact Bending Moment (ft-k) − | Total Bending Moment (ft-k) + | Total Bending Moment (ft-k) − |
|---|---|---|---|---|---|---|---|---|---|---|---|
| 1 | +960 | 22.71 | | 6.62 | | 974 | | 217 | | 2151 | |
| 2 | +1318 | 32.65 | | 10.88 | | 1415 | | 316 | | 3049 | |
| 3 | +1044 | 32.12 | | 12.13 | | 1406 | | 314 | | 2764 | |
| 4 | +147 | 24.25 | 16.25 | 7.17 | 7.78 | 1133 | 728 | 253 | 151 | 1533 | 732 |
| 5 | −1376 | 9.6 | 20.06 | 1.07 | 9.76 | 395 | 900 | 88 | 187 | | 2463 |
| 6 | −3508 | | 24.38 | | 11.50 | | 1090 | | 227 | | 4825 |
| 7 | −1280 | | 18.97 | | 7.85 | | 838 | | 174 | | 2292 |
| 8 | +223 | 23.95 | 15.33 | 7.48 | 6.35 | 1033 | 675 | 201 | 140 | 1457 | 592 |
| 9 | +1123 | 33.77 | | 12.13 | | 1472 | | 287 | | 2882 | |
| 10 | +1466 | 36.35 | | 14.95 | | 1604 | | 313 | | 3383 | |

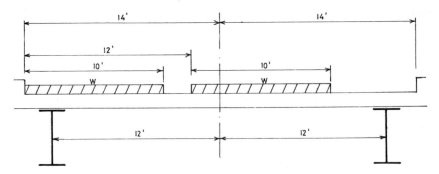

**Figure 5.3** Lane loading position.

8-k axle and the other 32-k axle are 14 feet to each side of the middle axle. The moments due to the truck then are the summation of the magnitudes of the axle loads times their respective ordinates of the influence diagram. This value of moment is also multiplied by the 1.25 distribution factor to obtain the total live-load moments (see Fig. 5.3 for position of vehicles).

The impact percentages are computed using the standard impact equation (equation 3.1) The AASHTO Specification for the definition of L for a continuous beam is followed. The Specification states that the value of L is the length of span under consideration for a positive moment and the average of the two adjacent.spans for a negative moment.

The total moment is the sum of the dead-load, live-load, and impact moment. The total moment is plotted in Fig. 5.4.

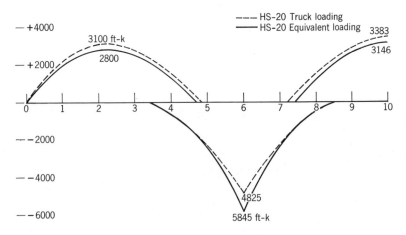

**Figure 5.4** Bending moment envelope.

## Table 5.5A    Girder Shears—Equivalent Lane Loading

| | | Live-Load Shear | | | | | | | | | | |
| | | Distribution | | | | | | Impact | Impact | | Total | |
| | Dead-Load | Load | | Conc. Load | | Total | | Percentage | | Shear (k) | Shear (k) | |
| Point | Shear | + | − | + | − | + | − | + | − | + | − | + | − |
|---|---|---|---|---|---|---|---|---|---|---|---|---|---|
| End Reaction | +77.9 | 35.4 | 8.1 | 32.5 | 4.1 | 67.9 | 12.2 | 22.3 | 22.3 | 15.1 | 2.7 | 160.9 | |
| 0–1 | +59.1 | 28.8 | 8.1 | 25.5 | 4.1 | 54.3 | 12.2 | 22.3 | 22.3 | 12.1 | 2.7 | 125.5 | |
| 1–2 | +19.1 | 18.5 | 11.8 | 18.7 | 7.0 | 37.2 | 18.8 | 22.3 | 22.3 | 8.3 | 4.2 | 64.6 | 3.9 |
| 2–3 | −16.3 | 10.9 | 16.6 | 12.5 | 13.8 | 23.4 | 30.4 | 22.3 | 22.3 | 5.2 | 6.8 | 12.3 | 53.5 |
| 3–4 | −54.0 | 5.8 | 24.7 | 7.2 | 20.0 | 13.0 | 44.7 | 22.3 | 22.3 | 2.9 | 10.0 | | 108.7 |
| 4–5 | −91.6 | 2.9 | 35.0 | 2.9 | 25.4 | 5.8 | 60.4 | 22.3 | 22.3 | 1.3 | 13.5 | | 165.0 |
| 5–6L | −129.3 | 1.7 | 47.6 | 1.1 | 29.6 | 2.8 | 77.2 | 22.3 | 22.3 | 0.6 | 17.2 | | 223.7 |
| 6R–7 | +131.9 | 51.9 | 5.7 | 29.4 | 3.6 | 81.3 | 9.3 | 19.5 | 19.5 | 15.8 | 1.8 | 229.0 | |
| 7–8 | +94.2 | 59.9 | 6.9 | 25.5 | 3.6 | 65.4 | 10.5 | 19.5 | 19.5 | 12.8 | 2.0 | 172.4 | |
| 8–9 | +56.5 | 29.6 | 9.8 | 21.0 | 7.0 | 50.6 | 16.8 | 19.5 | 19.5 | 9.9 | 3.3 | 117.0 | |
| 9–10 | +18.8 | 21.0 | 14.4 | 16.3 | 11.4 | 37.3 | 25.8 | 19.5 | 19.5 | 7.3 | 5.0 | 63.4 | 12.0 |
| 10 | 0 | 17.8 | 17.8 | 16.3 | 16.3 | 34.1 | 34.1 | 19.5 | 19.5 | 6.6 | 6.6 | 40.7 | 40.7 |

Tables 5.5A and 5.5B give the values of dead load, live load, and impact shear. The procedure used to obtain values for these tables is similar to that used for moment. The impact percentages were computed somewhat differently than those for the moment, since AASHTO does not make any definition for $L$ for shear in continuous beams. The value of $L$ used in the impact equation for shear was taken as the end-span

## Table 5.5B    Girder Shears—HS20 Truck Loading

| | | Summation of Ordinates | | | | | | | | | |
| | Dead Load | 32-k Axles | | 8-k Axle | | Live-Load Shear (k) | | Impact Shear (k) | | Total Shear (k) | |
| Point | Shear | + | − | + | − | + | − | + | − | + | − |
|---|---|---|---|---|---|---|---|---|---|---|---|
| End Reaction | +77.9 | 1.818 | 0.243 | 0.640 | 0.113 | 79.1 | 10.9 | 17.6 | 2.4 | 174.6 | |
| 0–1 | +59.1 | 1.395 | 0.243 | 0.443 | 0.113 | 60.3 | 10.9 | 13.4 | 2.4 | 132.8 | |
| 1–2 | +19.1 | 0.990 | 0.243 | 0.270 | 0.113 | 42.3 | 10.9 | 9.4 | 2.4 | 70.8 | |
| 2–3 | −16.3 | 0.697 | 0.670 | 0.129 | 0.065 | 29.2 | 27.5 | 6.5 | 6.1 | 19.4 | 49.9 |
| 3–4 | −54.0 | 0.240 | 1.068 | 0.027 | 0.278 | 9.9 | 45.5 | 2.2 | 10.1 | | 109.6 |
| 4–5 | −91.6 | 0.102 | 1.420 | −0.050 | 0.482 | 3.6 | 61.6 | 0.8 | 13.7 | | 166.9 |
| 5–6L[a] | −129.3 | 0.066 | 1.711 | 0.027 | 0.665 | 2.9 | 75.1 | 0.6 | 16.7 | | 221.1 |
| 6R–7[b] | +131.9 | 1.708 | 0.219 | 0.689 | 0.095 | 75.2 | 9.7 | 14.7 | 1.9 | 221.8 | |
| 7–8 | +94.2 | 1.453 | 0.219 | 0.545 | 0.095 | 63.6 | 9.7 | 12.4 | 1.9 | 170.2 | |
| 8–9 | +56.5 | 1.169 | 0.328 | 0.398 | 0.029 | 50.7 | 13.4 | 9.9 | 2.6 | 117.1 | |
| 9–10 | +18.8 | 0.875 | 0.588 | 0.257 | 0.131 | 37.6 | 24.8 | 7.3 | 4.8 | 63.7 | 10.8 |

[a] $L$ indicates left of point.
[b] $R$ indicates right of point.

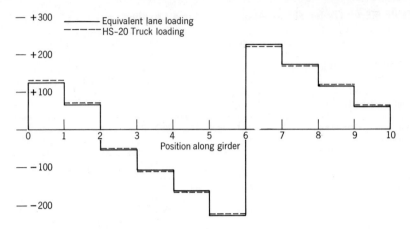

**Figure 5.5**   Shear envelope.

length for points in the end span and the length of middle span for points in the middle span. Figure 5.5 is a plot of the maximum value of shear at all points along the girder.

It should be noted that the toal moments and shears using either the truck loading or the equivalent load were nearly equal. For longer spans than those used in this particular design, the equivalent loading would

**Table 5.6   Design Bending Moments and Shears**

| Point | Bending Moment (ft-k) + | Bending Moment (ft-k) − | Shear (k) + | Shear (k) − | Point |
|---|---|---|---|---|---|
| 0 | | | 133 | | 0–1 |
| 1 | 2151 | | 71 | | 1–2 |
| 2 | 3049 | | | 50 | 2–3 |
| 3 | 2764 | | | 110 | 3–4 |
| 4 | 1533 | 732 | | 167 | 4–5 |
| 5 | | 2584 | | 221 | 5–6L[a] |
| 6 | | 5845 | 222 | | 6R–7[b] |
| 7 | | 2438 | 171 | | 7–8 |
| 8 | 1457 | 592 | 117 | | 8–9 |
| 9 | 2882 | | 64 | | 9–10 |
| 10 | 3383 | | | | |

[a] $L$ indicates left of point.
[b] $R$ indicates right of point.

produce the greatest stresses and for shorter spans, the truck loading would give greater values. The girder is designed for the greatest moment or shear irrespective of whether caused by truck or lane loading. Table 5.6 lists final design bending moments and shears.

## 5.4 GIRDER SELECTION

Once the total moments and shears have been determined, the selection of the girder cross section can begin. The first step in this action is to select the depth and thickness of the web plate. The maximum depth of the web plate is based on the magnitude of the maximum moment, and the thickness is dependent on the maximum shear, as well as the limiting depth-to-thickness ratio as specified in the AASHTO Specification.

Before the decision can proceed any further, it is necessary to determine whether the depth of the web plate should be variable or constant throughout the length of the bridge. The variable-depth girder is commonly referred to as a haunched girder. The straight girder is preferred for shorter spans, while the haunched girder proves more economical for longer spans. The straight girder is more easily shipped and handled. The haunched girder has flanges that are more nearly uniform in cross section throughout the girder length, which results in fewer flange splices.

After selection of the type of web plate, the depth of the web plate is chosen. The deeper the web plate, the greater the web thickness required to prevent web buckling. AASHTO limits the minimum web thickness to

$$t = \frac{D\sqrt{f_b}}{23,000} \tag{5.1}$$

where $D$ = depth of web plate

$f_b$ = calculated compressive bending stress in the flange

If the compressive bending stress is equal to the allowable bending stress for the various steels, the thickness of the web plate cannot be less than:

$$\frac{D}{165} \quad \text{for yield point} = 36,000 \text{ psi}$$

$$\frac{D}{150} \quad \text{for yield point} = 42,000 \text{ psi}$$

$$\frac{D}{145} \quad \text{for yield point} = 46,000 \text{ psi}$$

$$\frac{D}{140} \quad \text{for yield point} = 50,000 \text{ psi}$$

If longitudinal stiffeners are used, the web-plate thickness can be reduced to one-half of the above values.

As the yield point increases the thickness increases, because higher allowable bending stresses are permitted for the stronger steels; however, the buckling stress of the web plate is almost constant because all the steels have essentially the same modulus of elasticity.

In Design Example 5.1, a depth of 72 in. is selected. This is the maximum depth using a $\frac{7}{16}$-in. web plate without longitudinal stiffeners. An economic study should include alternate designs using longitudinal stiffeners, as well as the possibility of different depths and thicknesses of web plates. The depth of 72 in. gives a reasonable size of flange so the design is continued with this size web plate.

The novice bridge designer should here note that as the depth of the girder is increased, the thickness of the web increases, but the area of the flange decreases. The objective is to arrive at the proper depth so that the minimum cost is achieved; however this is not an easy task and requires the investigation of many possible alternatives. It should be realized also that minimum girder weight is not necessarily minimum girder cost. Minimum girder cost must include such items as fabrication and erection costs. Not only should the cost of the girder be included in the study, but the objective should be the effect of varying the girder depth on the cost of the total project. The depth of the girder affects the cost of the approach highway, as well as the substructure of the bridge. It should now be evident that a few extra dollars spent on engineering may result in many dollars saved in the total cost of the bridge.

The area of the flange plate varies along the length of the girder, since the total design moment is a different quantity at positions along the girder. The area of the flange plate is selected first at the point of maximum moment. This is at the interior support. The next selections of the flange plate are made at the points of maximum positive moment in each span. Once the flange is determined for these three points, the resisting moment of the girder with several selected flange plate areas is calculated. These values of the resisting moment can then be superimposed on the graph of the total design moments (see Fig. 5.6). This plot is then used to determine the required length of each size flange plate.

In selecting the size of the compression flange plate, the width ($b$) to thickness ($t$) ratio should be equal to or less than $3250/\sqrt{f_b}$, but in no case should $b/t$ exceed 24. The value of $f_b$ is the actual maximum compressive bending stress. This limitation of the $b/t$ ratio is to prevent buckling of the compression flange.

The bridge designer is faced with the decision of how many changes in the flange plate should be made in the length of the girder. Many

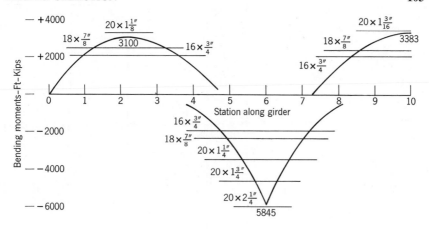

**Figure 5.6**  Flange plate lengths.

different sizes of flange plates result in the minimum weight girder, but not usually the minimum cost. A point of diminishing returns is reached when the cost of a butt welded splice is more than the money saved in plate weight.

Figure 5.7 shows four different variations of the flange plates. In trial no. 1 the size of the flange plate is changed to meet the required bending moment. There are nine flange shop splices between the end bearing of the girder and the center line of the bridge. The total weight of steel in one flange over this distance is 12,340 lb. If the cost of a butt weld is taken as equivalent to 800 lb of steel, then the equivalent amount of steel is 19,540 lb.

In trial no. 2 the minimum size of flange plate was increased from $16 \times \frac{3}{4}$ in. (trial no. 1) to $18 \times \frac{7}{8}$ in. and the short $20 \times 1\frac{1}{4}$ in. plates were eliminated. In this design the steel in the flanges increases by nearly 500 lb but the reduction in number of welds reduces the net by 1110 lb. If the equivalent cost of a flange weld was 600 lb instead of 800 lb, the designs would be closer in cost.

Trial no. 3 is based on using the maximum plate size required in the positive bending moment regions for as long a length as possible and then using the maximum size of plate in the negative moment region over the remaining length. The equivalent plate weight for welds is based on unlimited available plate lengths. Although the plate weight is higher than in the first two trial designs, the net weight is the lowest because theoretically there are only two welds. This unlimited availability of plate lengths is not very realistic, so trial no. 4 is the same as no. 3 with maximum plate lengths at 40 ft. This length introduces two more butt

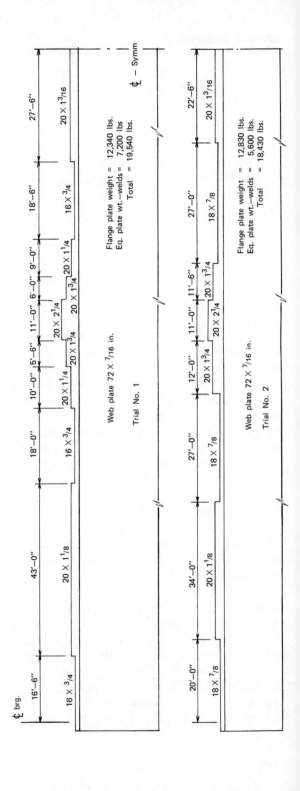

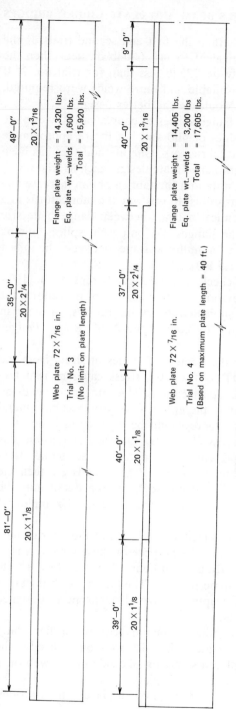

**Figure 5.7** Flange plate design: Trials 1–4.

welds. The net weight is still less than the first and second trials. If the cost of the butt welds were equivalent to just 600 lb of steel, then the weight for trial no. 4 would only be 225 lb less than for trial no. 2. If plates of 40 ft length were not available and more welds were required, then more changes in plate size would be the better design.

The equivalent weight cost of flange splices of course varies with plate thickness and weight and the values of 600 and 800 lb are approximate and should be checked with local fabricators. One method that has been used is to show a minimum-weight flange plate design, such as trial no. 2, on the design drawings and then note that the fabricator can increase any flange plate size, to eliminate butt welds. In this way the fabricator is in a position to determine the economical balance between flange plate size and number of butt welds in the flanges. At butt weld splices joining material of different thicknesses, a uniform slope between the offset surfaces of not more than 1 in 1.5 is required.

The girder design must be checked for fatigue strength. In 1974 AASHTO made considerable revision and clarification of procedures for the design of steel structures subjected to cyclic stresses. A help in understanding the fatigue design provisions is the AISC *Bridge Fatigue Guide* (2).

The first step is to determine the number of cycles of stress from the average daily truck traffic (ADTT). Then the stress category is determined from Table 1.7.2A2 in the Specifications. The category depends on the type of physical discontinuity in the girder. Category A is base metal only, while Category E is where there would be high stress concentrations. The allowable range of stress ($F_{sr}$), either tension or tension plus compression when reversal is present, is compared to the actual stress range ($f_{sr}$) due to live load plus impact stress. If $f_{sr}$ exceeds $F_{sr}$, then the girder section must be increased to decrease $f_{sr}$ or the detail must be changed to change the category to a higher $F_{sr}$. Design Example 5.1 shows the detail procedure.

After the girder cross section is selected, the design of the intermediate and bearing stiffeners is performed. The purpose of these stiffeners is to restrain the web from buckling. By using stiffeners fastened to the web, the thickness of the web plate can be reduced. The limiting web thickness values given previously in this chapter are based on properly designed transverse intermediate stiffeners.

In welded girders, the intermediate stiffeners consist of flat plates welded perpendicular to the web plate with a fillet weld on each side. The intermediate stiffeners can be placed singly on one side of the web or in pairs, one on each side of the web. When placed on one side only, the stiffener is welded to the compression flange. When placed on both sides, they need not be welded to the compression flange but must be tightly fitted to it. The usual practice is to place the stiffeners only on the inside of welded girders. They are not welded to the tension flange.

From a standpoint of fatigue crack resistance it is recommended that the transverse intermediate stiffener be terminated short of the web-to-flange weld by a distance of four to six times the web thickness. The stiffener-to-web fillet weld should not be returned around the ends of the stiffeners. The background for these requirements is contained in the previously mentioned AISC Guide.

The thickness, width, and spacing of the intermediate stiffeners have to be determined. The maximum spacing of the stiffeners is the depth of the girder between flanges or that given by the following formula, whichever is the lower value.

$$d = \frac{11,000t}{\sqrt{f_v}} \qquad (5.2)$$

where   $t$ = web thickness

$$f_v = \frac{\text{shear}}{\text{area of web}}$$

From the above equation, it is seen that the calculated value of $d$ varies with the value of shear. The usual practice is to use only a few different spacings along the girder.

### 5.4.1 Stiffener Design

The intermediate stiffeners can be omitted where the thickness of the web plate is greater than that given by the equation

$$t = \frac{D\sqrt{f_v}}{7500} \qquad (5.3)$$

The design example shows that for $t = \frac{7}{16}$ in. the stiffeners could be omitted where the shear is less than 65.4 k. However, the Specifications also state that if web stiffeners are omitted, then

$$t \not< \frac{D}{150} = \frac{72}{150} = 0.48 > \tfrac{7}{16} \text{ in.}$$

Therefore the intermediate web stiffeners cannot be omitted anywhere on the girder.

The girder has stiffeners at the floor beams for attachment (bolts) of the floor beam to the stringer. The stiffeners between the floor beams should have uniform spacing. There can be one, two, or three stiffeners in any panel between the floor beams. With one stiffener $d$ is 99 in. which is greater than the depth of web plate. Therefore two or three stiffeners are used. The maximum values of shear at these spacings are calculated. In a none-floor-beam bridge, three to four different spacings of stiffeners would be used throughout the length of the girder. Detailing would be similar to the detail spacing of stud shear connectors, as shown in Chapter 4.

The minimum width and thickness are selected as prescribed by the Specifications. The moment of inertia provided by the stiffener is checked against the required $I$ as given by equation 5.4.

$$I = \frac{d_0 t^3 J}{10.92} \qquad (5.4)$$

where $J = 25\left(\frac{D}{d}\right)^2 - 20 \not< 5.0$

$d_0$ = actual spacing of stiffeners

$d$ = minimum required spacing

In most cases the minimum size of stiffener meets the moment of inertia requirement.

The bearing stiffeners are designed to resist the reaction force at the piers and abutments. For welded girders, the bearing stiffeners can be either a single pair or a double pair of plates extending the full depth of the girder and welded to the web plate. The larger girders may have two pairs of plates spaced at a sufficient distance to enable a fillet weld being placed on both sides of each plate.

The bearing stiffeners are designed as columns carrying the total reaction. Since the load is applied to the stiffener as shear from the web, it would be very conservative to take the web-plate depth as the column length. A column length equal to three-quarters of the web-plate depth is considered satisfactory. In the design example, the column section consists of the two stiffener plates plus a width of web plate equal to 18 times the thickness of the web plate.

The stiffener plates must now be checked for bearing stress where they are in contact with the bottom flange. Since a bevel cut is made in the stiffener plate to clear the flange-to-web fillet weld, only the actual bearing area of the stiffener can resist the reaction force.

The stiffener plates should extend out to near the edge of the flange and the thickness should not be less than

$$\frac{b'}{12} \sqrt{\frac{Fy}{33,000}} \qquad (5.5)$$

where $b'$ = width of stiffeners

$F_y$ = yield point stress of stiffener material

The fillet welds fastening the bearing stiffeners to the web plate are designed to resist the full reaction. The maximum size weld based on thickness of the plates being attached is usually adequate.

### 5.4.2 Summary of Girder Design

Design Example 5.1 gives the detail design procedures for one type of plate girder, however, the resultant design is not necessarily the optimum design. An optimum design would require precise knowledge of material and fabrication costs.

In addition to the calculations shown, the complete design requires maximum live-load and impact deflections as a check against the allowable permitted by the Specifications. In hand calculations the conjugate-beam method is the most expeditious. The live load is placed in the middle span for maximum deflections. Space does not allow these detailed computations. Unless the girder is too shallow, a continuous bridge rarely has trouble meeting the deflection limitation.

Dead-load deflections are needed for establishing the camber of the girder. The camber is usually set at 1 to 1.5 times the dead-load deflection.

At the conclusion of the design the carry-over factor, the relative stiffnesses, and the fixed-end moments as assumed at the beginning of the design should be checked. These values can be readily calculated for the variable-section girder using the principles of moment area or conjugate beam.

Studies have shown that the bending moments in a three-span continuous-bridge girder are only slightly changed by a change of carry-over factor in the middle span. In fact, in changing from 0.52 to 0.60 (the likely maximum variation) the changes in peak bending moments were less than 5%.

The sensitivity of bending moments to changes in relative stiffness is also low. The bending moments are more sensitive to changes in the fixed-end-moment factor used. Some adjustment may be required if the first choice of fixed-end moments appear to be somewhat different from those determined from the shape of the designed girder. The values of frame constants used in Design Example 5.1 are very typical of actual bridge designs and will give results that require only slight modifications.

### Design Example 5.1

CONTINUOUS WELDED PLATE GIRDER BRIDGE (Fig. 5.8)

*Given:*

1. Three-span continuous with spans of 99, 132, and 99 ft.
2. Deck: as shown in Fig. 5.8.
3. Loading: two lanes of HS20 traffic.
4. Floor systems and girders: A36 steel.
5. Weight of steel handrail: 65 lb/ft.

DESIGN OF STRINGERS (TWO-SPAN CONTINUOUS) (Fig. 5.9)

Dead load:

Slab: $\dfrac{7.5}{12} \times 8 \times 0.15$: 0.75 k/ft

Stringer (estimated): $\dfrac{0.05}{0.80 \text{ k/ft}}$

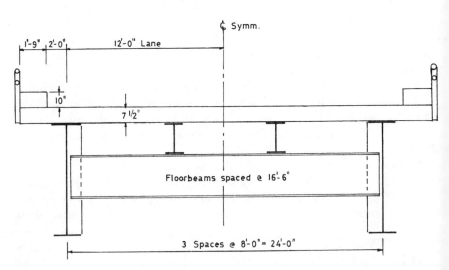

**Figure 5.8** Bridge cross section.

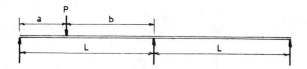

**Figure 5.9**  Stringer live loading.

Positive dead-load moment: $0.07wL^2 = 0.07 \times 0.80(16.5)^2$
$$= 15.25 \text{ ft-k (max at } 0.375L)$$
Negative dead-load moment: $0.125wL^2 = 0.125 \times 0.80(16.5)^2$
$$= 27.2 \text{ ft-k}$$

Positive live-load moment:   $\dfrac{Pab}{4L^3}(4L^2 - a(L+a))$

Maximum moment occurs when $a = 0.43L$ and $L = 16.5$ ft
Positive live-load moment: $0.2074PL = 0.2074 \times 16 \times 16.5 = 54.8$ ft-k
Distribution factor: $\dfrac{8}{5.5} = 1.454$
Positive moment per stringer: 79.7 ft-k
Negative live-load moment: $0.0964PL$ (when a = $0.58L$)
    ($a = 0.58L = 9.57$ ft; distance between loads in each span = $2(16.5 - 9.57) = 13.86$ ft, use 14)
Negative live-load moment: $0.964 \times 16 \times 16.5 \times 2 = -50.8$ ft-k
Negative live-load moment per stringer: 73.9 ft-k
Impact: 0.30

*Total positive moment*

Dead load:    15.2 ft-k
Live load:    79.7
Impact:       23.9
    $\Sigma = \overline{118.8 \text{ ft-k}}$

*Total negative moment*

Dead load:    27.2 ft-k
Live load:    73.9
Impact:       22.2
    $\Sigma = \overline{123.3 \text{ ft-k}}$

Required $S$: $\dfrac{123.3 \times 12}{20} = 74 \text{ in}^3$.

Use $W18 \times 45$, $S = 78.9 \text{ in}^3$.

Maximum live load $\Delta$: $\dfrac{0.015 PL^3}{EI} = \dfrac{0.015 \times 16 \times (16.5)^3 (12)^3}{704.5 \times 29 \times 10^3}$ (1.454)

$$= 0.132 \text{ in}.$$

Live load plus impact $\Delta = 0.132 \times 1.3 = 0.172 < \dfrac{16.5 \times 12}{800} = 0.248 \text{ in}.$

### DESIGN OF FLOOR BEAMS

The floor beams are designed for reactions ($W$) from stringers. For maximum load on one floor beam, one wheel is placed over a floor beam (Fig. 5.10) and the other two wheels are out on the span. To determine the load on a floor beam the slab is considered as a continuous beam supporting four wheel loads in a line (Fig. 5.11).

$R$ is now determined in terms of $P$. From influence lines for reactions (reference 1, p. 106):

$$R = P(1 + 0.2r_1 - 1.8r_1^2 + 0.6r_1^3) + P(0.8r_2 + 1.2r_2^2 - r_2^3)$$
$$+ P(0.8r_3 + 1.2r_3^2 - r_3^3) + P(-0.8r_4 + 1.2r_4^2 - 0.4r_4^3)$$

for

$$r_1 = \tfrac{4}{8} = 0.5 \quad r_2 = \tfrac{6}{8} = 0.75 \quad r_3 = \tfrac{2}{8} = 0.25 \quad r_4 = \tfrac{4}{8} = 0.5$$
$$W = R = P(0.725 + 0.8531 + 0.2594 - 0.150) = 1.69P$$

The value of $P$ is a contribution from all three axles. This value is now used in evaluating the reaction (Fig. 5.12) of the two-span continuous stringer on the floor beam.

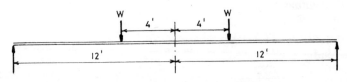

**Figure 5.10** Floor beam live loading.

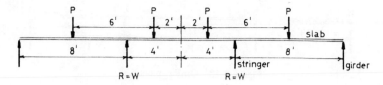

**Figure 5.11** Slab live loading.

The reaction at support $b$ is computed as

$$R_b = 16 + 20\left[1 - 1.5\left(\frac{14}{16.5}\right)^2 + 0.5\left(\frac{14}{16.5}\right)^3\right] = 20.5 \text{ k}$$

Live load $W$: $1.69 \times 20.5 = 34.6$ k
Impact $W$: $34.6 \times 0.3 = 10.4$ k
Dead load: $wL = 0.80 \times 16.5 \times 1.25 = 16.5$ k
Total load: $61.5$ k

Total bending moment: $61.5 \times 8 = 492$ ft-k.

Dead-load bending moment of floor beam: $\dfrac{0.11}{8}(24)^2 = 8$ ft-k.

Required $S$: $\dfrac{500 \times 12}{19.5} = 308 \text{ in}^3$.

Try $W30 \times 116$, $S = 329$, $b = 10.5$.
Compression flange, restrained at stringer location only.

Allowable $F_b$: $20,000 - 7.5\left(\dfrac{96}{10.5}\right)^2 = 19,370$ psi

Actual $F_b$: $\dfrac{500 \times 12}{329} = 18,240 \text{ psi} < 19,370$

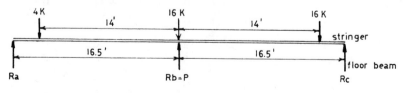

**Figure 5.12** Floor beam reactions.

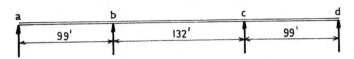

**Figure 5.13**  Girder spans.

Live load plus impact $\Delta$: $\dfrac{0.0355PL^3}{EI} = \dfrac{0.0355 \times 45 \times (24)^3 \times (12)^3}{4930 \times 29 \times 10^3}$

$$= 0.26 \text{ in}$$

Allowable $\Delta$: $\dfrac{24 \times 12}{800} = 0.36 \text{ in} > 0.26$

Web shear stress: $\dfrac{61.5}{30 \times 0.564} = 3.63 \text{ ksi} < 12 \text{ ksi}$

Use $W30 \times 116$

GIRDER MOMENTS (Fig. 5.13)

From Fig. 5.13:

$$K_{BA} = \left(\frac{0.75I}{99}\right) 0.9 = 0.00682I$$

$$K_{BC} = \left(\frac{I}{132}\right) 1.1 = 0.00833I$$

Distribution factors (DF): for BA = 0.45;   for   BC = 0.55

Assuming a fixed-end moment of 100 at the end of each member successively, the internal support moments are computed as follows:

| co | A | | B ←——— 0.56 ——→ C | | | D |
|---|---|---|---|---|---|---|
| DF | | 0.45 | 0.55 | 0.55 | 0.45 | |
| FEM | 0 | −100 | 0 | 0 | 0 | 0 |
| | 0 | +45 | +55 | 0 | 0 | 0 |
| | 0 | 0 | 0 | +31 | 0 | 0 |
| | 0 | 0 | 0 | −17 | −14 | 0 |
| | 0 | 0 | −9 | 0 | 0 | 0 |
| | 0 | +4 | +5 | 0 | 0 | 0 |
| | 0 | 0 | 0 | +3 | 0 | 0 |
| | 0 | 0 | 0 | −2 | −1 | 0 |
| Final $M$ | 0 | −51 | +51 | +15 | −15 | 0 |

| DF |  | 0.45 | 0.55 |  | 0.55 | 0.45 | 0 |
|---|---|---|---|---|---|---|---|
| FEM | 0 | 0 | +100 |  | 0 | 0 | 0 |
|  | 0 | −45 | −55 |  | 0 | 0 | 0 |
|  | 0 | 0 | 0 |  | −31 | 0 | 0 |
|  | 0 | 0 | 0 |  | +17 | +14 | 0 |
|  | 0 | 0 | 9 |  | 0 | 0 | 0 |
|  | 0 | −4 | −5 |  | 0 | 0 | 0 |
|  | 0 | 0 | 0 |  | −3 | 0 | 0 |
|  | 0 | 0 | 0 |  | +2 | +1 | 0 |
| Final $M$ | 0 | −49 | +49 |  | −15 | +15 |  |

This table shows the balancing of the fixed-end moments at $BA$ and $BC$, using the moment distribution technique.

From the final moment values the end moments for each span can be determined for actual fixed-end moments:

$$M_{BA} = 0.51M_{F_{BA}} - 0.49M_{F_{BC}} + 0.15M_{F_{CB}} + 0.15M_{F_{CD}}$$
$$M_{BC} = -0.51M_{F_{BA}} + 0.49M_{F_{BC}} - 0.15M_{F_{CB}} - 0.15M_{F_{CD}}$$
$$M_{CB} = -0.15M_{F_{BA}} - 0.15M_{F_{BC}} + 0.49M_{F_{CB}} - 0.51M_{F_{CD}}$$
$$M_{CD} = +0.15M_{F_{BA}} + 0.15M_{F_{BC}} - 0.49M_{F_{CB}} + 0.51M_{F_{CD}}$$

End moments are determined for unit loads at the panel points shown in (Fig. 5.14). Using

$$M_{F_{BA}} = \frac{Pab}{2L^2}(a+L)$$

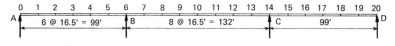

**Figure 5.14** Girder panel points.

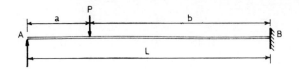

**Figure 5.15**  Exterior span A–B.

and $P = 1$, $L = 99$ ft (Fig. 5.15) gives:

| Point | $a$ | $b$ | $M_{F_{BA}}$ (ft-k) | $1.2\,M_{F_{BA}}$ (ft-k) |
|-------|------|------|------|------|
| 1 | 16.5 | 82.5 | −7.78 | −9.3 |
| 2 | 33.0 | 66.0 | −14.67 | −17.6 |
| 3 | 49.5 | 49.5 | −18.56 | −22.3 |
| 4 | 66.0 | 33.0 | −18.33 | −22.0 |
| 5 | 82.5 | 16.5 | −12.60 | −15.1 |

Using

$$M_{F_{BC}} = \frac{Pab^2}{L^2} \text{ and } M_{F_{CB}} = \frac{Pa^2b}{L^2}$$

and $P = 1$, $L = 132$ ft (Fig. 5.16) gives:

| Point | $a$ | $b$ | $M_{F_{BC}}$ (ft-k) | $M_{F_{CB}}$ (ft-k) | $1.10\,M_{F_{BC}}$ (ft-k) | $1.10\,M_{F_{CB}}$ (ft-k) |
|-------|------|------|------|------|------|------|
| 7 | 16.5 | 115.5 | 12.63 | −1.80 | 13.9 | −2.0 |
| 8 | 33.0 | 99.0 | 18.56 | −6.19 | 20.4 | −6.8 |
| 9 | 49.5 | 82.5 | 19.34 | −11.60 | 21.3 | −12.8 |
| 10 | 66.0 | 66.0 | 16.50 | −16.50 | 18.2 | −18.2 |

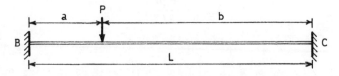

**Figure 5.16**  Center span B–C.

The end moments are tabulated below. The plus sign indicates a clockwise moment on the joint and the minus sign indicates a counterclockwise moment

| Point of Load | $M_{BA}$ (ft-k) | $M_{BC}$ (ft-k) | $M_{CB}$ (ft-k) | $M_{CD}$ (ft-k) |
|---|---|---|---|---|
| 1 | −4.7 | +4.7 | +1.4 | −1.4 |
| 2 | −9.0 | +9.0 | +2.6 | −2.6 |
| 3 | −11.4 | +11.4 | +3.3 | −3.3 |
| 4 | −11.2 | +11.2 | +3.3 | −3.3 |
| 5 | −7.7 | +7.7 | +2.3 | −2.3 |
| 7 | −7.1 | +7.1 | −3.1 | +3.1 |
| 8 | −11.0 | +11.0 | −6.4 | +6.4 |
| 9 | −12.4 | +12.4 | −9.5 | +9.5 |
| 10 | −11.6 | +11.6 | −11.6 | +11.6 |

*Dead loads per girder per panel point*

| | |
|---|---|
| Handrail: $0.065 \times 16.5$ | 1.07 k |
| Curb: $1.75 \times \dfrac{10}{12} \times 0.15 \times 16.5$ | 3.61 |
| Deck slab: $\dfrac{7.5}{12} \times 15.75 \times 0.15 \times 16.5$ | 24.36 |
| Stringer: $0.045 \times 16.5$ | 0.74 |
| Floor beam: $0.108 \times 12.0$ | 1.30 |
| Girder (estimated): $0.40 \times 16.5$ | 6.60 |
| $\sum = $ | 37.68 k |

Figure 5.3 can be used to determine the live loads per girder per panel point.

Number of lane loads to one girder: $\dfrac{W(9+21)}{24} = 1.25\,W$

Panel point load for uniform equivalent lane load: $0.64 \times 16.5 \times 1.25 = 13.2$ k

Concentrated load
  Moment: $18 \times 1.25 = 22.5$ k
  Shear: $26 \times 1.25 = 32.5$ k

Design moments are determined as in Tables 5.1 through 5.6.

*Impact percentages*

Positive moment

End span: $\dfrac{50}{125+99}(100) = 22.3$

Center span: $\dfrac{50}{125+132}(100) = 19.5$

Negative moment: $\dfrac{50}{125+(99+132)/2}(100) = 20.8$

DESIGN OF GIRDER CROSS SECTION

Minimum girder depth is $\frac{1}{25}$ of span where span is the distance between dead-load points of contraflexure. This span is approximately 75 ft, and $\frac{1}{25}$ (75) = 3 ft. This depth would result in flange plates too large for economy in A36 steel. A web plate depth of 6 ft is tried. The minimum web thickness is

$$\frac{D}{165} = \frac{72}{165} = 0.436 \text{ in.} \qquad \text{Use } 72 \times \tfrac{7}{16} \text{ in. web}$$

The maximum shear is 222 k at position $6R$-7 and the average web shear stress is $222/72(\tfrac{7}{16}) = 7.0 < 12$ ksi. The resisting moments of the girder with various flange plates are calculated in the table on page 121. The minimum $b/t$ ratio is 23(A36 steel).

CHECK OF FATIGUE STRESSES

From the plot of the bending moment diagram (Fig. 5.10) it is seen that there is a possible stress reversal in the girder from location 3.4 ft to 4.9 ft and from 7.2 to 8.5. The strength in fatigue must be checked in these positions along the girder. It is considered that the ADTT is less than 2500. Therefore, for truck loading, the number of cycles is 500,000. The maximum range of live load plus impact bending moment between locations 3.4 ft and 4.9 ft is about 2300 ft-k. If the flange is $20 \times 1\frac{1}{8}$ in. in this region, the range of stress is equal to

$$f_{sr} = \frac{2300(37.125)(12)}{73,760} = 13.9 \text{ ksi}$$

| Section | | $I$ of Web | $I$ of Flange | Total $I$ | Resisting Moment |
| Web | Flange | | | | |
|---|---|---|---|---|---|
| | $20 \times 2\frac{1}{4}$ | 13,610 | 124,040 | 137,650 | $\dfrac{20(137,650)}{12(38.25)} = 5998$ ft-k |
| | $20 \times 1\frac{3}{4}$ | 13,610 | 95,180 | 108,790 | $\dfrac{19.265^a(108,790)}{12(37.75)} = 4625$ |
| | $20 \times 1\frac{1}{4}$ | 13,610 | 67,070 | 80,670 | $\dfrac{19.265^a(80,670)}{12(37.75)} = 3477$ |
| | $18 \times \frac{7}{8}$ | 13,610 | 41,820 | 55,430 | $\dfrac{19.09^a(55,430)}{12(36.875)} = 2392$ |
| | $16 \times \frac{3}{4}$ | 13,610 | 31,750 | 45,360 | $\dfrac{18.85^a(45,360)}{12(36.75)} = 1940$ |
| | $20 \times 1$ | 13,610 | 53,290 | 66,900 | $\dfrac{20(66,900)}{12(37.0)} = 3013$ |
| | $20 \times 1\frac{1}{8}$ | 13,610 | 60,150 | 73,760 | $\dfrac{20(73760)}{12(37.125)} = 3310$ |
| | $20 \times 1\frac{3}{16}$ | 13,610 | 63,610 | 77,220 | $\dfrac{20(77,220)}{12(37.19)} = 3461$ |
| | $18 \times \frac{7}{8}$ | | | 55,430 | $\dfrac{20}{19.265}(2392) = 2483$ |
| | $16 \times \frac{3}{4}$ | | | 45,360 | $\dfrac{20}{18.85}(1940) = 2058$ |

(left margin, rotated: $72 \times \frac{7}{16}$ in.)

$^a$ $F_b = 20,000 - 7.5(16.5 \times 12/b)^2$ (in the negative moment region).

From the AASTO Specifications (Sec. 7) the allowable range of stress for the various categories is:

| Category | $F_{sr}$ (ksi) |
|---|---|
| A | 36 − 24* |
| B | 27.5 − 18* |
| C | 19.0 − 13* |
| D | 16.0 − 10* |
| E | 12.5 − 8* |

* Non-redundant load path structures.

The actual $f_{sr}$ of 13.9 ksi is below that allowable for categories A through D. The groove weld at the flange splice is category B, as is the flange-to-web fillet weld. The weld on the transverse stiffeners is category C. The $f_{sr}$ at the end of the stiffener would also be less than 13.0 ksi. All $f_{sr}$ stresses are below the allowable since there are no category stresses below C in the girder. See Fig 5.7 for design plate sizes.

### DESIGN OF INTERMEDIATE STIFFENERS

The transverse intermediate stiffeners can be omitted where thickness of web plate is not less than

$$t = \frac{D\sqrt{Fv}}{7500} \qquad Fv = \frac{V}{Dt}$$

For $t = \frac{7}{16}$ in. and $D = 72$ in.,

$$V = \frac{7}{16}(72)\frac{(7500 \times \frac{7}{16})^2}{72} = 65,420 \text{ lb}$$

If $t < D/150 = 0.48$ in. $> \frac{7}{16}$ (stiffeners cannot be omitted), spacing of stiffeners is given by

$$d = \frac{11,000t}{\sqrt{Fv}}$$

At the end support $V = 133$ k. Then

$$d = \frac{11,000(\frac{7}{16})}{\sqrt{133,000/72(\frac{7}{16})}} = 74 \text{ in.}$$

The first two stiffeners must be at $74/2 = 37$ in. Vertical stiffeners are required for bolted attachment to the floor beam. Other stiffeners should be equally spaced between floor beams. (see table below).

| No. of Stiffeners Between Floor Beams | $d$ (in.) |
|---|---|
| 1 | 99 > 72 (cannot use) |
| 2 | 66 |
| 3 | 49.5 |

$$\text{Maximum spacing} = \frac{11,000}{\sqrt{fv}} = \frac{4810}{\sqrt{V/31.5}}$$

$$V = \frac{31.5(4810)^2}{d^2}$$

Therefore for $d = 66$ in.

$$V = 167.3\,k$$

and for $d = 49.5$ in.

$$V = 297.5\,k$$

| Panel | Stiffener Spacing (in.) |
|-------|-------------------------|
| 0–1 | 66 (place first at 33) |
| 1–2 | 66 |
| 2–3 | 66 |
| 3–4 | 66 |
| 4–5 | 66 |
| 5–6L | 49.5 |
| 6R–7 | 49.5 |
| 7–8 | 49.5 |
| 8–9 | 66 |
| 9–10 | 66 |

A single stiffener on one side only is used, the dimensions of which are given below (see Fig. 5.17):

Minimum width: $2 + \frac{1}{30}(D) = 4.4$ in. but preferably not less than $20/4 = 5$ in.

Minimum thickness: $\frac{5}{16}$ in.

Actual $I_{1-1}$: $\frac{1}{3}(\frac{5}{16})(5)^3 = 13.0$ in$^4$.

Minimum $I_{1-1}$: $\frac{d_o t^3 J}{10.92}$; then $J = 25\left(\frac{D^2}{d^2}\right) - 20 \nless 5.0$

Smallest required $d$: $\dfrac{4810}{\sqrt{222{,}000/31.5}} = 56.4$; then $J = 25\left(\dfrac{72}{56.4}\right)^2 - 20 = 20.7$

Minimum $I_{1-1}$: $\dfrac{49.5(7/16)^3(20.7)}{10.92} = 7.9 < 13.0$

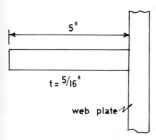

5″

$t = \frac{5}{16}''$

web plate

**Figure 5.17** Web plate stiffener detail.

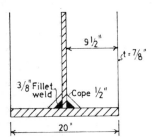

**Figure 5.18**   Bearing stiffener—center pier.

For $d_o$ 66 in. $J$ is less than 20.7 and $I_{1-1} < 13.0$. Use $5 \times \frac{5}{16}$ in. intermediate stiffener.

DESIGN OF BEARING STIFFENERS (Fig. 5.18)

At pier, $R = 221 + 222 = 443$ k using a single stiffener on each side, the maximum width is computed as:

Maximum width: $\dfrac{20}{2} - \dfrac{0.4375}{2} = 9.78$ in.; $9\frac{1}{2}$-in. width should be tried.

Minimum thickness: $\dfrac{b}{12}\sqrt{\dfrac{F_y}{33,000}} = \dfrac{9.9}{12} = 0.83$ in.; $\frac{7}{8}$-in. thickness should be tried.

Bearing area on flange: $(9.5 - 0.5)(0.875)(2) = 15.75$ in.$^2$

$$f_b = \frac{443}{15.75} = 28.1 < 29.0 = F_b$$

The stiffeners are checked for column strength Fig. (5.19):

$$I = \frac{0.875(19.44)^3}{12} = 536 \text{ in.}^4$$

$A = 19(\frac{7}{8}) + 7.9(0.44) = 20.1$ in.$^2$

$f_a = \dfrac{443}{20.1} = 22.0$ ksi (too large)     approximate required $A = \dfrac{443}{15.8} = 28.0$ in.

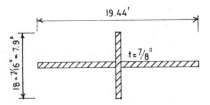

**Figure 5.19**   Bearing stiffener—end pier.

Approximate $t = \dfrac{28.0}{19.5} = 1.4$ in.

$t = 1\frac{3}{8}$ in. is tried. Therefore $A = 19(1.375) + 7.9(0.44) = 29.6$ in.$^2$

$$I = \frac{1.375(19.44)^3}{12} = 842 \text{ in.}^4$$

$$r = \sqrt{\frac{842}{29.6}} = 5.33 \text{ in.}$$

$$F_a = 16,000 - 0.3 \left(\frac{72 \times 0.75}{5.33}\right)^2 = 15,960 \text{ psi}$$

$$f_a = \frac{443}{29.6} = 15.0 < 15.97 \text{ ksi}$$

Use $9\frac{1}{2}$ in. $\times 1\frac{3}{8}$ in. bearing stiffener at pier locations.

At end of bridge, $R = 175$ k. Stiffener plate $8\frac{3}{4}$ in. is tried $\times \frac{3}{4}$; minimum $t = \frac{3}{4}$ in.

$$A = 17.5(0.75) + 7.9(0.44) = 16.6 \text{ in.}^2$$

$$I = \frac{0.75(17.9)^3}{12} = 358 \text{ in.}^4$$

$$r = \sqrt{\frac{358}{16.6}} = 4.65$$

$$F_a = 15.96 \text{ ksi} \qquad f_a = \frac{175}{16.6} = 10.5$$

Use $8\frac{3}{4}$ in. $\times \frac{3}{4}$ in. in bearing stiffener at end bearings.

DESIGN OF WELDING

Fillet welds—flange to web

Maximum material thickness: $2\frac{1}{4}$ in.
Minimum size of fillet weld: $\frac{1}{2}$ in.
Shear per inch: $\dfrac{VQ}{I}$; maximum $V = 222$ k

$$Q = 20 \times 2\frac{1}{4}(36 + 1.125) = 1671 \text{ in.}^3$$

$$I = 137,650 \text{ in.}^4$$

$$v = \frac{222(1671)}{137,650} = 2.69 \text{ k/in.}$$

Strength of two $\frac{1}{2}$-in. fillet welds: $2 \times \frac{1}{2} \times 0.707 \times 15.7 = 11.1$ k/in. (1977 revision)

Welds of $\frac{1}{2}$-in. are quite adequate but cannot be reduced because of $2\frac{1}{4}$-in. plate. At the end of the bridge the weld can be reduced to $\frac{5}{16}$ in. (maximum thickness $= 1\frac{1}{8}$ in.). Then

$$Q = 20 \times 1\frac{1}{8}(36 + 0.56) = 823 \text{ in}^3$$

$$V = 133 \text{ k}$$

$$I = 73,760 \text{ in}^4$$

$$v = 1.5 \text{ k/in.} < 2 \times \frac{5}{16} \times 0.707 \times 15.7 = 6.9 \text{ k/in.}$$

All web-to-flange fillet welds can be minimum size consistent with maximum material thickness.

Fillet welds—stiffener to web
End bearing: $R = 175$ k

$$\text{Shear per inch of weld: } \frac{175}{2(72)} = 1.21 \text{ k}$$

Maximum material thickness: $\frac{3}{4}$ in.
Minimum fillet weld: $\frac{5}{16}$ in.
Strength of $\frac{5}{16}(0.707)(15.7) = 3.5$ k/in. $> 1.21$; use $\frac{5}{16}$-in. weld

For the bearing at the pier; $R = 443$ k,

$$\text{shear per inch of weld} = \frac{443}{2(72)(2)} = 1.54 \text{ k/in.}$$

The $\frac{5}{15}$-in. weld is adequate for strength and since maximum material thickness is $1\frac{3}{8}$ in., $\frac{5}{16}$-in. weld is satisfactory.

## 5.5 HYBRID GIRDERS

A hybrid girder is one that has the main load-carrying plates fabricated from more than one type of steel. The common example has the web of a plate girder made from A36 steel, while the flange plates are made of a higher-strength steel. In most plate girder bridges, A36 is adequate to carry the shear stresses, however for flange plates the use of a higher-strength steel results in a more reasonable size. A question does exist in a hybrid girder in that the extreme fiber of the web plate has stresses in bending that are only slightly smaller (especially for a deep girder) than the stresses in the flange plate. If the actual bending stress in a flange plate of steel with a yield stress of 50 ksi were near the allowable value (27 ksi), then the bending stress in the extreme fiber of the web plate would very likely be greater than 20 ksi, the allowable tensile bending stress for A36 steel. The Specification does allow the stress in the lower-strength plate to exceed the allowable bending stress when it is attached to a higher-strength plate. The procedure in the Specification is not, however, to increase the allowable bending stress in the web, but to reduce the allowable bending stress in the flange so that the bending stress in the web is within a reasonable value.

The Specification states that the stress in the web steel can exceed the allowable stress provided the actual stress in the flange steel does not exceed the allowable stress for the flange steel reduced by the factor $R$, which is given by

$$R = 1 - \frac{\beta\psi(1-\alpha)^2(3-\psi+\psi\alpha)}{6+\beta\psi(3-\psi)}$$

where    $\alpha$ = the minimum specified yield strength of the web divided by the minimum specified yield strength of the tension flange

$\beta$ = the area of the web divided by the area of the tension flange

$\psi$ = the distance from the outer edge of the tension flange to the neutral axis (of the transformed section for composite girders) divided by the depth of the steel section

As an example the girder in Design Example 5.1 is redesigned at the section of maximum bending moment (5845 ft-k) using A572 Grade 50 steel for flange plates but with the web of A36 steel.

The normal allowable bending stress for A572 grade 50 for thicknesses up to 2 in. is 27 ksi.

Checking on a girder with web of 72 in. × $\frac{7}{16}$ in. (A36) and flanges of 20 in. × $1\frac{5}{8}$ in. (A572) gives

$$I = 13,610 + 32.5(2710) = 101,685 \text{ in.}^4$$

$$f_b = \frac{5845(12)(37.625)}{101,685} = 26.0 \text{ ksi}$$

$$\alpha = \frac{f_{yw}}{f_{yf}} = \frac{36}{50} = 0.72$$

$$\beta = \frac{A_w}{A_f} = \frac{31.5}{32.5} = 0.97$$

$$\psi = \frac{c}{d} = 0.5$$

$$R = 1 - \frac{0.97(0.72)(1-0.72)^2(3-0.5-0.5\times0.72)}{6+0.97(0.5)(3-0.5)}$$

$$= 0.984$$

allowable bending (A572) = 27(0.984) = 26.6 > 26.0 ksi

The above calculations show that the girder can have a 20 in. $\times 1\frac{5}{8}$ in. flange of A572 steel in place of a 20 in. $\times 2\frac{1}{4}$ in. (A36) flange. Although the unit price of A572 plate is about 15% greater, the reduction in area of flange plate is 27.8%; thus this hybrid design would likely be cheaper than the all-A36 design.

## 5.6   ALTERNATE DESIGNS

The bridge selected for Design Example 5.1 consists of two girders and a floor system. An alternative choice is four welded plate girders without a floor system. This probably would not result in any saving of steel but it could result in a lower fabrication cost and thus a lower total cost. The former scheme is used here to show the procedure for designing floor system, as well as a plate girder. The design example procedure for stringers and floor beams is similar for long-span steel bridges such as continuous trusses and arches.

The girders can be made composite with the deck, however, this would considerably increase the design calculations. Since a composite design is covered in Chapter 4 and because of the size limitation of this book, such a design is not covered here.

## 5.7   CURVED GIRDERS

During the past five years there has been a trend toward the design and construction of horizontally curved highway bridges, as shown in Figs. 5.20 and 5.21. In the past alignment of these curved bridges was provided by straight girders or chords that met the curvature. This solution, however, usually necessitated additional piers and an unattractive structure. The actual curving of girders has permitted greater span lengths, fewer piers, and more aesthetically pleasing structures.

This type of structure, however, has inherently created new design problems for the engineer. Due to the curve of the girder and the eccentricity of the midspan with respect to the supports, the girder under load will twist in addition to distorting vertically. This twisting action induces additional stresses that must be considered along with the normal bending stresses. If one considers the twisting of an isolated curved girder under load, the effects are quite severe. However if that same girder is interacting, by means of transverse diaphragms, with a series of curved girders, the twisting effects are minimized, thus affording a more economical design. Therefore the engineer should consider the system response

**Figure 5.20** Continuous-span curved I girders.

in the design of curved-plate girder bridges. The analyses and subsequent design of such a highly indeterminate system have generally required the use of specialized computer programs (3–7). However, in all instances these programs require girder stiffness and thus preliminary examination of the system and estimation of girder forces is needed. The following, kindly suggested by Mr. D. Nettleton of FHWA, will describe a simplified prodedure for estimating the dead- and live-load forces on curved bridges (8).

DEAD-LOAD RESPONSE: BENDING EFFECTS

*Bending Moment.* The bridge system (Fig. 5.22) that is subjected to a dead load $W$ is considered here. The total overturning moment in the system due to this dead load can be estimated by considering the effective moment to be developed about the chord of each girder of the bridge as shown in Fig. 5.23. The average moment arm $Y_{av}$ as related to the

**Figure 5.21** Erection of curved I girders.

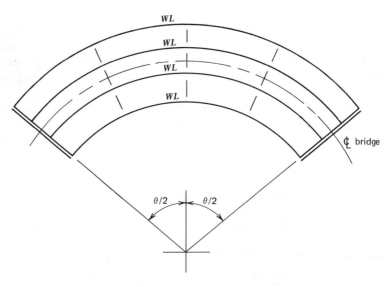

**Figure 5.22** Curved-girder dead-load system.

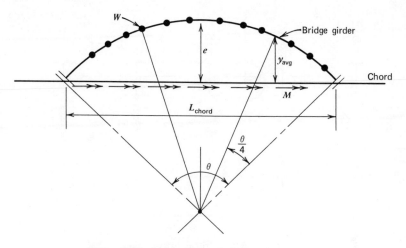

**Figure 5.23**   Isolated curved girder—dead load.

moment arm at the center line $e$ is related by

$$Y_{av} = e\left(1 - \frac{1 - \cos \theta/4}{1 - \cos \theta/2}\right) \tag{5.6}$$

For ranges of $\theta$ applicable to highway bridges equation 5.6 gives

$$Y_{av} = \tfrac{3}{4}e \tag{5.6a}$$

which is within less than 1% for any value of $\theta$. The resulting overturning moment per girder is then

$$M_G = (\tfrac{3}{4}e)\,WL \tag{5.7}$$

The total combined effect for all girders $(N)$ is

$$M_T = N(\tfrac{3}{4}e)WL \tag{5.8}$$

The effect of this total torque on the system can now be determined by examining a cross section of the bridge (Fig. 5.24). Assuming the system rotates rigidly about the bridge center line, then the induced forces imposed by the torque $M_T$ are

$$M_T = \tfrac{3}{2}SG_1 + S/2\,G_2 + S/2\,G_3 + 3/2SG_4 + \cdots + XG_N \tag{5.9}$$

With the assumption of linear distribution of forces induced, which is similar to the rivet problem,

$$\frac{G_1}{\tfrac{3}{2}S} = \frac{G_2}{S/2} = \frac{G_3}{S/2} = \frac{G_4}{\tfrac{3}{2}S} = \frac{G_N}{X} \tag{5.10}$$

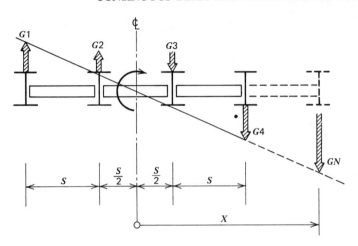

**Figure 5.24** Bridge cross section.

The forces relative to $G_N$ and $X$ are therefore

$$G_N = \frac{G_N X}{X}; \qquad G_1 = \frac{\frac{3}{2}S}{X} G_N, \qquad G_2 = \frac{S/2}{X} G_N \cdots$$

Substituting these values into equation 5.9 gives

$$M_T = (\tfrac{3}{2}S)\left(\frac{\frac{3}{2}S}{X}\right) G_N + S/2 \frac{(S/2)}{X} G_N + \cdots \frac{X^2}{X} G_N$$

$$M_T = (\tfrac{3}{2}S)^2 \frac{G_N}{X} + (S/2)^2 \frac{G_N}{X} + \cdots X^2 \frac{G_n}{X}$$

$$M_T = \frac{G_n}{X} \left[(\tfrac{3}{2}S)^2 + (S/2)^2 + \cdots X^2\right]$$

In general, therefore,

$$M_T = \frac{G_n}{X} \sum X^2 \qquad (5.11)$$

The forces on each girder can now be written as

$$G_N = \frac{M_T X}{\sum X^2} \qquad (5.12)$$

$$G_1 = \frac{M_T(\tfrac{3}{2}S)}{\sum X^2} \cdots$$

where $\sum X^2$ is the sum of the squares of the distance to each girder. In terms of load, therefore, using equation 5.8 the girder force $G_N$ is

$$G_N = \frac{M_T X}{\sum X^2} = (N(\tfrac{3}{4}e)\ WL) \frac{X}{\sum X^2} \tag{5.13}$$

which represents the additional induced vertical load to girder $N$. This load is then added to the effective dead load $WL$. The girder moments are computed using

$$M_G = \frac{(WL + G_N)L}{8} \tag{5.14}$$

*Bending Stress.*   The dead-load bending stress is then determined using the conventional equation

$$\sigma_b = \frac{M_G}{S} \tag{5.15}$$

where $S =$ section modulus of girder.

DEAD-LOAD RESPONSE: WARPING EFFECTS

*Warping Moment.* As mentioned previously, when a girder is subjected to torsional forces normal stresses may be developed in the cross section. These stresses occur as a result of warping or lateral bending of the flanges. This type of distortion is shown in Fig. 5.25. Also illustrated are the normal warping stress distributions in an I-shaped girder, in addition to the normal bending stresses.

These lateral bending or warping stresses can be approximated by examining the effects the bending stress has on the curved flange. The effect of the bending stresses on the curved flange located between transverse diaphragms is shown in Fig. 5.26. These resultant stresses give forces $F$ or

$$F = \frac{M_G}{h} \tag{5.16}$$

Examination of the equilibrium of a piece of this flange (Fig. 5.27) shows that the vertical forces $F_V$ are not balanced by external loading, thus an internal moment must be present (warping moment). However, for ease of development a given load $q$ to be applied to the element, is considered; this load will balance the forces $F$. Thus we have

$$F_V = 0$$
$$2F \sin \Delta = S \cdot q$$

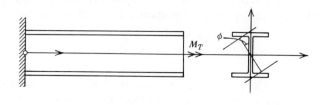

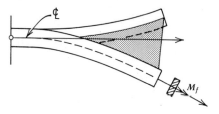

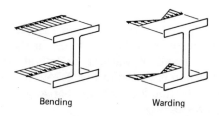

Bending                    Warding

**Figure 5.25**  Warping of *W*-shape section.

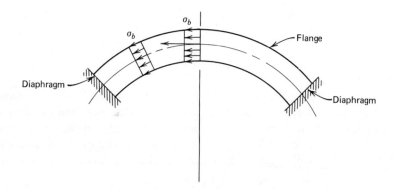

**Figure 5.26**  Bending stresses in curved flange.

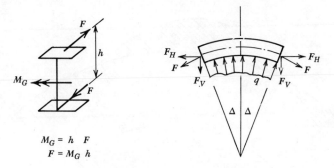

$$M_G = h \ F$$
$$F = M_G \ h$$

**Figure 5.27** Flange equilibrium.

However,

$$S = R(2\Delta)$$
$$2F \sin \Delta = R(2\Delta)$$

and for a small angle $\sin \Delta = \Delta$; therefore

$$2F\Delta = R2\Delta q$$

and

$$q = \frac{F}{R} \tag{5.17}$$

Substituting in equation 5.16 gives

$$q = \frac{M_G}{h} \frac{1}{R} \tag{5.18}$$

This load is now applied to the girder flange, acting as a beam as shown in Fig. 5.28. The induced moments are then determined from

$$M_f(\text{end}) = \frac{q(kL)^2}{12} \tag{5.19}$$

where $kL = $ diaphragm spacing, $M_G$ is assumed constant, and the flange is fixed at the diaphragm. Substituting equation 5.18 into 5.19 gives the resulting warping moment:

$$M_w = \frac{M_G(kL)^2}{12hR} \tag{5.20}$$

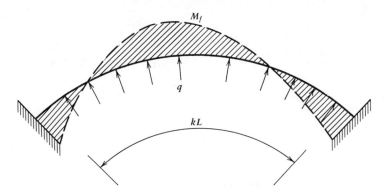

**Figure 5.28**   Equivalent flange load.

*Warping Stress.*    The warping stresses are now evaluated by examining a section of the flange, as shown in Fig. 5.29

$$\sigma_w = \frac{M_w}{S} \tag{5.21}$$

where

$$S = \frac{b_f^2 t_f}{6},$$

Therefore $\sigma_w$ is given by

$$\sigma_w = \frac{6M_w}{b_f^2 t_f} \tag{5.22}$$

where  $b_f$ = flange width
$T_F$ = flange thickness

### DEAD-LOAD RESPONSE: DISTORTIONAL EFFECTS

The general response of thin-walled structural elements considers that the elements will remain at right angles during their deformation. If the

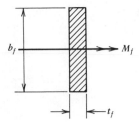

**Figure 5.29**   Flange moment.

elements, which comprise the girder and systems, are subjected to variations in shears, distortion of the system takes place and thus induced normal stresses occur. For the distortional normal stresses to be minimum, lateral bracing or the use of diaphragms is required. In the case of curved girders, distortion effects are more severe because of the curvature. The complete analysis of distortional effects is quite complicated (10), however, recent research (11) has resulted in direct application of this theory for design use. The induced distortional normal stress, due to dead load is approximated by the following equation:

$$\frac{(\sigma_d)}{\sigma_b + \sigma_w} = \frac{[A(X)^3 + B(X)^2 + C(X) + D]}{100} \qquad (5.23)$$

where
$\sigma_d$ = induced normal distortional stress (ksi)
$\sigma_b$ = normal bending stress (ksi) (equation 5.15)
$\sigma_w$ = normal warping stress (ksi) (equation 5.22)
$X$ = diaphragm spacing/span length
$A, B, C, D$ = constants as given in Table 5.7.

**Table 5.7 Dead-Load Distortional Coefficients**

| Girder Length (ft) | Radius (ft) | A | B | C | D |
|---|---|---|---|---|---|
| | 250 | 0 | 19.75 | 4.722 | 0.3027 |
| 50 | 500 | 0 | 44.38 | −2.619 | 0.9935 |
| | 750 | 0 | 61.38 | −8.986 | 1.667 |
| | 100 | 0 | 73.11 | −13.67 | 2.175 |
| | 250 | 0 | −15.71 | 18.20 | −1.196 |
| 75 | 500 | 0 | −1.732 | 17.13 | −1.221 |
| | 750 | 0 | 8.400 | 15.32 | −1.129 |
| | 1000 | 0 | 15.72 | 13.70 | −1.025 |
| | 250 | −14.93 | 4.022 | 10.89 | −0.5629 |
| 100 | 500 | −24.60 | 28.10 | 7.729 | −0.4398 |
| | 750 | −9.880 | 32.48 | 6.910 | −0.4023 |
| | 1000 | 3.445 | 33.40 | 6.564 | −0.3824 |
| | 250 | −181.2 | 70.24 | 0.9687 | −0.06552 |
| 125 | 500 | −208.6 | 101.3 | −2.562 | 0.05658 |
| | 750 | −203.7 | 111.0 | −3.913 | 0.1090 |
| | 1000 | −187.4 | 111.8 | −4.163 | 0.1200 |
| | 250 | −254.1 | 108.2 | −3.548 | 0.1287 |
| 150 | 500 | −258.5 | 131.4 | −6.327 | 0.2254 |
| | 750 | −233.5 | 132.8 | −6.709 | 0.2411 |
| | 1000 | −217.3 | 133.2 | −6.918 | 0.2508 |

For Table 5.7 interpolation may be used for data between the limiting parameters.

*Bending Moment.*   In considering live loads, the standard HS20 truck which governs for spans of $L \le 140$ ft. is used. If lane loading governs, then the combination of the dead-load and live-load techniques can be emphasized.

The truck loading, when applied to the curved bridge system, is assumed to initiate two conditions:

1. Torsional moments about the chords of the respective loaded curved girder, which may be designated as *local* loading.
2. Torsional moments about the centerline of the entire bridge system, due to the combined truck loading or center of gravity of these loads. This effect may be designated as *system* loading.

These two combined effects are now explained, by examining Figs. 5.30 through 5.32.

*Local Loading.*   This effect, as noted, takes into account the curvature of the girder and the induced torsional moments due to this curvature. It is assumed in this development that the vehicle axle weight is concentrated along the curved girder, as shown in Fig. 5.30. The effective induced

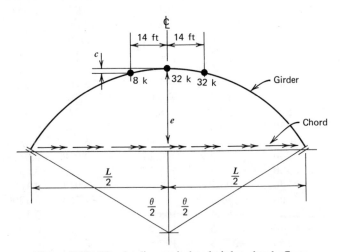

**Figure 5.30**   Live loading on isolated girder—local effects.

moment, as in the case of dead-load effects is of the form

$$M = yP \qquad (5.24)$$

where $y$ is the moment arm to the respective axle loads. In the case of the 32-k load at the centerline $y = e$ and for the other two loads $y = (e - c)$. The $c$ distance is found by assuming the curve to have the form $y = AX^2 + B$. Evaluation of these constants gives $y = e[1 - (X/0.5L)^2]$. The $y$ value at $X = 14\,ft$ ft is therefore $y = e[1 - (14/0.5L)^2]$ and the offset $c = e - y$, or $c = (14/0.5L)^2 e$ or $c = \bar{K}e$. Examination of this $c$ value for various span lengths gives the following data:

| $L$ (ft) | $c/e = \bar{K}$ |
|---|---|
| 140 | 0.040 |
| 100 | 0.0785 |
| 70 | 0.160 |

The total moment about the chord is now evaluated as

$$M_G = (e - c)8 + (e - c)32 - e32$$

$$M_G = 72e\left[1 - \frac{40}{72}\bar{K}\right] \qquad (5.25)$$

where $\bar{K}$ is given in the table above and is dependent on the span length. The term in brackets in equation 5.25 gives the following values for these span lengths:

| $L$ (ft) | $[1 - (40/72)\bar{K}]$ |
|---|---|
| 140 | 0.98 |
| 100 | 0.96 |
| 70 | 0.91 |

An average value of 0.95 is obtained; therefore

$$M_G = 72e[0.95] \qquad (5.26)$$

If there are numerous vehicles on the structure, assuming each is concentrated on a given girder, the total induced moment is

$$M_T = (0.95e)\,72N_T \qquad (5.27)$$

where $N_T =$ number of trucks.

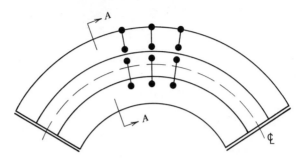

**Figure 5.31** Live loading on girders—system effects—plan.

*System Loading.* The effect of all vehicles on the bridge system may now be determined by examining the general plan and cross section of the bridge, as shown in Figs. 5.31 and 5.32. The response of the vehicles is examined relative to the bridge center line and the resultant of all of these vehicles. As shown in Fig 5.32,

$$M_T = \bar{X}x(2 \times 72) \tag{5.28}$$

or if there are numerous trucks ($N_T$) then

$$M_T = \bar{X}N_T72 \tag{5.29}$$

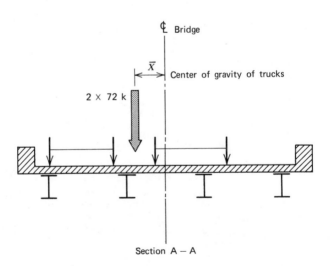

**Figure 5.32** Live loading on girders—systems effects—cross section.

*Combined Loading.* The combined induced torsional moment is now determined by summing equations 5.22 and 5.24, which gives

$$M_T = 72N_T(0.95e + \bar{X}) \qquad (5.30)$$

where $N_T$, $e$, and $\bar{X}$ are as defined previously. Equation 5.30 is similar to the dead-load torque expression equation 5.8. The forces to each girder, due to this live load torque, are similarly assumed distributed as according to the dead-load effects as shown in Fig. 5.24 and given by equation 5.13. Therefore the additional force on a given girder, at some distance $X$ from the bridge center line is

$$G_N = \frac{M_T X}{\sum X^2}$$

or

$$G_N = 72N_T(0.95e + \bar{X})\frac{X}{\sum X^2} \qquad (5.31)$$

This additional force effect is then added to the existing line of AASHTO truck wheels (36 k) on the span.

BENDING STRESS

The normal bending stress is

$$\sigma_b = \frac{M_B}{S} \qquad (5.32)$$

where $\quad M_b = M_{\text{AASHTO}}(1 + CE)$

$$CE = 2N_T(0.95e + \bar{X})\frac{X}{\sum X^2} \qquad (5.33)$$

$M_{\text{AASHTO}}$ = induced moment due to a line of
$\qquad\qquad$ AASHTO truck wheels (36 k)
$\qquad S$ = section modulus

WARPING STRESS

The value for warping stress is determined according to

$$\sigma = \frac{6M_w}{b_f^2 t_f} \qquad (5.34)$$

where $\; M_w = M_b(KL)^2/12hR$
$\qquad M_h = q$

**Table 5.8   Live-Load Distortional Coefficients**

| Girder Length (ft) | Radius (ft) | A | B | C | D |
|---|---|---|---|---|---|
| | 250 | 0 | −7.115 | 12.88 | 4.155 |
| 50 | 500 | 0 | 6.160 | 14.17 | 4.573 |
| | 750 | 0 | 16.75 | 14.30 | 4.482 |
| | 1000 | 0 | 26.65 | 12.53 | 4.632 |
| | 250 | 0 | −78.09 | 40.65 | −0.3318 |
| 75 | 500 | 0 | −90.80 | 54.60 | −1.585 |
| | 750 | 0 | −88.91 | 58.57 | −2.004 |
| | 1000 | 0 | −84.77 | 59.83 | −2.168 |
| | 250 | 662.8 | −418.8 | 94.61 | −3.479 |
| 100 | 500 | 663.6 | −407.8 | 100.1 | −3.944 |
| | 750 | 699.2 | −411.3 | 102.9 | −4.164 |
| | 1000 | 710.8 | −405.2 | 102.8 | −4.181 |
| | 250 | −2067.0 | 761.1 | −79.58 | 4.989 |
| 125 | 500 | −2709.0 | 1027.0 | −107.4 | 6.038 |
| | 750 | −2807.0 | 1085.0 | −113.5 | 6.269 |
| | 1000 | −3072.0 | 1193.0 | −125.6 | 6.727 |
| | 250 | 4783.0 | −1949.0 | 265.5 | −9.159 |
| 150 | 500 | 5209.0 | −2077.0 | 283.4 | −9.873 |
| | 750 | 5474.0 | −2159.0 | 293.8 | −10.28 |
| | 1000 | 5632.0 | −2207.0 | 299.8 | −10.51 |

DISTORTIONAL STRESS

Distortional stress is given by

$$\left(\frac{\sigma_d}{\sigma_b + \sigma_w}\right) = \frac{[A(X)^3 + B(X)^2 + C(X) + D]}{100} \tag{5.35}$$

where
$\sigma_d$ = induced normal distortional stress (ksi)
$\sigma_b$ = normal bending stress (ksi), equation 5.32
$\sigma_w$ = normal warping stress (ksi), equation 5.34
$X$ = diaphragm spacing per span length
$A, B, C, D,$ = constants as given in Table 5.8.

## Design Example 5.2

STEEL CURVED COMPOSITE DESIGN PLATE GIRDER BRIDGE (Figs. 5.33 and 5.34)

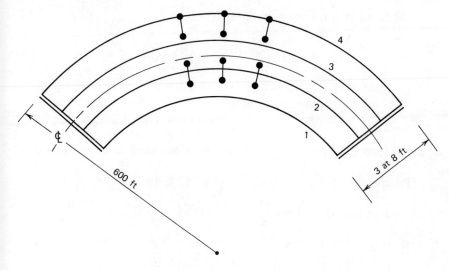

**Figure 5.33**  Plan of curved girder bridge.

In this example, the design of an exterior curved composite girder is presented. A description of the bridge and loading are given in Fig. 5.33. Given

1. Span: 100 ft 0 in. center-to-center bearings outside girder.
2. Structural slab: $7\frac{1}{2}$ in.
3. Girder spacing: 8 ft (four girders).
4. Outside girder radius: 600 ft.
5. Future paving: 25 psf.
6. Load: HS20, no sidewalk live load.
7. A36 = Steel girder.
8. Bridge constructed without shoring.
9. Diaphragms spaced at 20 ft.

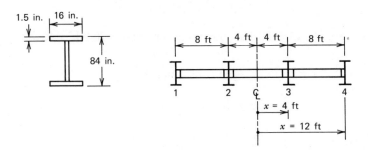

**Figure 5.34**  Dead-load system details.

Dead Load A (no composite action)

| | |
|---|---|
| Slab: $\dfrac{7.5}{12} \times 8.0 \times 0.150$ | 0.750 k/ft |
| Haunch: $2 \times \frac{3}{12} \times \frac{1}{12} \times 0.150$ | 0.006 |
| Steel beam, diaphragms assumed | $\underline{0.250}$ |
| | 1.006 k/ft |

Dead load B (composite action equally distributed to all girders)

| | |
|---|---|
| Parapet: $2 \times 1.0 \times 1.5 \times 0.150/4$ | 0.112 k/ft |
| Asphalt paving: $0.025 \times \dfrac{28}{4}$ | 0.175 |
| Railing $(15 \text{ lb/ft}) \times 2/4$ | $\underline{0.007}$ |
| | 0.294 k/ft |

DEAD LOAD A—BARE STEEL GIRDER (outside)

If the chord and arc lengths are assumed to be essentially the same, the dead-load moment is computed from equations 5.13 and 5.14;

$$G_N = N(\tfrac{3}{4}e)\,WLX/\bar{Z}X^2$$

where  $G_n$ = total force on the girder due to curvature
$N = 4.0$
$e = 2.0$ ft
$W = 1.01$ k/ft,
$L = 100.0$ ft

The term $\sum X^2$ is computed as follows, referring to Fig. 5.34.

$$\sum X^2 = 2[(4)^2 + (12)^2] = 320 \text{ ft}^2$$

The total force $G_N$ on the outside girder, where $X = 12.0$ ft, is

$$G_4 = 4.0(3/4 \times 2)(1.01) \times \frac{12}{320} L$$

$$G_4 = 0.227 \text{ k/ft} \times L$$

$$G_4 = 0.227 \text{ k/ft}$$

The total load per foot of the outside girder is therefore

$$W_T = (W + G_4)$$
$$W_T = (1.01 + 0.227)$$
$$W_T = 1.237 \text{ k/ft}$$

when equation 5.14 is applied, the dead-load $A$ moment is

$$M_{G4} = \frac{W_T L^2}{8}$$

$$M_{G4} = \frac{1.237(100)^2}{8}$$

$$M_{G4} = 1546.3 \text{ k-ft}$$

DEAD LOAD B—COMPOSITE STEEL GIRDER (outside, $n = 30$)

The dead load $B$, as given previously, is 0.294 k/ft. Following the same procedure as given for dead load $A$, the additional load on the outside girder due to curvature is

$$G_4 = 0.227 \text{ k/ft} \times \frac{0.294}{1.01}$$

$$G_4 = 0.0650 \text{ k/ft}$$

The total load $W_T$ is therefore

$$W_T = (0.246 + 0.0650)$$
$$W_T = 0.311 \text{ k/ft}$$

When equation 5.14 is applied, the dead-load $A$ moment is

$$M_{G4} = \frac{W_T L^2}{8}$$

$$M_{G4} = \frac{0.311(100)^2}{8}$$

$$M_{G4} = 388.0 \text{ k-ft}$$

LIVE LOAD

When the HS20 truck is positioned as shown in Fig. 5.35, equation 5.31 can be applied to determine the additional load caused by girder curvature. Thus equation 5.31

$$G_N = 72 N_T (0.95 e + \bar{X}) \frac{X}{\sum X^2}$$

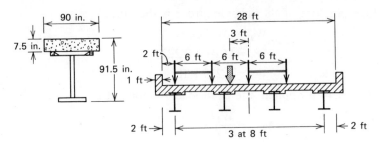

**Figure 5.35** Live-load system details.

where    $N_T = 2$ (number of trucks)

$e = 2.0$ ft

$\bar{X} = 3.0$ (center of gravity of all trucks W.R. to center-to-center of bridge)

$x = 12.0$ ft

$\sum X^2 = 320.0$ ft$^2$

Substituting these values into the equation for $G_4$ gives

$$G_4 = 72 \times 2(0.95 \times 2 + 3)\frac{12}{320}$$

$$G_4 = 26.46 \text{ k}$$

This load represents a 74% increase in a set of truck wheels (36 k) positioned on the outside girder, that is $26.46/36 \times 100 = 74\%$. A set of these wheels (4, 16, and 16 k) positioned on a 100.0 ft girder to induce maximum moment is shown in Fig. 5.36. The value for $M_{max}$ is given by

$$M_{max} = \left[\frac{36}{100}(47.67)^2 - 14 \times 4\right]$$

$$= (818.07 - 56)$$

$$= 762.1 \text{ k-ft}$$

This value is increased by 74%, or

$$M_{LL} = 1.74 \times 762.1 = 1326.05 \text{ k-ft}$$

Thus

$$\text{impact} = \frac{50}{125 + 100} = 0.222$$

$$M_I = 0.222 \times 1326.05 = 294.37 \text{ k-ft}$$

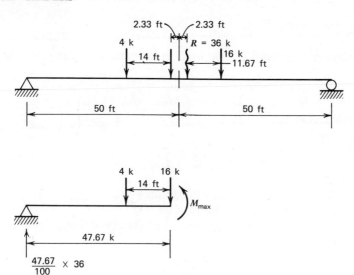

**Figure 5.36** Live-load girder moment.

The total live load + impact moment is given by

$$M_{G4} = 1620.50 \text{ k-ft}$$

Alternately, equation 5.33 may be applied directly, which gives

$$M_{LL} = M_{AASHTO}(1 + CE)$$

$$CE = 2N_T(0.95e + \bar{X})\frac{X}{\sum X^2}$$

$$CE = 2 \times 2(0.95 \times 2 + 3)\frac{12}{320}$$

$$CE = 0.74$$

$$M_{AASHTO} = 762.1 \text{ k-ft } x(1 + 0.222) = 931.3$$

Therefore

$$M_{LL+I} = 931.3(1 + 0.74)$$
$$M_{LL+I} = 1620.5 \text{ k-ft}$$

which agrees with the moment $M_{G4}$ previously calculated.

SECTION PROPERTIES

The following equation can be used to estimate the required flange area (9)

$$A_{sb} = \frac{12}{F_b} \left( \frac{M_{DL}}{d_{cg}} + \frac{M_{SDL} + M_{LL} + M_I}{d_{cg} + t} \right)$$

$$A_{st} = A_{sb} \left( \frac{50}{190 - L} \right)$$

where   $A_{sb}$ = estimated bottom flange area (in.$^2$)
   $A_{st}$ = estimated top flange area (in.$^2$)
   $d_{cg}$ = distance between centers of gravity of flanges of steel shape (in.)
   $t$ = thickness of concrete slab (in.)
   $M_{DL}$ = dead-load moment (k-ft)
   $M_{SDL}$ = superimposed dead-load moment (k-ft)
   $M_{LL}, M_I$ = static-live-load and impact moments (k-ft)
   $F_b$ = allowable flange stress (ksi)

For this design the moments are

$$M_{DL} = 1546.3 \text{ k-ft}$$
$$M_{SDL} = 376.3 \text{ k-ft}$$
$$M_{LL+I} = 1620.5 \text{ k-ft}$$

Assuming a girder depth $d_{cg} \cong 84.0$ in., the estimated flange area is

$$A_{sb} = \frac{12}{20} \left( \frac{1546.3}{84} + \frac{1997.0}{91.5} \right)$$
$$A_{sb} = 24.1 \text{ in.}^2$$

Assuming a flange area of $16.0 \text{ in} \times 1.5 \text{ in.}$ ($A = 24.0 \text{ in.}^2$) and a web thickness of 0.50 in. The properties of this section are computed.

*Steel Section.*

| Element | $A$ | $y$ | $y^2$ | $Ay^2$ | $I_0$ |
|---------|-----|-----|-------|--------|-------|
| Web | 40.5 | — | — | — | 22,1143.4 |
| Flanges (2) | 48.0 | 41.25 | 1701.6 | 81,675 | 13.5 |
| $\Sigma$ | 88.5 | | | 81,675 | 22,156.9 |

Thus

$$I_T = 81,675 + 22,156.9 = 103,831.9 \text{ in.}^4$$

$$S_T = S_B = \frac{103,831.9}{42} = 2472.2 \text{ in.}^3$$

*Composite Section* (n = 30). The effective flange width for composite action is

$$100 \times \tfrac{1}{4} = 25 \text{ ft}$$

$$\text{Center-to-center beams} = 8.0 \text{ ft}$$

$$12 \times \frac{7.5}{12} = 7.5 \leftarrow \text{governs}$$

$$\text{Transformed width of flange} = \frac{7.5 \times 12}{30} = 3.0 \text{ in.}$$

1. *Centroid location* (*reference at center of gravity of slab*).

| Element | A | y | Ay |
|---|---|---|---|
| Slab | $3 \times 7.5 = 22.5$ | — | — |
| Girder | 88.5 | 45.75 | 4048.9 |
| Total | 111.0 | | |

Using the above data gives

$$\bar{y} = \frac{4048.9}{111} = 36.5 \text{ in.}$$

2. *Section Modulus.*

| Element | $I_0$ | A | y | $y^2$ | $Ay^2$ |
|---|---|---|---|---|---|
| Steel section | 103,831.9 | 88.5 | 9.25 | 85.6 | 7,575.6 |
| Slab | 805.5 | 22.5 | 36.5 | 1332.3 | 29,975.6 |
| | 103,937.4 | | | | 37,551.2 |

$I_T$ can now be calculated:

$$I_T = 103,937.4 + 37,551.2$$

$$I_T = 141,488.5 \text{ in.}^4$$

Then

$$S_T = \frac{141,488.5}{40.25} = 3515.2 \text{ in.}^3 \qquad S_B = \frac{141,488.5}{51.25} = 2760.8 \text{ in.}^3$$

*Composite Section* (n = 10).

$$\text{Transformed width of flange} = \frac{7.5 \times 12}{10} = 9.0 \text{ in.}$$

1. *Centroid location (reference about center of gravity of slab).*

| Element | A | y | Ay |
|---|---|---|---|
| Slab | $9 \times 7.5 = 67.5$ | — | — |
| Girder | 88.5 | 45.75 | 4048.9 |
| $\Sigma$ | 156.0 | | |

Using the above data gives

$$\bar{y} = \frac{4048.9}{156} = 26.0 \text{ in.}$$

2. *Section Modulus.*

| Element | $I_0$ | A | y | $y^2$ | $Ay^2$ |
|---|---|---|---|---|---|
| Steel section | 103,831.9 | 88.5 | 19.75 | 390.1 | 34,528.9 |
| Slab | 316.4 | 67.5 | 26.0 | 676.0 | 45,630.0 |
| | 104,148.3 | | | | 80,153.9 |

*Induced stresses on bottom flange.* The bending stresses for a dead load A are

$$\sigma_{DL} = \frac{1546.3 \times 12}{2472.2} = 7.51 \text{ ksi}$$

For dead load B they are

$$\sigma_{DL} = \frac{388.0 \times 12}{2760.8} = \frac{1.68 \text{ ksi}}{9.19 \text{ ksi}}$$

The dead-load warping stress is given by

$$\sigma_w = \frac{6M_w}{b_f^2 t_f}$$

where

$$M_w = \frac{M_G(KL)^2}{12\,hR} = \frac{(1546.3 \times 12)(20)^2}{12 \times 82.5 \times 600} = 12.5 \text{ k-ft}$$

Thus

$$\sigma_w = \frac{12.5 \times 12 \times 6}{(16)^2 \times 1.5} = 2.3 \text{ ksi}$$

Dead-load distortion is calculated from

$$\sigma_d = (\sigma_b + \sigma_w)\frac{[A(X)^3 + B(X)^2 + C(X) + D]}{100}$$

$$\sigma_b + \sigma_w = 9.19 + 2.3 = 11.5 \text{ ksi}$$

$$X = \frac{20}{100} = 0.2$$

From Table 5.7:

$$A = -18.71 \qquad B = 29.95 \qquad C = 7.40 \qquad D = 0.425$$

Substituting into the above equation gives

$$\sigma_d = \frac{11.5}{100.0}[-18.71(0.2)^3 + 29.85(0.2)^2 + 7.4(0.2) - 0.425]$$

$$\sigma_d = 0.23 \text{ ksi}$$

Thus the total dead-load stress is

$$\sigma_T = \sigma_b + \sigma_w + \sigma_d$$

$$\sigma_T = 9.2 + 2.3 + 0.23$$

$$\sigma_T = 11.73 \text{ ksi}$$

Live-load bending stresses are given by

$$\sigma_{LL+I} = \frac{1620.5 \times 12}{2984.7} = 6.52 \text{ ksi}$$

Live-load warping stress is

$$\sigma_w = \frac{6M_w}{b_f^2 t_f}$$

where $M_w = \dfrac{M_b(KL)^2}{12h} = \dfrac{(1620.5 \times 12)(20)^2}{12 \times 82.5 \times 600}$

$$M_w = 13.1 \ \ k\text{-ft}$$

$$\sigma_w = \frac{6 \times 13.1 \times 12}{(16)^2 \times 1.5} = 2.5 \text{ ksi}$$

Live-load distortion stresses are calculated from;

$$\sigma_d = (\sigma_b + \sigma_w)\frac{[A(X)^3 + B(X)^2 + C(X) + D]}{100}$$

$$\sigma_b + \sigma_w = 6.52 + 2.5 = 9.02 \text{ ksi}$$

$$X = \frac{20}{100} = 0.2$$

From Table 5.8

$$A = 649.36 \qquad B = -409.2 \qquad C = 101.2 \qquad D = 4.03$$

Substituting into the above equation gives

$$\sigma_d = \frac{9.02}{100}[649.36(0.2)^3 - 409.2(0.2)^2 + 101.22(0.2) - 4.03]$$

$$\sigma_d = 0.45 \text{ ksi}$$

Thus total live-load stress is

$$\sigma_T = \sigma_b + \sigma_w + \sigma_d$$
$$\sigma_T = 6.52 + 2.5 + 0.45$$
$$\sigma_T = 9.47 \text{ ksi}$$

Combining the stresses total dead-load and live-load gives

$$\sigma = 11.73 + 9.47$$
$$\sigma = 21.2 \text{ ksi}$$

which slightly exceeds the allowable value of 20 ksi.

Design of the other girder follows a similar procedure, with due consideration of girder length, radius, and diaphragm spacing. The design for the web stiffeners and shears connectors follows the designs given in Section 5.5. Also, if the curved girder bridge is continuous, the approximate procedure must be modified by using an effective span length, between supports, when considering torsional effects.

## REFERENCES

1   D. Allan Firmage, *Fundamental Theory of Structures*, Krieger, Huntington, New York, 1971.
2   J. W. Fisher, *Bridge Fatigue Guide*, American Institute of Steel Construction, 1977.

3  K. R. Spates and C. P. Heins, "The Analysis of Single Curved Girder with Various Loadings and Boundary Conditions," *Civil Engineering Report* No. 20, University of Maryland, College Park, Maryland, June 1968.

4  S. Shore and S. Chaudhuri, "Statical Dynamic Analysis of Curved I Bridges," Report No. T0374, Department of Civil Engineering, University of Pennsylvania, Philadelphia, Pennsylvania, May 1974.

5  L. C. Bell and C. P. Heins, "The Solution of Curved Bridge Systems Using the Slope-Deflection Fourier Series Method," *Civil Engineering Report No. 19*, University of Maryland, College Park, Maryland, June 1968.

6  L. C. Bell and C. P. Heins, "Curved Girder Computer Manual," *Civil Engineering Report No. 30*, University of Maryland, College Park, Maryland, September 1969.

7  F. H. Lavelle and J. S. Boick, "A Program to Analyze Curved Girder Bridges," Engineering Bulletin No. 8, Division of Engineering Research and Development, University of Rhode Island, December 1965.

9  J. C. Hacker, "A Simplified Design of Composite Bridge Structures," *Struct. Div. Am. Soc. Civil Eng.*, ASCE Proc. Paper No. 1432, Vol. 93, ST11, November 1957.

10  R. Dabrowski, *Curved Thin-Walled Girders*, Cement and Concrete Association, London, England, 1972.

11  C. P. Heins and T. L. Martin, "Distortional Effects of Curved I Girder Bridges," *Civil Engineering Report* No. 65, University of Maryland, College Park, Maryland, February 1977.

# 6

# Design of Orthotropic Deck Bridges

## 6.1 INTRODUCTION

### 6.1.1 General

An orthotropic steel deck bridge, generally called an "orthotropic bridge," employs a stiffened steel plate to support the vehicle wheel loads instead of a reinforced concrete slab as used in conventional bridge construction. The word orthotropic is derived from the conjunction of two words, *ortho*gonal aniso*tropic*, and means material properties having differences at right angles. The general definition of an orthotropic plate, which has a constant plate thickness $t$, can be described by the following material terms:

$$Dx = Ex \cdot I$$
$$Dy = Ey \cdot I$$
$$H = GI$$

where $Ex$, $Ey$, and $G$ are material properties and $I$ is the plate stiffness per unit of width or $I = \frac{1}{12}t^3$, as shown in Fig. 6.1. Such a plate might be fabricated of wood (plywood).

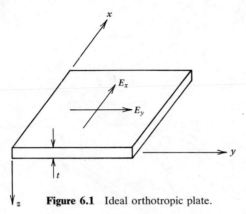

**Figure 6.1**  Ideal orthotropic plate.

In an orthotropic bridge deck, as shown in Fig. 6.2, the deck plate is stiffened in one direction by open ribs or closed ribs. Therefore this type of construction creates a larger inertia ($I$) per unit of width at right angles to the stiffeners than it does on a section that is parallel to the stiffeners. Thus the general stiffness equations can be written as

$$Dx = EIx$$
$$Dy = EIy$$
$$H = GK_T$$

where $Ix$ and $Iy$ are the plate stiffnesses (including stiffeners in two directions) and $K_T$ is the torsional plate stiffness. The evaluation of the plate stiffnesses, for various stiffener configurations, is presented later. With these plate stiffnesses known, the behavior of the deck plate interacting with transverse floor beams and main longitudinal girders of the bridge can then be determined.

In designing bridge structures, it is useful to know their general details. A study (1) of existing orthotropic bridges indicates the following trends:

1. *Open Cellular Decks*
   Size of rib: $\frac{3}{8}$ in. $\times$ 8 in. to 1 in. $\times$ 12 in.
   Rib spacing: 12 in. to 16 in.
   Floor-beam spacing: 4 ft to 7 ft
2. *Closed Cellular Decks*
   Cell spacing: 24 in. to 28 in.
   Size of cell: 12 in. $\times$ 12 in. $\times \frac{5}{16}$ in.
   Floor-beam spacing: 4 ft to 15 ft

A survey of typical properties of orthotropic bridges in existence is given in Tables 6.1 and 6.2.

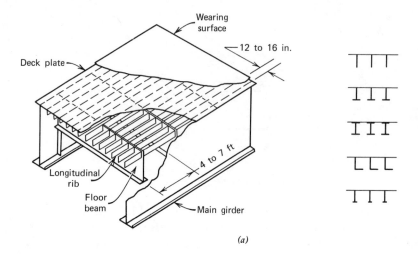

*(a)*

**Figure 6.2** Orthotropic deck bridges:
(*a*) open ribs, (*b*) closed ribs, (*c*) typical
system, (*d*) rib splice.

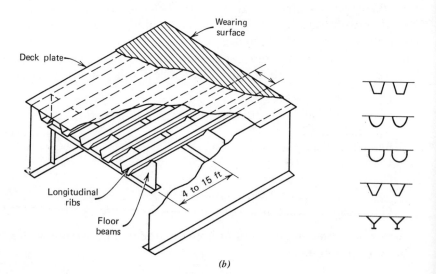

*(b)*

(c)

(d)

**Figure 6.2**   *Continued*

157

## Table 6.1 Properties of the Open-Deck System

| Bridge | Rib Space (in.) | Rib Type | Rib Size (in.) | Deck Plate (in.) | $s/l$ | Depth Ratio[a] | Spacing Ratio[b] | Thickness Ratio[c] |
|---|---|---|---|---|---|---|---|---|
| Save River | 12 | Open | $4\frac{3}{4}\times\frac{7}{16}$ $10\frac{1}{2}\times1$ | $\frac{3}{8}$–1 | 0.133 | 1:9 to 1:3 | 0.025 | 0.079 0.095 |
| Duesseldorf Neuss | 17 | Split I-beam | 7 to 11 deep | $\frac{9}{16}$–$1\frac{3}{32}$ | 0.250 | 1:2 | 0.0575 | 0.080 0.099 |
| Duesseldorf North | 16 | Angles | $8\times4\times\frac{7}{16}$ | $\frac{9}{16}$ | 0.125 | 1:4 | 0.0272 | 0.070 |
| Severin | 12 | Open | $9\frac{1}{2}\times\frac{5}{16}$ $11\frac{1}{2}\times\frac{3}{4}$ | $\frac{3}{8}$ | 0.105 | 1:3.5 | 0.0161 | 0.040 0.030 |
| Cologne Muelhein | 12 | Bulb | $7\times\frac{5}{16}$ | $\frac{1}{2}$ | 0.095 | 1:4 | 0.0178 | 0.0715 |
| AISC Example | 12 | Open | $8\frac{1}{2}\times\frac{1}{2}$ | $\frac{3}{8}$ | 0.120 | 1:3 to 1:4 | 0.02 | 0.044 |

[a] Ratio of depth of rib to depth of floor beam.
[b] Ratio of rib spacing to space width $l$.
[c] Ratio of deck-plate thickness to depth of plate stiffener.

## Table 6.2 Properties of Closed-Deck System

| Bridge | Rib Space (in.) | Rib Type | Rib Size (in.) | Deck Plate Size (in.) | $s/l$ | Depth Ratio[a] | Floor-Beam Properties[b] (in.) |
|---|---|---|---|---|---|---|---|
| Weser | 24 | U 12 in. / 12 in. | $12\times12\times\frac{1}{4}$ | $\frac{1}{2}$ to $\frac{3}{4}$ | 0.40 | 1:2.4 | |
| Duisburg | $23\frac{1}{2}$ | trapezoid 10 in. | $\frac{5}{16}$ | $\frac{9}{16}$ to $\frac{11}{16}$ | 0.133 | 1:8.4 | |
| Port Mann | 24 | U 12 in. / 11 in. | $11\times12\times\frac{5}{16}$ | $\frac{7}{16}$ | 0.093 | 1:4 | $36\times\frac{5}{16}$ $12\times\frac{5}{8}$ |
| AISC Exp. | 24 | trapezoid 13 in. / 9½ in. / 6 in. | $\frac{1}{4}$ | $\frac{3}{8}$ | 0.30 | 1:3–1:4 | $48$–$60\times\frac{5}{16}$ $10\times\frac{9}{16}$ |
| California | 24 | trapezoid 12 in. / 8½ in. / 6 in. | $\frac{5}{16}$ | $\frac{3}{8}$–$\frac{7}{16}$ | 0.65 | 1:3 | $24\times\frac{1}{2}$ $10\times\frac{7}{8}$ |
| St. Louis | 26 | trapezoid 12 in. / 10 in. / 6 in. | $\frac{1}{2}$ | 0.50 | 1:4 | |
| Bethlehem Steel | 6 | trapezoid 5½ in. / 2½ in. / 3 in. | $\frac{3}{16}$ | $\frac{3}{16}$ | 1.0 | 1:3 | $34.5\times0.28$ $11\times1$ |

[a] Ratio of depth of rib to depth of floor beam.
[b] Floor-beam properties; depth and thickness of web and size of flange.

## 6.1.2  History

Orthotropic decks were first conceived in the early 1930s, for use in movable bridges, and were called battledecks. The battledeck had a steel plate supported by longitudinal rolled stringers. These stringers spanned transverse floor beams. The advantage of this system was that the original concrete deck slab was replaced by a light steel deck plate. However, the interactions of the stringers, floor beams, and deck were not considered and thus the real advantage of welded orthotropic bridge design was not realized.

The real advantage was not perceived until the 1940s, in Germany, after World War II. During that time steel was in short supply and thus bridges had to be made as light as possible. These bridges were constructed of open ribs with closely spaced floor beams (4 to 5 ft) and were analyzed by considering the deck as a grid. With continued construction, new analytic techniques were devised and in the late 1950s the decks were being analyzed as continuous orthotropic plates. Such analytical techniques were refined and in 1957 Pelikan and Esslinger (2) presented a simplified method, which has been widely used and is given herein.

Orthotropic decks were first used for medium-span girder bridges of 200 to 300 ft. They were then used for suspension bridges and cable-stiffened girder bridges and now have been applied to varied bridge systems. It has been found that the orthotropic deck is applicable for spans of 80 to 120 ft, as well as for long-span structures.

## 6.1.3  Economy

As mentioned previously, the main advantage of the orthotropic deck is its reduction in dead weight in comparison to the concrete deck. It also has the ability to act compositely with the main girders, floor beams, and stiffeners.

In long-span bridges, the bending moment due to the dead load comprises a large portion of the design moment. Therefore a reduction in the dead load can significantly affect the economy of the bridge. As the bridge span decreases, the influence of the dead-load moment on the design, and thus economy, also decreases, until conventional bridge designs start to govern. Also to be considered is the fabrication of these steel deck bridges, which require a great deal of welding. A study of the costs of conventional concrete deck bridges and orthotropic deck bridges, has resulted in typical trends as shown in Fig. 6.3. This study (3) recommends the following:

Orthotropic deck systems with closed ribs will be the most economical. A $\frac{3}{8}$ in. deck plate supported at 12 in. centers by trapezoidal trough stiffeners $\frac{1}{4}$ in. thick

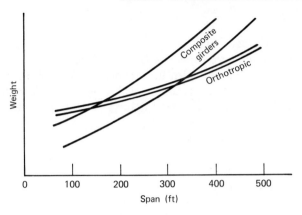

**Figure 6.3** Economy of orthotropic and conventional bridges.

and spanning between floorbeams at 15 ft centers provides the minimum recommended deck section. For short and medium span bridges the ribs can be 8 in. deep, increasing to 12 in. deep for spans of 600 ft or more. The grade of steel to be used for the deck plate, ribs and floorbeam webs, should be normalized to provide for notch toughness and a minimum shaping of 20 ft. lb. The trough stiffeners should pass through notches in the floorbeam webs and field splices should be all welded and predominantly in the longitudinal direction. Fatigue will not be a governing design consideration and radiographic inspection should be kept to the minimum which would be necessary to secure good workmanship.

## 6.2 DECK-PLATE BEHAVIOR

The response of such a plate is a function of the plate stiffness. Such a consideration therefore requires a study of the load deformation of such a plate element. To develop an equation, the intersecting girder system shown in Fig. 6.4 is considered. One girder has a bending stiffness of $EIx$ and torsional stiffness $GK_{Tx}$ and the other has values of $EIy$ and $GK_{Ty}$. These intersecting girders are in equilibrium with the external load as a result of the action of their internal forces $Vx$, $M_{Tx}$ and $Vy$, $My$, $M_{Ty}$. A change in these forces results from moving a distance $dx$ along the $x$ axis and $dy$ along the $y$ axis; the new forces have differential changes as shown in Fig. 6.4.

Since this system is in equilibrium,

$$\sum F \text{ vertical} = 0$$
$$\sum Mx \text{ axis} = 0$$
$$\sum My \text{ axis} = 0$$

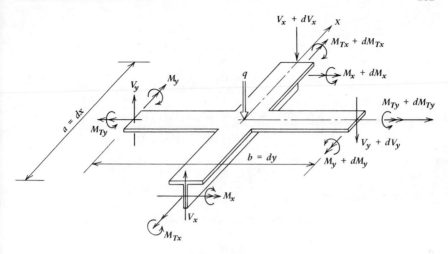

**Figure 6.4** Forces on intersecting beam elements.

Applying these equations gives

$$V_x + V_y - V_x - dV_x - V_y - dV_y - q\,dx\,dy = 0$$

or

$$+ dV_x + dV_y + q\,dx\,dy = 0$$

If the shear forces are considered to have units of k/ft then the total force is

$$dV_x\,dy + dV_y\,dx + q\,dx\,dy = 0$$

or

$$\frac{dVx}{dx} + \frac{dVy}{dy} = q \tag{6.1}$$

Assuming bending and torsion moments are also per unit of width, that is k-in./in., then each must be multiplied by $dx$ or $dy$. Now $\sum M = 0$ about the $x$ and $y$ axes gives

$$\sum Mx = 0 = My\,dx - (My + dMy)\,dx$$
$$+ M_{Tx}\,dy - (M_{Tx} + dM_{Tx})\,dy$$
$$+ Vy\,dx\frac{dy}{2} + (Vy + dVy)\,dx\frac{dy}{2} = 0$$

Collecting terms gives

$$-dMy\, dx - dM_{Tx}\, dy + V_y\, dx\, dy = 0$$

or

$$-\frac{dMy}{dy} - \frac{dM_{Tx}}{DX} + Vy = 0$$

$$\sum My = 0 = Mx\, dy - (Mx + dMx)\, dy \qquad (6.2)$$
$$+ M_{Ty}\, dx - (M_{Ty} + dM_{ty})\, dx$$
$$+ V_x\, dy\frac{dx}{2} + (V_x + dV_x)\, dy\frac{dx}{2} = 0$$

Collecting terms gives

$$-dMx\, dy - dM_{Ty}\, dx + Vx\, dy\, dx = 0$$

or

$$-\frac{dMx}{dx} - \frac{dM_{Ty}}{dy} + Vx = 0 \qquad (6.3)$$

Now taking the derivatives of equation 6.2, $d/dy$, and equation 6.3, $d/dx$, and substituting these into equation 6.1 gives

$$-q = \frac{d}{dx}\left(\frac{dMx}{dx} + \frac{dM_{Ty}}{dy}\right) + \frac{d}{dy}\left(\frac{dM_y}{dy} + \frac{dM_{Tx}}{dx}\right)$$

$$-q = \frac{d^2M_x}{dx^2} + \frac{d^2M_{Ty}}{dy\, dx} + \frac{d^2My}{dy^2} + \frac{d^2M_{Tx}}{dx\, dy}$$

Collecting terms,

$$\frac{d^2Mx}{dx^2} + \left(\frac{d^2M_{Ty}}{dy\, dx} + \frac{d^2M_{Ty}}{dx\, dy}\right) + \frac{d^2My}{dy^2} = -q \qquad (6.4)$$

which is the basic equilibrium equation of the intersecting beam elements. Note that if only one beam exists, that is $M_{T_y} = My = 0$, and torsion is neglected, then

$$\frac{d^2Mx}{dx^2} = -q$$

which is the basic strength of materials equation.

The general moment–load equation (equation 6.4) can be related to deformations by relating moments to curvature.

1. $Mx$. The general relationship between moment and deformation is given by

$$\frac{Mx}{EIx} = -\frac{d^2w}{dx^2} \qquad (6.5a)$$

Taking the second derivative of equation 6.5a gives

$$\frac{d^2}{dx^2}(Mx) = \frac{d^2}{dx^2}\left[-EIx\frac{d^2w}{dx^2}\right]$$

or

$$\frac{d^2Mx}{dx^2} = -EIx\frac{d^4w}{dx^4} \qquad (6.5b)$$

2. *My*. The moment–deformation relationship is

$$\frac{My}{EIy} = \frac{d^2w}{dy^2} \qquad (6.6a)$$

The second derivative of equation 6.6a is

$$\frac{d^2(My)}{dy^2} = -EIy\frac{d^4w}{dy^4} \qquad (6.6b)$$

3. $M_{Tx}$, $M_{Ty}$. The general torsional response of a beam is given by the familiar equation

$$\phi = \frac{TL}{GK_T}$$

or

$$T = M_T = GK_T\left(\frac{\phi}{L}\right) \qquad (6.7a)$$

Noting that $\phi/L$ is the rate of change in slope per unit length,

$$M_T = GK_T\phi' \qquad (6.7b)$$

Equation 6.7b can now be expressed in terms of $M_{Tx}$ and $M_{Ty}$ given below.

4. $M_{Tx}$. The rotation of the girder along the $x$ axis is $(-dw/dy)$ and the rate of change is $d/dx(-dw/dy)$. The torque is therefore given by

$$M_{Tx} = -GK_{Tx}\frac{d^2w}{dx\,dy} \qquad (6.8a)$$

and

$$\frac{d^2M_{Tx}}{dx\,dy} = -GK_{Tx}\frac{d^4w}{dx^2\,dy^2} \qquad (6.8b)$$

Similarly, the rotation of the girder along the $y$ axis is $(-dw/dx)$ and the change per unit of length is $d/dy(-dw/dx)$. The torque therefore is

$$M_{Ty} = -GK_{Ty}\frac{d^2w}{dy\,dx} \qquad (6.9a)$$

$$\frac{d^2M_{Tx}}{dx\,dy} = -GK_{Ty}\frac{d^4w}{dy^2\,dx^2} \qquad (6.9b)$$

Now, substituting equations 6.5b, 6.6b, 6.8b, and 6.9b into the basic equilibrium equation 6.4 gives

$$EIx\frac{d^4w}{dx^4}+(GK_{Tx}+GK_{Ty})\frac{d^2w}{dx^2\,dy^2}+EIy\frac{d^4w}{dy^4}=q \qquad (6.10)$$

Assuming the stiffnesses per unit of width are

$$Dx=\frac{EIx}{b}$$

$$Dy=\frac{EIy}{c} \qquad (6.11)$$

$$2H=\frac{GK_{Tx}}{b}+\frac{GK_T}{c}$$

equation 10 can be reduced to the following general orthotropic plate equation, considering partial rather than exact derivatives and that the beam elements are continuous plates.

$$Dx\frac{\partial^4w}{\partial x_4}+2H\frac{\partial^4w}{\partial x^2\,\partial y^2}+Dy\frac{\partial^4w}{\partial y^4}=q \qquad (6.12)$$

This equation represents the load-displacement response of a continuous orthotropic plate, which is the bridge deck. It should be noted that equation 6.12 reduces to the conventional beam equation $d^4w/dx^4=q/Dx$ if $Dy=H=0$. What is now required is to solve this equation, relative to bridge decks, when it is applied of bridge loadings.

### 6.3  PELIKAN-ESSLINGER METHOD

### 6.3.1  General

The response of continuous orthotropic plates on flexible supports, as applied to bridge structures, has been solved by such techniques as Fouries series (1), finite differences (1, 4) and finite elements (5, 6). In all these methods, the numerical computations are quite tedious and lengthy, and thus they all require use of the computer. However, a procedure has been developed that does not require the use of a computer. This method, which employs some simplifying assumptions, is called the Pelikan-Esslinger method (2) after its originators. It has had extensive application in Europe and has been employed in the United States and Canada for major orthotropic bridge designs (7–11).

This method first considers the bridge to be divided into three systems, whose individual actions are added to yield the final bridge response. The three systems are discussed below.

1. *System I.* The local response of the deck plate spanning the distance between the supporting ribs.
2. *System II.* The response of the deck plate, ribs, and transverse floor beams, which form the bridge deck. This system is considered as a continuous orthotropic plate on flexible supports.
3. *System III.* The response of the main longitudinal girder, acting with the deck plate and longitudinal ribs. This system is assumed to be a large beam, spanning the distance between the main supports.

The general response of systems II and III is shown in Fig. 6.5. System I is generally not included in this total stress summary because of the excessive load-carrying capacity of the plate, which cannot be predicted by first-order theory.

The major contribution of Pelikan and Esslinger is in the prediction of the behavior of system II, the continuous orthotropic plate on flexible supports. They assume that the continuous orthotropic plate is supported rigidly by the main longitudinal girders and elastically by the floor beams. This system is then designed in two steps: (1) design of continuous orthotropic deck plate on rigid supports and (2) correction to this system by considering the floor beams as elastic supports. In addition to this analytical procedure, Pelikan and Esslinger have modified the general plate equation (equation 6.12), depending on the type of deck-plate stiffener.

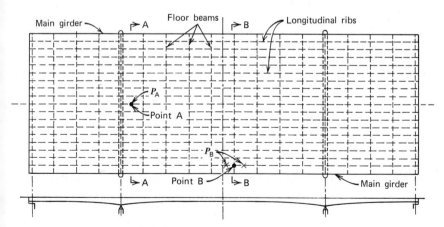

**Figure 6.5**  Stresses on bridge system.

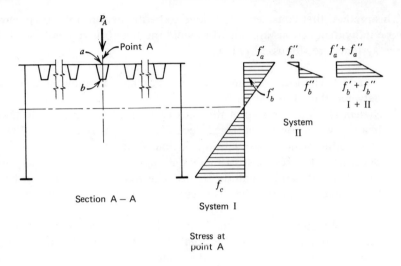

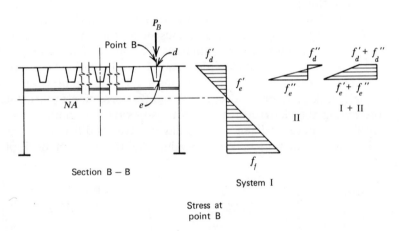

**Figure 6.5** *Continued.*

1. *Deck Plate with Open Ribs.* Deck plates stiffened by open ribs, as shown in Fig. 6.2, have minimal stiffnesses $Dx$ and $H$ relative to the primary longitudinal stiffness $Dy$. Therefore, in the design of such a plate, $Dx = H = 0$ and the general plate equation (equation 6.12) becomes

$$Dy\frac{\partial^4 w}{\partial y^4} = q \qquad (6.13)$$

This equation represents the load-deflection response of a beam of stiffness $Dy$, which is continuous over the transverse floor beams.

2. *Deck Plate with Closed Ribs.* Deck plates stiffened by closed ribs, as shown in Fig. 6.2, have minimum transverse flexural rigidity $Dx$ relative to $Dy$ and $H$. Therefore, letting $Dx = 0$, equation 6.12 reduces to

$$2H\frac{\partial^4 w}{\partial x^2 \partial y^2} + Dy\frac{\partial^4 w}{\partial y^4} = q \qquad (6.14)$$

In the analysis of either the open or closed stiffened-rib deck plates, the following design information is required:

1. Effective width of stiffened deck plate.
2. Effective wheel-load distribution on the deck plate.
3. Properties of stiffened deck plate.
4. Influence lines of loaded deck plate.
5. Effects of flexible floor beams on deck response.

These design data have been developed (12, 13) and are now presented for each deck rib type.

## 6.3.2 Effective Plate Width and Load Distribution of Ribs and Floor Beams

1. *Rigid Supports.* The effective plate width of longitudinal open or closed stiffened plates is defined as $a_0$. This dimension accounts for that portion of the top plate that acts with the stiffener when examined as an isolated element. That portion of the plate that acts with the floor beam is designated as $l$. Evaluation of these dimensions can be obtained by using Fig. 6.6.

As seen in Fig. 6.6, the evaluations of $a$ and $l$ are dependent on the parameters:

$$\beta = \frac{\pi a^*}{l} \quad \text{for rib stiffeners}$$

$$\beta = \frac{\pi l^*}{b} \quad \text{for floor beam}$$

where $\quad l_1 = 0.70l$, $l = $ spacing of floor beams
$\qquad b = $ spacing between main girders
$\quad a^*, l^* = $ effective rib and floor-beam spacing, which accounts for unequal load distribution.

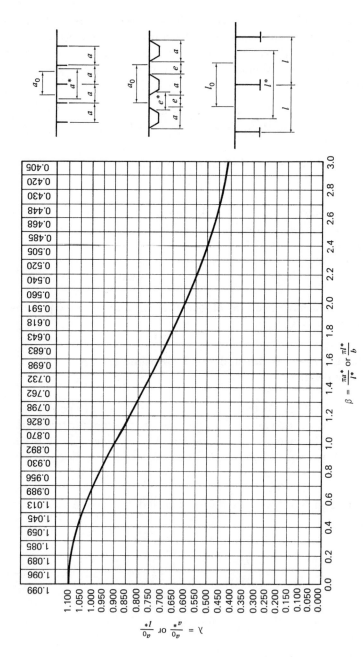

**Figure 6.6** Effective width of orthotropic deck plate, $a$ and $l_0$ = effective width of the deck plate $a^*$ and $l^*$ = ideal span of T-ribs, box ribs, or floor beams.

In general the effective floor-beam spacing $l^*$ is equal to $l$, the actual floor-beam spacing. The evaluation of $a^*$, however, depends on the load application. The two basic loading cases (13), given in Fig. 6.7, are considered here.

1. Wheel load over one open stiffener.
2. Wheel load over several open stiffeners.

These two cases relate the load distribution to the effective rib spacing $(a^*)$ to the actual spacing $(a)$ as a function of $B/a$. $B = 2g$, which is the tire width (24 in.).

The evaluation of the equivalent loads to each rib can also be obtained by using Fig. 6.8. It should be noted that Figs. 6.7 and 6.8 are applicable to open-ribbed decks.

In evaluating the effective width of closed cells, Fig. 6.6 can be used assuming

$$a = \lambda_1 a + \lambda_2 e$$

where $\lambda_1$ is found from $\beta_1 = \pi a / l_1$
$\lambda_2$ is found from $\beta_2 = \pi e / l_1$
$l_1 = 0.70l$
$e =$ spacing between cells
$a =$ width of cell

2. *Flexible Supports.* Yielding of the floor beam has considerable effect on the effective span of the ribbed plate, which in turn influences the effective plate width. When the floor beams are flexible, the effective span is greater than the floor-beam spacing; therefore

$$l_1 = \infty \text{ (rather than } 0.7l) \qquad \therefore \beta = \frac{\pi a}{l_1} = 0$$

From Fig. 6.6,

$$\lambda = \frac{a_0}{a^*} = 1.10$$

and because of the flexible floor beams $a^* = a$, the effective plate width is

$$a_0 = 1.10a \qquad (6.15)$$

The effective floor-beam span $l = l^*$, and $b = b^*$ in computing the floor-beam plate width.

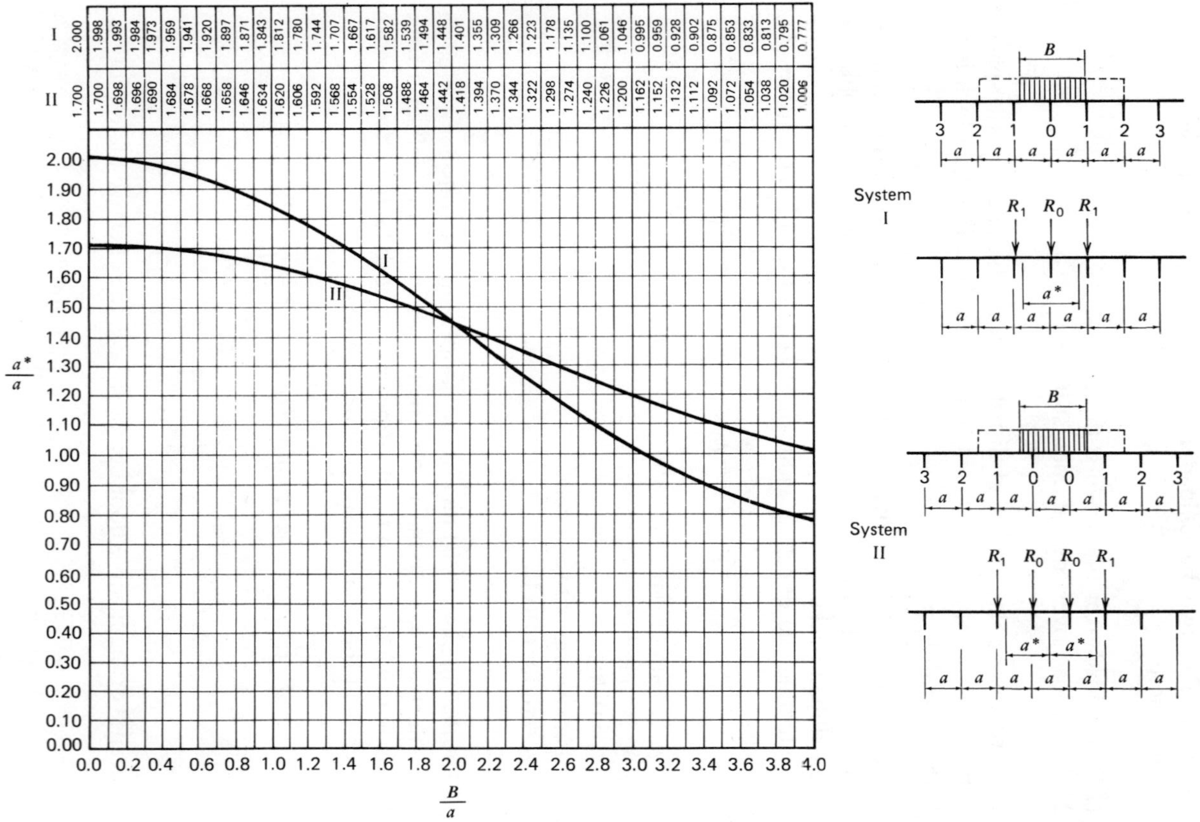

System I

System II

Figure 6.7 Ideal spacing of flexible ribs.

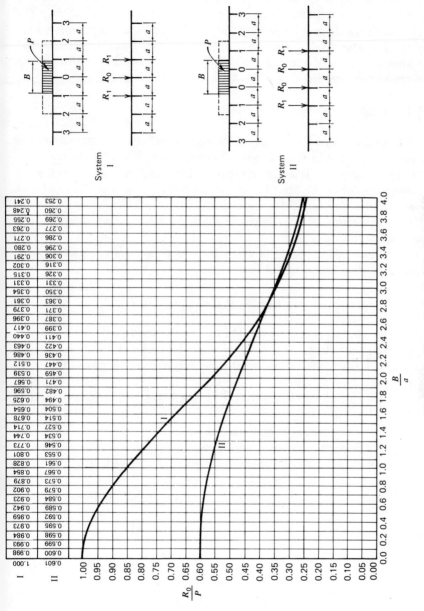

**Figure 6.8** Open-rib loading distribution.

171

### 6.3.3   Effective Plate Width of Main Longitudinal Girder

The complete cross section of the deck plus longitudinal stiffeners can be taken as effective with the main girders provided the ratio of the length to the width of the deck is large. Generally, if $L \leq 3B$, the full plate is active, where $B$ is the total bridge width and $L$ is the span between simple supports or points of contraflexure. The AASHTO code is more restrictive than this criterion.

### 6.3.4   Solution for Open-Rib Deck Plate

1. *Influence Lines.* The behavior of an open-rib deck plate, supported by transverse flexible floor beams, is predicted by use of influence lines (12). These influence lines are relative to the locations on the deck plate (1) over support of the floor beam ($M_s$) and (2) midspan of deck ($M_c$), as shown in Figs. 6.9 and 6.10 respectively.

The coordinates of the influence lines, assuming rigid supports, are given below.

The moment at midspan is given by two expressions; equation 6.16 considers a distributed wheel load ($2c$) at $y = l/2$ and is

$$\left[\frac{M_c}{Pl}\right]_{00'} = \left[0.1708 - 0.250\left(\frac{c}{l}\right) + 0.1057\left(\frac{c}{l}\right)^2\right] \qquad (6.6a)$$

where $2c =$ contact wheel length (12 in.). If the loads are applied in any other span, the effects of the wheel load dimensional width $2c$ is neglected and a point load is assumed; thus the moment is given by

$$\left[\frac{M_c}{Pl}\right]_m = \left[-0.183\frac{y}{l} + 0.317\left(\frac{y}{l}\right)^2 - 0.134\left(\frac{y}{l}\right)^3\right](-0.268)^m \qquad (6.16b)$$

where     $P =$ any concentrated wheel load

$l =$ spacing between floor beams

$m =$ the smaller of the two support numbers enclosing the span under consideration

$y =$ location of the level with respect to left support of loaded span

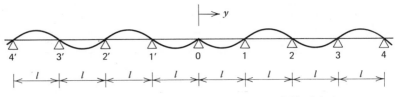

**Figure 6.9**   Influence lines for support moment ($Ms$) of deck.

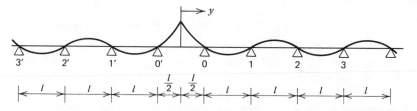

**Figure 6.10**  Influence lines for midspan moment (*Mc*) of deck.

The moment over the support is given by equation 6.17:

$$\left[\frac{M_s}{Pl}\right]_m = \left[-0.5\left(\frac{y}{l}\right) + 0.866\left(\frac{y}{l}\right)^2 - 0.366\left(\frac{y}{l}\right)^3\right](-0.268)^m \quad (6.17)$$

The effect of the contact wheel length 2*c* is neglected and only the point wheel load is used. When the load is between 0–1′, and 0–1 the term $(-0.268)^m$ is set equal to 1.0 and the remaining equation (equation 6.17) is applied.

Also required in the bridge system design are the forces (12) on the floor beams due to the action of the load through the deck. The reaction at support 0 below due to a load applied in span 0–1 and in any other span is given below.

For Span 0–1:

$$\left[\frac{F_0}{p}\right] = \left[1 - 2.19\left(\frac{y}{l}\right)^2 + 1.106\left(\frac{y}{l}\right)^3\right] \quad (6.18a)$$

For any other span:

$$\left[\frac{F_0}{p}\right] = \left[-0.804\left(\frac{y}{l}\right) + 1.292\left(\frac{y}{l}\right)^2 - 0.5885\left(\frac{y}{l}\right)^3\right](-0.268)^{m-1} \quad (6.18b)$$

where $F_0$ = support reaction and $y$, $m$, $P$, and $l$ are as defined previously.

2. *Relaxation of Floor Beams.* The resulting moments and reactions, using the above influence line equations (equations 6.16 through 6.18), assume rigid support. To account for the floor-beam flexibility, the interactions of deck and floor beam must be considered. Such interaction is first related by the stiffness according to

$$\gamma = \frac{I_r b^4}{I_f l^3 a \pi^4} \quad (6.19)$$

where  $I_r$ = inertia of T-rib

$I_f$ = inertia of floor beam, including effective plate width

$b$ = spacing of open ribs

$l$ = spacing of floor beams

$a$ = flange width of T-rib

Assuming this relative stiffness, influence line tables have been developed (12), for $M$, $M_c$, and $F_0$ as given in Tables 6.3 through 6.5. Using these tables, in conjunction with the influence line equations (equations 6.16 through 6.18), the modified moments in the deck and floor beams can be determined.

(a)  Bending Moment Modification at Center of Rib ($\Delta M_c$)

Because of flexibility of the floor beam, the rigid-support induced moment $M_c$ (equation 6.16), is modified by an amount given by

$$\Delta M_c = Q_0 la \frac{Q_{ix}}{Q_0} \sum \frac{F_m}{P} \frac{\eta_m}{l} \qquad (6.20)$$

where  $Q_0 = P/2g$

$P$ = wheel load intensity

$g$ = contact width of tires

$F_m$ = reaction due to load $P$ at support $m$ of continuous beam on rigid supports (equation 6.18)

$\eta_m$ = influence line ordinates at flexible support $m$ for the bending moment at midspan (Table 6.3)

$l$ = floor-beam spacing

$a$ = flange width of T-rib

$$\frac{Q_{ix}}{Q_0} = \frac{8}{\pi} \cos \frac{\pi g}{b} \left( \sin \frac{\pi d_1}{b} + \cdots + \sin \frac{\pi d_n}{b} \right) \sin \frac{\pi}{b}$$

where  $d_1, \ldots, dn$ = distance from support to center of gravity of wheel set

$e$ = spacing between wheels divided by 2.0

$b$ = spacing between main floor beams

as shown in Fig. 6.11. The moment $\Delta M_c$ increases the previously computed moment and thus is added to the original moment.

Ordinates for Bending Moment $M/l$ at Section $m\text{-}m$ due to a Load $P = 1$ at Any Arbitrary Support

| $\gamma$ | 0 | 1 | 2 | 3 | 4 | 5 | 6 | 7 | 8 |
|---|---|---|---|---|---|---|---|---|---|
| 0.02 | +0.01321 | −0.01465 | +0.00125 | +0.00024 | −0.00006 | 0 | | | |
| 0.04 | +0.02313 | −0.02333 | −0.00087 | +0.00102 | −0.00004 | −0.00004 | 0 | | |
| 0.06 | +0.03130 | −0.02917 | −0.00383 | +0.00150 | −0.00029 | −0.00010 | +0.00001 | 0 | |
| 0.08 | +0.03839 | −0.03344 | −0.00700 | +0.00149 | −0.00068 | −0.00002 | −0.00005 | −0.00001 | |
| 0.1 | +0.04468 | −0.03674 | −0.01015 | +0.00122 | +0.00155 | +0.00009 | −0.00007 | −0.00002 | 0 |
| 0.2 | +0.06902 | −0.04538 | −0.01628 | +0.00495 | −0.00198 | +0.00100 | +0.00010 | −0.00009 | 0 |
| 0.3 | +0.08713 | −0.04835 | −0.03450 | −0.00816 | −0.00148 | +0.00187 | +0.00062 | 0 | −0.00004 |
| 0.4 | +0.10188 | −0.04903 | −0.04268 | −0.01384 | +0.00005 | +0.00237 | +0.00122 | +0.00023 | −0.00010 |
| 0.5 | +0.11450 | −0.04858 | −0.04932 | −0.01935 | −0.00196 | +0.00248 | +0.00180 | +0.00057 | −0.00002 |
| 0.6 | +0.12559 | −0.04743 | −0.05473 | −0.02454 | −0.00428 | +0.00226 | +0.00227 | +0.00096 | +0.00023 |
| 0.7 | +0.13555 | −0.04591 | −0.05593 | −0.02941 | −0.00681 | +0.00178 | +0.00262 | +0.00136 | +0.00034 |
| 0.8 | +0.14459 | −0.04417 | −0.06311 | −0.03395 | −0.00943 | +0.00107 | +0.00286 | +0.00175 | +0.00052 |
| 0.9 | +0.15297 | −0.04198 | −0.06640 | −0.03819 | −0.01210 | +0.00019 | +0.00297 | +0.00211 | +0.00077 |
| 1.0 | +0.16073 | −0.04018 | −0.06923 | −0.04216 | −0.01478 | −0.00080 | +0.00297 | +0.00242 | +0.00110 |
| 1.5 | +0.19322 | −0.02944 | −0.07881 | −0.05860 | −0.02771 | −0.00715 | +0.00166 | +0.00340 | +0.00243 |
| 2.0 | +0.21895 | −0.01876 | −0.08383 | −0.07094 | −0.03933 | −0.01436 | −0.00115 | +0.00333 | +0.00340 |
| 3.0 | +0.25930 | +0.00107 | −0.08748 | −0.08821 | −0.05869 | −0.02881 | −0.00889 | +0.00092 | +0.00396 |
| 4.0 | +0.29104 | +0.01881 | −0.08705 | −0.09957 | −0.07298 | −0.04215 | −0.01764 | −0.00328 | +0.00300 |
| 5.0 | +0.31756 | +0.03484 | −0.08469 | −0.10751 | −0.08641 | −0.05413 | −0.02651 | −0.00839 | +0.00099 |
| 6.0 | +0.34052 | +0.04947 | −0.08130 | −0.11315 | −0.09665 | −0.06485 | −0.03512 | −0.01395 | −0.00174 |
| 7.0 | +0.36139 | +0.06347 | −0.07733 | −0.11722 | −0.10526 | −0.07448 | −0.04335 | −0.01970 | −0.00497 |
| 8.0 | +0.37926 | +0.07552 | −0.07302 | −0.12014 | −0.11257 | −0.08317 | −0.05117 | −0.02549 | −0.00849 |
| 9.0 | +0.39604 | +0.08728 | −0.06855 | −0.12221 | −0.11855 | −0.09103 | −0.05856 | −0.03123 | −0.01223 |
| 10.0 | +0.41154 | +0.09835 | −0.06394 | −0.12362 | −0.12430 | −0.09820 | −0.06555 | −0.03688 | −0.01610 |

Courtesy of Lincoln Arc Welding Foundation.

175

**Table 6.4  Influence Lines for Support Moments; Continuous Beam Over Infinite Number of Elastic Supports**

Beam diagram with elastic supports numbered (left to right): 9  8  7  6  5  4  3  2  1  0  1  2  3  4  5  6  7  8  9, each span = $l$.

Ordinates for Bending Moment $M/l$ at Support 0 due to a Load $P = 1$ at Any Arbitrary Support

| $\gamma$ | 0 | 1 | 2 | 3 | 4 | 5 | 6 | 7 | 8 | 9 |
|---|---|---|---|---|---|---|---|---|---|---|
| 0.02 | +0.05763 | −0.03121 | +0.00191 | +0.00060 | −0.00011 | 0 | | | | |
| 0.04 | +0.08919 | −0.04294 | −0.00361 | +0.00187 | +0.00017 | −0.00009 | 0 | | | |
| 0.06 | +0.11111 | −0.04850 | −0.00985 | +0.00219 | +0.00080 | −0.00023 | +0.00003 | 0 | | |
| 0.08 | +0.12800 | −0.05128 | −0.01566 | +0.00166 | +0.00133 | −0.00004 | −0.00009 | −0.00002 | | |
| 0.1 | +0.14183 | −0.05257 | −0.02091 | +0.00061 | +0.00182 | +0.00027 | −0.00010 | −0.00004 | 0 | |
| 0.2 | +0.18859 | −0.05056 | −0.04019 | −0.00764 | +0.00226 | +0.00170 | −0.00031 | −0.00010 | −0.00007 | 0 |
| 0.3 | +0.21870 | −0.04445 | −0.05224 | −0.01665 | +0.00034 | +0.00262 | +0.00112 | +0.00011 | −0.00012 | −0.00008 |
| 0.4 | +0.24142 | −0.03766 | −0.06040 | −0.02496 | −0.00272 | +0.00281 | +0.00192 | +0.00052 | −0.00006 | −0.00013 |
| 0.5 | +0.25989 | −0.03090 | −0.06625 | −0.03238 | −0.00631 | +0.00240 | +0.00256 | +0.00103 | +0.00011 | −0.00014 |
| 0.6 | +0.27556 | −0.02438 | −0.07049 | −0.03897 | −0.01010 | +0.00154 | +0.00298 | +0.00155 | +0.00037 | −0.00008 |
| 0.7 | +0.28924 | −0.01814 | −0.07368 | −0.04486 | −0.01395 | −0.00034 | +0.00321 | +0.00203 | +0.00068 | |
| 0.8 | +0.30143 | −0.01225 | −0.07609 | −0.05013 | −0.01776 | −0.00109 | +0.00323 | +0.00249 | +0.00100 | +0.00014 |
| 0.9 | +0.31245 | −0.00651 | −0.07791 | −0.05489 | −0.02149 | −0.00271 | +0.00309 | +0.00286 | +0.00135 | +0.00031 |
| 1.0 | +0.32254 | −0.00108 | −0.07927 | −0.05919 | −0.02512 | −0.00444 | +0.00280 | +0.00314 | +0.00169 | +0.00050 |
| 1.5 | +0.36346 | +0.02297 | −0.08184 | −0.07577 | −0.04142 | −0.01399 | −0.00031 | +0.00363 | +0.00317 | +0.00168 |
| 2.0 | +0.39470 | +0.04319 | −0.08071 | −0.08695 | −0.05492 | −0.02373 | −0.00498 | +0.00268 | +0.00397 | +0.00283 |
| 3.0 | +0.44218 | +0.07641 | −0.07428 | −0.10068 | −0.07573 | −0.04165 | −0.01596 | −0.00181 | +0.00364 | +0.00428 |
| 4.0 | +0.47857 | +0.10351 | −0.06588 | −0.10822 | −0.09092 | −0.05704 | −0.02726 | −0.00803 | +0.00148 | +0.00453 |
| 5.0 | +0.50847 | +0.12664 | −0.05697 | −0.11241 | −0.10261 | −0.07021 | −0.03805 | −0.01496 | −0.00182 | −0.00379 |
| 6.0 | +0.53406 | +0.14698 | −0.04804 | −0.11456 | −0.11174 | −0.08156 | −0.04814 | −0.02209 | −0.00580 | −0.00231 |
| 7.0 | +0.55656 | +0.16522 | −0.03927 | −0.11538 | −0.11905 | −0.09146 | −0.05751 | −0.02920 | −0.01019 | −0.00025 |
| 8.0 | +0.57672 | +0.18180 | −0.03075 | −0.11529 | −0.12499 | −0.10015 | −0.06618 | −0.03616 | −0.01482 | −0.00217 |
| 9.0 | +0.59502 | +0.19706 | −0.02250 | −0.11456 | −0.12986 | −0.10785 | −0.07422 | −0.04291 | −0.01956 | −0.00490 |
| 10.0 | +0.61185 | +0.21122 | −0.01452 | −0.11335 | −0.13389 | −0.11471 | −0.08168 | −0.04941 | −0.02434 | −0.00785 |

Support labels (left to right): 8  7  6  5  4  3  2  1  0  1  2  3  4  5  6  7  8 — spacing $l$

| | Ordinates for Reaction at Support 0 due to a Load $P = 1$ at Any Arbitrary Support | | | | | | | | |
|---|---|---|---|---|---|---|---|---|---|
| $\gamma$ | 0 | 1 | 2 | 3 | 4 | 5 | 6 | 7 | 8 |
| 0.02 | +0.82232 | +0.12196 | -0.03443 | +0.00060 | +0.00082 | -0.00011 | 0 | 0 | 0 |
| 0.04 | +0.73574 | +0.17146 | -0.03385 | -0.00718 | +0.00144 | +0.00035 | -0.00009 | 0 | 0 |
| 0.06 | +0.68078 | +0.19826 | -0.02661 | -0.01343 | +0.00036 | +0.00129 | -0.00029 | +0.00003 | 0 |
| 0.08 | +0.64144 | +0.21490 | -0.01830 | -0.01765 | -0.00096 | +0.00116 | +0.00020 | -0.00005 | -0.00002 |
| 0.1 | +0.61120 | +0.22606 | -0.01013 | -0.02032 | -0.00276 | +0.00118 | +0.00043 | -0.00002 | -0.00004 |
| 0.2 | +0.52170 | +0.24951 | +0.02219 | -0.02266 | -0.01046 | -0.00083 | +0.00098 | +0.00044 | +0.00003 |
| 0.3 | +0.47368 | +0.25537 | +0.04337 | -0.01853 | -0.01471 | -0.00379 | +0.00050 | +0.00078 | +0.00027 |
| 0.4 | +0.44183 | +0.25634 | +0.05819 | -0.01321 | -0.01670 | -0.00642 | -0.00050 | +0.00082 | +0.00051 |
| 0.5 | +0.41842 | +0.25555 | +0.06916 | -0.00777 | -0.01736 | -0.00855 | -0.00169 | +0.00060 | +0.00067 |
| 0.6 | +0.40013 | +0.25382 | +0.07763 | -0.00265 | -0.01723 | -0.01019 | -0.00288 | +0.00025 | +0.00074 |
| 0.7 | +0.38524 | +0.25184 | +0.08436 | +0.00290 | -0.01662 | -0.01108 | -0.00405 | -0.00017 | +0.00067 |
| 0.8 | +0.37264 | +0.24998 | +0.08980 | +0.00641 | -0.01571 | -0.01235 | -0.00506 | -0.00075 | +0.00062 |
| 0.9 | +0.36208 | +0.24756 | +0.09443 | +0.01038 | -0.01462 | -0.01298 | -0.00602 | -0.00129 | +0.00047 |
| 1.0 | +0.35276 | +0.24543 | +0.09827 | +0.01400 | -0.01340 | -0.01344 | -0.00690 | -0.00178 | +0.00026 |
| 1.5 | +0.31901 | +0.23569 | +0.11087 | +0.02829 | -0.00692 | -0.01375 | -0.00973 | -0.00441 | -0.00103 |
| 2.0 | +0.29700 | +0.22760 | +0.11766 | +0.03826 | -0.00084 | -0.01243 | -0.01110 | -0.00637 | -0.00243 |
| 3.0 | +0.26847 | +0.21507 | +0.12430 | +0.05135 | +0.00912 | -0.00838 | -0.01153 | -0.00871 | -0.00481 |
| 4.0 | +0.24988 | +0.20568 | +0.12705 | +0.05964 | +0.01657 | -0.00409 | -0.01056 | -0.00972 | -0.00646 |
| 5.0 | +0.23633 | +0.19823 | +0.12817 | +0.06523 | +0.02261 | -0.00025 | -0.00906 | -0.00996 | -0.00753 |
| 6.0 | +0.22584 | +0.19206 | +0.12849 | +0.06935 | +0.02736 | +0.00324 | -0.00737 | -0.00976 | -0.00819 |
| 7.0 | +0.21730 | +0.18686 | +0.12831 | +0.07244 | +0.03127 | +0.00636 | -0.00565 | -0.00929 | -0.00857 |
| 8.0 | +0.21016 | +0.18236 | +0.12802 | +0.07485 | +0.03454 | +0.00913 | -0.00395 | -0.00868 | -0.00870 |
| 9.0 | +0.20408 | +0.17840 | +0.12750 | +0.07676 | +0.03732 | +0.01161 | -0.00232 | -0.00796 | -0.00869 |
| 10.0 | +0.19875 | +0.17488 | +0.12691 | +0.07830 | +0.03971 | +0.01385 | -0.0076 | -0.00720 | -0.00858 |

Courtesy of Lincoln Arc Welding Foundation.

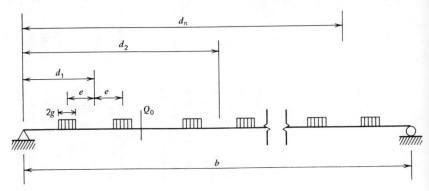

**Figure 6.11** Wheel loads across bridge width.

## (b) Bending Moment Modification at Support of Rib ($\Delta M_s$)

The modification of the support moment $M_s$, due to flexibility of the floor beams, generally reduces these moments and therefore the influence of floor-beam flexibility is generally neglected.

### (c) Modification of Floor-Beam Moment

The wheel loads applied to the continuous rib on rigid supports induce a reaction $F_0$, as given by equation 6.18. If the floor beams or supports are flexible those reactions are modified and, in turn, induce a modified floor-beam moment. As given in reference 12, this relief moment is

$$\Delta M_F = Q_0 \left(\frac{l}{\pi}\right)^2 \frac{Q_{1x}}{Q_0} \left[\frac{F_0}{P} - \sum \frac{F_0}{P} \gamma_m\right] \tag{6.21}$$

where $F_0/P$ = induced reactions, when floor beams are rigid, as given by
equation 6.18
$\gamma_m$ = influence lines for reaction of a flexible floor beam (Table 6.5)

The other terms $Q_0$, $l$, and $Q_{1x}/Q_0$ are as defined previously.

## 6.3.5 Deck Plate Solution for Closed-Rib—Influence Lines

The development of influence lines for the continuous closed-rib deck plate requires solution of the differential plate equation (equation 6.14). This equation has been solved (12), resulting in the moment equations (6.22) and (6.33) given below for the locations (1) over the support of the floor beam ($M_s$) and (2) midspan of deck ($M_c$).

As given in the open-rib case:

$$M_s = Q_0 l \sum \frac{Q_{nx}}{Q_0} \frac{n}{l} \tag{6.22}$$

where $\quad \dfrac{n}{l} = \dfrac{M_0^*}{l} k_n^m \left( C_1 \sin h\gamma_n y + C_2 \cos h\gamma_n y + C_3 \dfrac{y}{l} + C_4 \right)$

$$\frac{Q_{nx}}{Q_0} = \frac{4}{n\pi} \sin \frac{n\pi g}{b}$$

$b = 14c$

$2c =$ contact wheel length (12 ft)

$2g =$ dual tire width

$Q_0 = \dfrac{P}{2g}$

$P =$ wheel-load intensity

$l =$ spacing between floor beams

$$\gamma_n = \frac{n\pi}{b} \sqrt{\frac{2H}{Dy}}$$

$m =$ smaller of the two support numbers enclosing the plate panel under consideration

$$C_1 = \frac{-k_n + \cos \gamma_n l}{\sin h\gamma_n l}$$

$C_2 = -1$

$C_3 = k_n - 1$

$C_4 = 1$

$$\frac{M_0^*}{l} = \frac{1}{\gamma_n^*} \frac{k_n}{1 - k_n^2}$$

$$k_n = -C' + \sqrt{C'^2 - 1}$$

$$C' = \frac{\gamma_n l \cdot \coth \gamma_n l - 1}{\gamma_n^*}$$

$$\gamma_n^* = 1 - \frac{\gamma_n l}{\sinh \gamma_n l}$$

The midspan deck-plate moment is given for two cases: (1) load at midlength of central panel and (2) load in any other panel. The load in the central panel is given by

$$M_c = Q_0 l \sum \frac{Q_{nx}}{Q_0} \frac{M_{cn}}{Ql} \tag{6.23}$$

where

$$\frac{M_{cn}}{Ql} = \frac{1}{2\gamma_n lnc}\left[1 - \frac{\cosh \gamma_n(l/2-c)}{\cosh \gamma_n l/2}\right] + \frac{M_0^*}{l}\left(1 - \frac{\sinh \gamma_n c}{\gamma_n C' \cosh (\gamma_n l/2)}\right)$$

$$\frac{M_0^*}{l} = \frac{kn}{\gamma_n^*(1-kn)}\frac{1}{2\cosh \gamma_n(l/2)}$$

The other terms in this equation are as given previously for equation 6.22.

For a load in any other panel the moment is:

$$M_c = Q_0 l \sum \frac{Q_{nx}}{Q_0}\frac{n}{l} \qquad (6.24)$$

where

$$\frac{n}{l} = \frac{M_0^*}{l} k_n^m \left(C_1 \sinh \gamma_n y + C_2 \cosh \gamma_n y + C_3 \frac{y}{l} + C_4\right)$$

$$\frac{M_0^*}{l} = \frac{k_n}{\gamma_n^*(1-k_n)}\frac{1}{2\cosh (\gamma_n l/2)}$$

and $C_1$, $C_2$, $C_3$, and $C_4$ are as defined for the support bending moment case $M_s$, equation (6.22). The solutions of these various equations are obtained term by term, that is $n = 1$, 2, 3, etc. In the case of the floor-beam reactions, equation 6.18, used for open-deck stiffeners, can be applied.

## 6.4   DESIGN CRITERIA

### 6.4.1   Loading

The orthotropic bridge should be designed for those loads that are generally considered for most highway bridges: (1) dead load, (2) live load, and (3) impact.

The dead load consists of the weight of the bridge system including the wearing surface.

The live load is in accordance with the AASHTO Specification in Article 1.2.5 and is shown in Fig. 6.12 for the main bridge system design and in Fig. 6.13 for the deck system.

The live loading to be used for the deck design is truck wheel loads, rather than lane loads. Wheel loads are used because they are more representative of actual loading. These truck wheel loads are used for the design of the deck plating, ribs, and floor beams. For HS20 trucks,

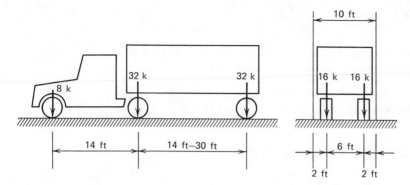

**Figure 6.12** AASHTO-HS 20 truck loading. Axle loads for main bridge system design.

the axial loading of 32 k consists of two random loads of 16 k each spaced at 4.0 ft as shown in Fig. 6.13 or one 24-k axle load, whichever produces the greatest effect.

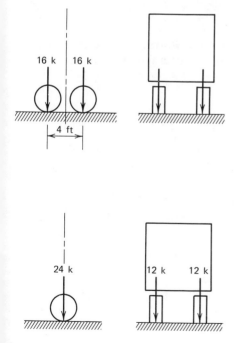

**Figure 6.13** AASHTO wheel loading for deck design. Axle loads for deck design.

## 6.4.2 Impact

The dynamic effect on the deck and main bridge system is considered by applying the general impact equation:

$$I = \frac{50}{L+125} \leq 30\%$$

## 6.4.3 Wheel-Load Distribution

The wheel loads just described should be uniformly distributed to the deck plate over the rectangular area defined below, as given in AASHTO Section 1.7.140

| Wheel Load (k) | Width (2g) Perpendicular to Traffic (in.) | Length (2c) in Direction of Traffic (in.) |
|---|---|---|
| 8 | $20+2t$ | $8+2t$ |
| 12 | $20+2t$ | $8+2t$ |
| 16 | $20+2t$ | $8+2t$ |

In the above table, $t$ is the thickness of the wearing surface in inches, $2g$ is the tire-width in inches, $2c$ is the tire contact length in inches, contact as shown in Fig. 6.16.

## 6.4.4 Effective Width of Deck Plate

The effective width of the deck plate acting as the top flange of a longitudinal rib or transverse beam may be considered by applying the methods given in this text, as outlined in AASHTO 1.7.141.

The main longitudinal girder deck plate may be considered fully active if

$$L \geq 5b \qquad L \geq 10b_1$$

where   $b =$ distance between girder webs
$b_1 =$ distance between edge of deck and nearest girder web
$L =$ effective girder span, that is, distance between simple supports or points of contraflexure (continuous spans)

## 6.4.5 Allowable Stresses: Article 1.7.142 of AASHTO

(A) *Local Bending Stresses in Deck Plate.* The term local bending stresses refers to the stresses caused in the deck plate as it carries a wheel load to the ribs and beams. The local transverse bending stresses caused in the deck plate by the specified wheel load plus 30 percent impact shall not exceed 30,000 psi unless a higher allowable stress is justified by a detailed fatigue analysis or by applicable fatigue-test results. For deck configurations in which the spacing of transverse beams is at least 3 times the spacing of longitudinal-rib webs, the local longitudinal and transverse bending stresses in the deck plate need not be combined with the other bending stresses covered in paragraphs (B) and (C) below.

(B) *Bending Stresses in Longitudinal Ribs.* The total bending stresses in longitudinal ribs due to a combination of (1) bending of the rib and, (2) bending of the girders may exceed the allowable bending stresses in Articles 1.7.1 and 1.7.3 by 25 percent. The bending stress due to each of the two individual modes shall not exceed the allowable bending stresses in Articles 1.7.1 and 1.7.3.

(C) *Bending Stresses in Transverse Beams.* The bending stresses in transverse beams shall not exceed the allowable bending stresses in Articles 1.7.1 and 1.7.3.

(D) *Intersections of Ribs, Beams, and Girders.* Connections between ribs and the webs of beams, holes in the webs of beams to permit passage of ribs, connections of beams to the webs of girders, and rib splices may affect the fatigue life of the bridge when they occur in regions of tensile stress. Where applicable, the number of cycles of maximum stress and the allowable fatigue stresses given in Section 1.7.3 shall be applied in designing these details; elsewhere, a rational fatigue analysis shall be made in designing the details. Connections between webs of longitudinal ribs and the deck plate shall be designed to sustain the transverse bending fatigue stresses caused in the webs by wheel loads where Article 1.7.1 pertains to "Allowable Stresses" for various grades of steel and designation and Article 1.7.3 relates to "Fatigue Stresses."

## 6.4.6 Thickness of Plate Elements

AASHTO Article 1.7.143 relates to the thicknesses of the longitudinal ribs, deck plate, girders, and transverse beams and is given below.

(A) *Longitudinal Ribs and Deck Plate.* Plate elements comprising longitudinal ribs, and deck-plate elements between webs of these ribs, shall meet the minimum thickness requirements of Article 1.7.88; $f_a$ may be taken as 75 percent of the sum of the compressive stresses due to (1) bending of the rib and, (2) bending of the girder, but not less than the compressive stress due to either of these two individual bending modes.

(B) *Girders and Transverse Beams.* Plate elements of box girders, plate girders, and transverse beams shall meet the requirements of Articles 1.7.69, 1.7.70, 1.7.71, 1.7.72, 1.7.73, and 1.7.105 where the referenced articles pertain to the following requirements:

Article 1.7.69       Flanges of Plate Girder
Article 1.7.70       Thickness of Web Plates of Plate Girders
Article 1.7.71       Transverse Intermediate Stiffeners of Plate Girders
Article 1.7.72       Longitudinal Stiffeners of Plate Girders
Article 1.7.73       Bearing Stiffeners of Plate Girders
Article 1.7.88       Compression Members—Thickness of Metal
Article 1.7.105      Design of Bottom Flange Plates—Box Girders

## 6.4.7   Maximum Slenderness of Longitudinal Ribs: AASHTO Article 1.7.144

The maximum slenderness of a longitudinal rib shall not exceed the following value:

$$\left(\frac{L}{r}\right)_{max} = 1000 \sqrt{\frac{1500}{Fy} - \frac{2700F}{Fy^2}}$$

where   $L$ = distance between transverse beams

$r$ = radius of gyration about the horizontal centroidal axis of rib including an effective width of deck plate

$F$ = maximum compressive stress (in psi) in the deck plate as a result of the deck acting as the top flange of the girders; this stress shall be taken as positive

$F_y$ = yield strength of rib material in psi

## 6.4.8   Stiffness Requirements

The deflections of ribs, beams, and girders due to live load plus impact may be the requirements given in Article 1.7.12, that is,

$$\Delta \le \frac{L}{800} \quad \text{or} \quad \Delta \le \frac{L}{1000} \text{ (urban areas)}$$

but preferably shall not exceed

$$\Delta \le \frac{L}{500}$$

The deflection of the deck plate due to the wheel load plus 30% impact shall be

$$\Delta \le \frac{L}{300}$$

where $L$ = distance between ribs.

## Design Example 6.1

A six-lane bridge, continuous over three spans with floor beams spaced at 12-ft intervals, (Figs. 6.14 and 6.15) is to be designed. The deck is composed of open ribs and the loading is HS20 in accordance with AASHTO. Given:

1. Roadway: 72 ft (6 to 12-foot lanes).
2. Floor-beam spacing: 12 ft.
3. Rib spacing: 18 in.
4. Main-girder spacing: 60 ft.
5. Material: A588, A572, and A36 steels.
6. Wearing surface: 2 in.-thick asphalt.

The design of this structure follows the procedure given in Section 6.3.4, which was presented by Dann Hall (14), at a lecture series to FHWA. His contribution is gratefully appreciated. This example design is also examined using the equations proposed by Heins and Perry (15). It is divided into the following sequence:

1. Deck plate design
2. Rigid support design
   a. Ribs
   b. Floor beams
3. Elastic support design
   a. Ribs
   b. Floor beams
4. Total stresses

DECK PLATE DESIGN

The minimum required deck-plate thickness is given by equation 6.25:

$$t_p = 0.0065a\sqrt[3]{P} \qquad (6.25)$$

where  $t$ = required plate thickness (in.)
$a$ = rib spacing (in.)
$P$ = wheel pressure (psi) for 12-k wheel load

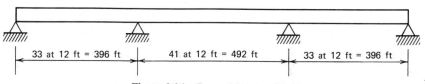

|← 33 at 12 ft = 396 ft →|← 41 at 12 ft = 492 ft →|← 33 at 12 ft = 396 ft →|

**Figure 6.14** General layout of bridge.

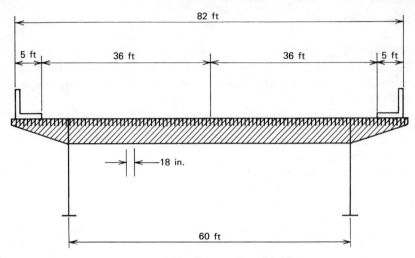

**Figure 6.15**   Cross section of bridge.

Referring to Fig. 6.16, the contact area is computed as

$$A = 12 \text{ in.} \times 24 \text{ in.} = 288 \text{ in.}^2$$

$$P = \frac{1.3 \times 12,000}{288} = 54.2 \text{ psi (including 30\% impact)}$$

The plate thickness for a rib spacing of $a = 18$ in., as shown in Fig. 6.17, is

$$t = 0.0065(18)\sqrt[3]{54.2}$$

$$t = 0.442 \text{ in.} \qquad \text{Use } \tfrac{1}{2} \text{ in. plate}$$

The local induced plate stress can be determined by analyzing the plate on a series of continuous rigid supports, as shown in Fig. 6.18. Such an analysis gives an area under the curve of

$$A = 0.4506 \times 2 = 0.9012$$

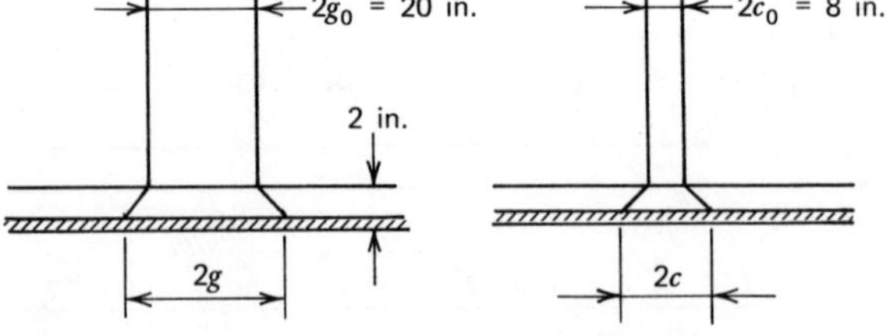

**Figure 6.16**   Wheel load on plate.

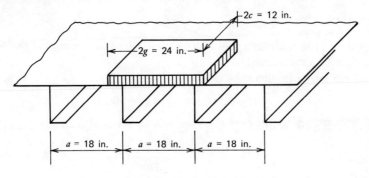

**Figure 6.17** Rip spacing.

The induced moment at the support is therefore

$$M_s = A \times W \times L$$
$$M_s = 0.9012 \times 54.2 \times 18$$
$$M_s = -879.2 \ 16\text{-in.}$$

The induced stress is therefore

$$f = \frac{M}{S} \qquad S = \tfrac{1}{6} t^2 b = \tfrac{1}{6} (\tfrac{1}{2})^2 (1) = \tfrac{1}{24}$$

$$f = \frac{879.2}{1000 \times 1/24} = 21.1 \ \text{ksi (transverse)}$$

Hence the longitudinal stress is

$$f = \mu \times f_t$$
$$f = 0.3 \times 21.2 = 6.3 \ \text{ksi (longitudinal)}$$

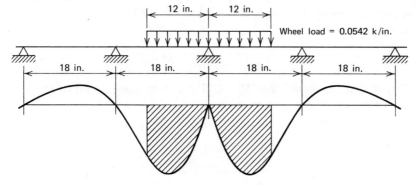

**Figure 6.18** Local loading on continuous plate.

To evaluate the moments in the deck plate, the cross-sectional properties must first be determined.

The rib length is given by

$$l_1 = 0.7(l) = 0.7(144) = 100.8$$

The effective rib width $(a_0)$ is

$$\frac{B}{a} = \frac{\text{tire width}}{18} = \frac{24}{18} = 1.33$$

From Fig. 6.7, for $B/a = 1.33$ and $a^*/a = 1.74$,

$$a^* = 1.74 \times 18 = 31.32 \text{ in.} \quad \text{(ideal rib span)}$$

$$\beta = \frac{\pi a^*}{l_1} = \frac{3.14 \times 31.32}{100.8} = 0.976$$

From Fig. 6.6, for $\beta = 0.976$ and $a_0/a^* = 0.90$, the effective plate width is

$$a_0 = 0.90a^*$$

$$a_0 = 0.90 \times 31.32$$

$$a_0 = 28.19 \text{ in.}$$

*Rib Geometry.* To estimate the size of the rib required, an approximate moment can be determined by estimating the induced forces $R_0$ and $R_0'$ on the rib, as shown in Fig. 6.19.

$$R_0 \times 18 = (0.5 \times 12)(6 + 6)$$

$$R_0 = \frac{6 \times 12}{18} = 4 \text{ k} \quad \text{and} \quad R_0' = 4 \text{ k}$$

The total reaction $R = 8$ k and $M = PL/4$ or $M = (8 \times 1.3 \times 12)/4 = 31.2$-ft per rib. Assuming the allowable live load stress is 11.0 ksi, the required section modulus is

$$S_{B_{req}} = \frac{31.2 \times 12}{11.0} = 34 \text{ in.}^3$$

Try a channel section $MC18 \times 42.7$ cut in half, as shown in Fig. 6.20. The total area of this section is given by

$$A_{\text{plate}} = 28.19 \times 0.50 = 14.10$$

$$A_{\text{web}} = 8.375 \times 0.45 = \;\; 3.77$$

$$A_{\text{flange}} = 3.95 \times 0.625 = \;\; 2.47$$

$$A_T = \overline{20.34} \text{ in.}^2$$

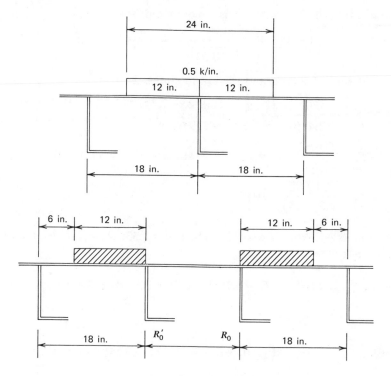

**Figure 6.19** Rib reactions.

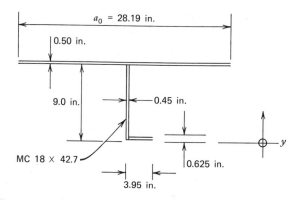

**Figure 6.20** Rib geometry.

The location of the neutral axis is computed:

$$\bar{y} = \frac{14.10(9.25) + 3.77(4.81) + 2.47(0.31)}{20.34} = 7.34 \text{ in.}$$

The moment of inertia and section modulus are therefore

$$I = \tfrac{1}{12}(0.45)(8.37)^3 + 14.10(9.25 - 7.34)^2 + 3.77(4.81 - 7.34)^2$$
$$+ 2.47(0.31 - 7.34)^2 = 219.62 \text{ in.}^4$$

$$S_B = \frac{219.62}{7.34} = 29.92 \text{ in.}^3 \qquad (\text{estimated } S_{\text{req}} = 34.0 \text{ in.}^3)$$

$$S_T = \frac{219.62}{9.5 - 7.34} = 101.7 \text{ in.}^3$$

*Live Loads on Ribs.* From Fig. 6.8, $B/a = 1.33$, $R_0/P_1 = 0.76$; therefore for an 8-k wheel

$$R_0 = \frac{R_0}{P_1} \times P_1 = 0.76 P_1 = 0.76 \times 1.3 \times 8.0 = 7.90 \text{ k}$$

and for a 12-k wheel

$$R_0 = \frac{R_0}{P_2} \times P_2 = 0.76 P_2 = 0.76 \times 1.3 \times 12.0 = 11.85 \text{ k}$$

*Live-Load $M_c$ Moment in Rib.* The moment at midspan of the rib deck for a distributed wheel load, as shown in Fig. 6.21, is given by equation 6.16a for a 12-k wheel load:

$$\left[\frac{M_c}{Pl}\right] = \left[0.1708 - 0.250\left(\frac{c}{l}\right) + 0.1057\left(\frac{c}{l}\right)^2\right] \tag{6.16a}$$

$$M_{c_{0-0'}} = 2R_0 cl\left[0.1708 - 0.25\frac{c}{l} + 0.1057\left(\frac{c}{l}\right)^2\right]$$

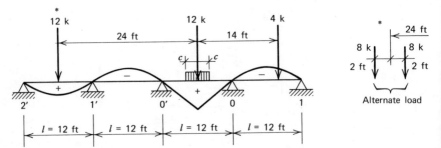

**Figure 6.21** Truck loading for evaluating $M_{\text{rib}}$ at midspan.

where $R_0 = 11.85$ k
$c = 0.5$ ft
$l = 12$ ft

at midspan of rib using a 12-k wheel load

$$M_{c_{0-0'}} = 2(11.85)(0.5)(12)\left[0.1708 - 0.25\left(\frac{0.5}{12}\right) + 0.1057\left(\frac{0.5}{12}\right)^2\right]$$

$$M_{c_{0-0'}} = 22.84 \text{ k-ft}$$

The contribution of the other two wheel loads, positioned on spans 1'–2' and 0–1', can now be evaluated using equation 6.16b. For the 4-k load on span 0–1,

$$M_{c_{0-1}} = R_0 l\left[-0.1830\frac{y}{l} + 0.3170\left(\frac{y}{l}\right)^2 - 0.1340\left(\frac{y}{l}\right)^3\right](-0.268)^m$$

where $R_0 = (1.3 \times 4.0 \times 0.76) = 3.95$ k
$l = 12.0$ ft
$y = 8.0$ ft

Then

$$M_{c_{0-1}} = 3.95(12.0)[0.1830(\tfrac{8}{12}) + 0.3170(\tfrac{8}{12})^2 - 0.134(\tfrac{8}{12})^3](-0.268)^0$$

$$M_{c_{0-1}} = -0.949 \text{ k/ft}$$

The bending moment in midspan due to a 12-k load in span 1'–2' is

$$M_{c_{0-1}} = 11.85(12.0)[-0.1830(\tfrac{6}{12}) + 0.3170(\tfrac{6}{12})^2 - 0.134(\tfrac{6}{12})^3](-0.268)^1$$

$$M_{c_{0-1}} = 1.10 \text{ k-ft}$$

The total bending moment at midspan is

$$M_c = 22.84 - 0.95 + 1.10 = 22.99 \text{ k-ft per rib} \quad \text{with 4-k wheel load}$$

$$M_c = 22.84 + 1.10 = 23.94 \text{ k-ft per rib} \quad \text{without 4-k wheel load}$$

*Live-Load $M_s$ Support Moment in Rib.* Using equation 6.17 the rib moment can now be calculated with the vehicle positioned as shown in Fig. 6.22:

$$\left[\frac{M_s}{Pl}\right] = \left[-0.5\frac{y}{l} + 0.866\left(\frac{y}{l}\right)^2 - 0.366\left(\frac{y}{l}\right)^3\right](-0.268)^m$$

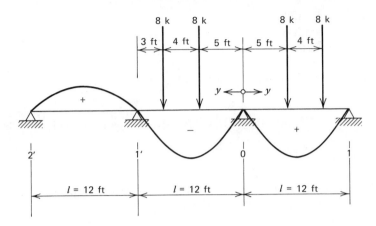

**Figure 6.22**  Truck loading for evaluating $M_{rib}$ at support.

For the loads in span 0–1 and 0–1′, the term $(-0.268)^m$ is set equal to 1.0. Applying the above equation gives for span 0–1 and span 0–1′

$$\text{8-k load, } y = 5 \text{ ft;} \qquad \frac{y}{l} = 0.42; \qquad \left(\frac{y}{l}\right)^2 = 0.174; \qquad \left(\frac{y}{l}\right)^3 = 0.073$$

$$M_s = (7.9 \times 12)[-0.5(0.42) + 0.866(0.174) - 0.366(0.073)]$$

$$= -7.9 \times 12(0.083)$$

$$\text{8-k load, } y = 9 \text{ ft;} \qquad \left(\frac{y}{l}\right) = 0.75; \qquad \left(\frac{y}{l}\right)^2 = 0.563; \qquad \left(\frac{y}{l}\right)^3 = 0.422$$

$$M_s = (7.9 \times 12)[-0.5(0.75) + 0.866(0.563) - 0.366(0.422)]$$

$$= -7.9 \times 12(0.043)$$

The combined effects give

$$M_s = -7.9 \times 12(0.083 + 0.083 + 0.043 + 0.043)$$

$$M_s = -23.9 \text{ k-ft}$$

*Live-Load Moment in Floor Beam*

1. *All Lanes Loaded.* Six trucks are positioned longitudinally on the bridge in their respective lanes and induce reactions $F_0$, as shown in Fig. 6.23. The reactions are computed below assuming the main longitudinal girders are the supports.

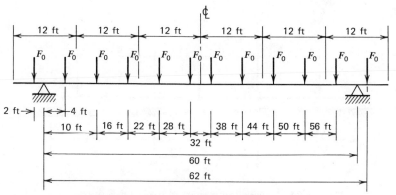

**Figure 6.23** Truck loading across bridge width.

The summation of the moments about left reaction (floor beam) is given by

$$60R_R = (-2 + 4 + 10 + 16 + 22 + 28 + 32 + 38 + 44 + 50 + 56 + 62)F_0$$

$$R_R = 6.0F_0$$

$$M_A = [28 \times 6 - (6 + 12 + 18 + 24 + 32)]F_0$$

$$M_A = 76F_0 \text{ k-ft} \quad (75\% \text{ for AASHTO})$$

$$M_A = 57.0F_0 \text{ k-ft}$$

where $F_0$ = one-wheel reaction ($R_0$) with impact.

2. *Three Lanes Loaded.* The summation for this case is given by

$$60R_R = (22 + 28 + 32 + 38 + 44 + 50)F_0$$

$$R_R = (3.57)F_0$$

$$M_A = [3.57(28) - (6 + 12 + 18)]F_0$$

$$M_A = [100 - 36]F_0 = 64F_0 \quad (90\% \text{ for AASHTO})$$

$$M_A = 0.90(64)F_0 = 57.6F_0 \text{ controls}$$

The total truck-wheel reaction on the floor beams is found by application of the influence line reaction equations (equations 6.18a and 6.18b) with all trucks positioned longitudinally, as shown in Fig. 6.24.

For span 0–1:

$$R_0 = \left[1 - 2.196\left(\frac{y}{l}\right)^2 + 1.196\left(\frac{y}{l}\right)^3\right]$$

$$y = 7 \text{ ft}, \quad \frac{y}{l} = 0.58, \quad \left(\frac{y}{l}\right)^2 = 0.34, \quad \left(\frac{y}{l}\right)^3 = 0.197$$

$$R_0 = [1 - 2.196(0.34) + 1.196(0.197)] = 0.489$$

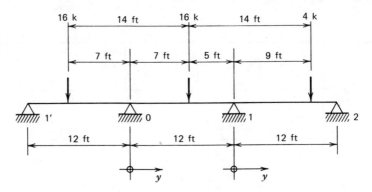

**Figure 6.24** Longitudinal position of trucks for floor-beam reaction.

For span 1–2:

$$R_0 = \left[ -0.804\frac{y}{l} + 1.392\left(\frac{y}{l}\right)^2 - 0.588\left(\frac{y}{l}\right)^3 \right](-0.268)^{m-1}$$

$$y = 9\,\text{ft}, \quad \frac{y}{l} = 0.75, \quad \left(\frac{y}{l}\right)^2 = 0.56, \quad \left(\frac{y}{l}\right)^3 = 0.42, \quad m = 1$$

$$R_0 = [-0.804(0.75) + 1.392(0.56) - 0.588(0.42)] = 0.065$$

Therefore total reaction including impact is given by

$$F_0 = \left[ 1 + \frac{50}{125+60} \right][2 \times 16 \times 0.489 + 4 \times 0.065]$$

$$F_0 = (1.27)(15.65 + 0.260)$$

$$F_0 = 20.23$$

Then the induced moment in the floorbeam is given by

$$M_A = 57.6F_0$$

$$M_A = 57.6 \times 20.23$$

$$M_A = 1165.2\,\text{k-ft}$$

*Dead-Load Bending Moments in Ribs.* The dead load of a typical rib, as shown in Fig. 6.25, is computed below.

| | |
|---|---|
| 2-in. asphalt: $\frac{2}{12} \times 1.0 \times 120 \times \frac{18}{12}$ | 30.0 lb per rib |
| Deck plate: $\frac{1}{2}$ in. $\times 18$ in. $\times 1.0$ ft $\times 3.4$ | 30.6 lb per rib |
| Rib $M_c$ 18 $\times$ 42.7 ($\frac{1}{2}$ section) $= \frac{1}{2}(42.7)$ | 21.4 lb per rib |

Weight ($w$) per rib for 18-in. spacing = 82.0 lb per ft

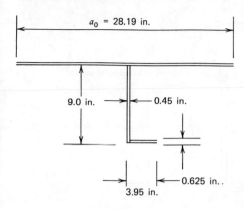

9.0 in.    0.45 in.

0.625 in.      **Figure 6.25**  Rigid rib section—dead-
3.95 in.       load analysis.

$a_0$ = 28.19 in.

The induced dead-load midspan and support moments respectively, are therefore

$$M_{sp} = \frac{wl^2}{24} = \frac{82.0 \times (12)^2}{24 \times 1000} = 0.49 \text{ k-ft per rib}$$

and

$$M_{sup} = \frac{wl^2}{12} = -0.98 \text{ k-ft per rib}$$

*Approximate Floor-Beam Section.*   Using the computed induced live-load moment, calculated as $M_A = 1165.2$ k-ft, and assuming $f_b = 27$ ksi, the required section modulus of the floor beam is

$$S = \frac{1165.2 \times 12}{27} = 518 \text{ in.}^3$$

The following values are tried.

Web:     48 in. $\times \frac{7}{16}$ in.

Flange:   10 in. $\times \frac{9}{16}$ in.

The section is described in Fig. 6.26.

*Dead-Load Bending Moment in Floor Beam.*

| | | |
|---|---|---|
| 2 in. Asphalt: $\frac{2}{12} \times 12.0 \times 120$ | 240 lb/ft | |
| Ribs: $21.4 \times 12.0 \times \frac{2}{3}$ | 171 | |
| Deck plate: $\frac{1}{2} \times \frac{1}{12} \times 12 \times 490$ | 245 | |
| Web and flange plate: $71.4 + 19.1$ | 90.5 | |
| | 746 lb/ft of floor beam | |

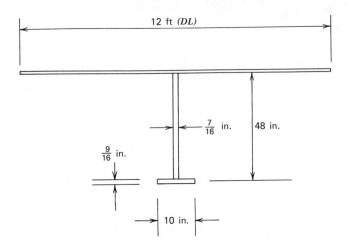

**Figure 6.26**  Rigid floor-beam section.

Then the resulting induced moment is

$$M = \frac{wl^2}{8} = \frac{746(60)^2}{8 \times 1000} = 337 \text{ k-ft per floor beam}$$

INFLUENCE OF ELASTIC SUPPORTS

As given in Section 6.3.2, the effective span $l_1 = \infty$ and ideal span $a^* = a = 18$ in.; therefore

$$\beta = \frac{\pi a^*}{l_1} = \frac{3.14 \times 18}{\infty} = 0$$

From Fig. 6.6, with $\beta = 0$, $\lambda = a_0/a^* = 1.10$, the effective width $a_0 = (a_0/a^*)a^* = 1.10$ in. $\times 18$ in. $\times 19.8$ in.

*Floor Beam.* From Section 6.3.2, the effective span $b^* = b = 60$ ft and $l^* = l = 12$ ft $= 144$ in.; therefore

$$\beta = \frac{\pi l^*}{C^*} = \frac{3.14 \times 144}{60 \times 12} = 0.628$$

From Fig. 6.6, for $\beta = 0.628$ and $\lambda = 1.02$, the effective width $l_0 = (l_0/l^*) = 1.02(144) = 147$ in. With these new effective widths, new properties are required and are determined.

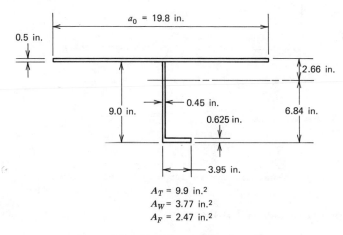

$A_T = 9.9$ in.$^2$
$A_W = 3.77$ in.$^2$
$A_F = 2.47$ in.$^2$

**Figure 6.27** Flexible support—rib section.

*Rib Section Properties.* Referring to Fig. 6.27, the location of the neutral axis of the rib, considering flexible supports, is computed as

$$\bar{y} = \frac{9.9(9.25) + 3.77(4.81) + 2.47(0.31)}{9.9 + 3.77 + 2.47} = 6.84 \text{ in.}$$

The moment of inertia $I$ is then

Web: $\dfrac{0.45(8.37)^3}{12} + 3.77(6.84 - 4.81)^2$      42.87

Flange: $2.47(6.84 - 0.31)^2$      105.32

Top plate: $9.90(9.25 - 6.84)^2$      57.50

$$I = 205.7 \text{ in.}^4$$

The section modulus is then computed as

$$S_T = \frac{205.7}{2.66} = 77.4 \text{ in.}^3$$

$$S_B = \frac{205.7}{6.84} = 30.1 \text{ in.}^3$$

*Floor-Beam Section Properties.* From Fig. 6.28, the location of the neutral axis, considering flexible supports, is computed as

$$\bar{y} = \frac{(73.5)(48.81) + (21)(24.56) + (5.62)(0.28)}{100.1} = 41.00 \text{ in.}$$

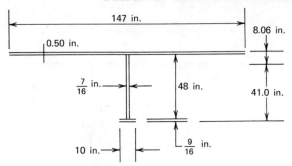

**Figure 6.28**   Flexible support—floor-beam section.

The moment of inertia $I$ is therefore

Top plate: $73.5(8.06 - 0.25)^2$                      $4483$ in.$^4$

Web: $\dfrac{7}{16} \times \dfrac{(48)^3}{12} + 21(41.00 - 24.56)^2$       $9708$

Flange: $5.62(41.00 - 0.28)^2$                        $\underline{9319}$

$$I = 23{,}510 \text{ in.}^4$$

The section modulus is then computed as

$$S_T = \frac{23{,}510}{8.10} = 2917 \text{ in.}^3$$

$$S_B = \frac{23{,}510}{41.00} = 573 \text{ in.}^3$$

With these properties, the floor-beam coefficient of flexural rigidity (equation 6.19) can be determined. This coefficient is used in conjunction with the influence line coefficients given in Tables 6.3 through 6.5 in the evaluation of moment relief or relaxation due to the flexible floor beams. The truck loading is as shown in Figs. 6.29 and 6.30. The flexural rigidity coefficient is computed as follows:

$$\gamma = \frac{I_r b^4}{I_f l^3 a \pi^4}$$

$$\gamma = \frac{205.9(60 \times 12)^4}{23{,}510(144)^3(18)(3.1416)^4} = 0.50$$

(6.19)

where $I_r$ = moment of inertia or rib
= main girder spacing
$I_f$ = moment of inertia of floor beam
$l$ = span of ribs
$a$ = spacing of ribs

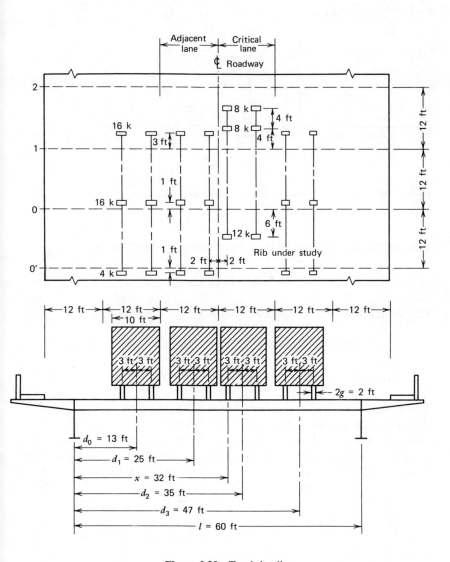

**Figure 6.29** Truck loading.

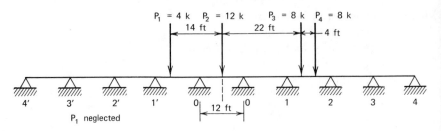

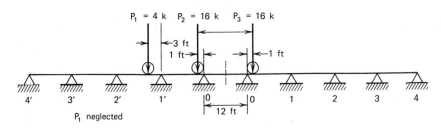

**Figure 6.30**  Floor-beam truck loading. (*Top*) Critical lane loads. (*Bottom*) Adjacent lane loads.

The additional moment in the ribs is given by equation 6.20 or

$$\Delta M_c = Q_0 la \frac{Q_{ix}}{Q_0} \sum \frac{F_m}{P} \frac{\eta_m}{l} \qquad (6.20)$$

The evaluation of this equation requires:

1. For $F_m/P$: Induced floor-beam reactions of a plate on rigid supports (equation 6.8).
2. For $\eta_m/l$: Influence line coefficients (Table 6.3) for $\gamma = 0.50$.

The loading required consists of four trucks positioned longitudinally on the bridge, as shown in Fig. 6.29. The position of the wheel loads in the lane enclosing the rib under study must be in accordance with the loading used under the rigid-floor-beam study, neglecting the 4 k load, as shown in Fig. 6.21. The loadings for the critical lane and adjoining lanes are shown in Fig. 6.30. Note that the two 8-k loads spaced at 4 ft are used in place of the 12-k load in Fig. 6.21, and the distance between these two 8 k axles is equal to 4 ft.

The floor-beam reactions are first computed by using equations 6.18 and 6.18b, which gives influence line ordinates (critical lane loaded) as shown below.

1. *Ordinates at* 0.

$$\frac{F_0}{P_2} = 1 - 2.1961\left(\frac{y}{l}\right)^2 + 1.1961\left(\frac{y}{l}\right)^3$$

$$\frac{F_0}{P_2} = 1 - 2.1961(\tfrac{6}{12})^2 + 1.1961(\tfrac{6}{12})^3 = 0.6005 = 0.601$$

$$\frac{F_0}{P} = \left[-0.8038\left(\frac{y}{l}\right) + 1.3923\left(\frac{y}{l}\right)^2 - 0.5885\left(\frac{y}{l}\right)^3\right](-0.268)^{m-1}$$

$$\frac{F_0}{P_3} = [-0.8038(\tfrac{4}{12}) + 1.3923(\tfrac{4}{12})^2 - 0.5885(\tfrac{4}{12})^3](-0.268)^0 = -0.135$$

$$\frac{F_0}{P_4} = [-0.8038(\tfrac{8}{12}) + 1.3923(\tfrac{8}{12})^2 - 0.5885(\tfrac{8}{12})^3](-0.268)^0 = -0.091$$

2. *Ordinates at* 1.

$$\frac{F_1}{P_2} = [-0.8038(\tfrac{6}{12}) + 1.3923(\tfrac{6}{12})^2 - 0.5885(\tfrac{6}{12})^3](-0.268)^0 = -0.1274$$

$$\frac{F_1}{P_3} = 1 - 2.1961(\tfrac{4}{12})^2 + 1.1961(\tfrac{4}{12})^3 = 0.8003$$

$$\frac{F_1}{P_4} = 1 - 2.1961(\tfrac{8}{12})^2 + 1.1961(\tfrac{8}{12})^3 = 0.3784$$

3. *Ordinates at* 2.

$$\frac{F_2}{P_2} = \frac{F_1}{P_2}(-0.268) = 0.0341$$

$$\frac{F_2}{P_3} = \frac{F_1}{P_4} = 0.3784$$

$$\frac{F_2}{P_4} = \frac{F_1}{P_3} = 0.8003$$

4. *Ordinates at* 3.

$$\frac{F_3}{P_2} = \frac{F_2}{P_2}(-0.268) = (0.0341)(-0.268) = -0.0091$$

$$\frac{F_3}{P_3} = \frac{F_0}{P_4} = -0.091$$

$$\frac{F_3}{P_4} = \frac{F_0}{P_3} = -0.135$$

5. *Ordinates at* 4.

$$\frac{F_4}{P_2} = \frac{F_2}{P_2}(-0.268)^2 = (0.0341)(-0.268)^2 = 0.0024$$

$$\frac{F_4}{P_3} = \frac{F_3}{P_3}(-0.268) = (-0.091)(-0.268) = 0.0243$$

$$\frac{F_4}{P_4} = \frac{F_3}{P_4}(-0.268) = (-0.135)(-0.268) = 0.0361$$

6. *Ordinates at* 1'.

$$\frac{F'_1}{P_3} = \frac{F_4}{P_4}(-0.268) = (0.0361)(-0.268) = -0.0097$$

$$\frac{F'_1}{P_4} = \frac{F_4}{P_3}(-0.268) = (0.0243)(-0.268) = -0.0065$$

1. *Ordinates at* 2'.

$$\frac{F'_2}{P_3} = \frac{F_4}{P_4}(-0.268)^2 = (0.0361)(-0.268)^2 = 0.0026$$

$$\frac{F'_2}{P_4} = \frac{F_4}{P_3}(-0.268)^2 = (0.0243)(-0.268)^2 = 0.0017$$

The adjacent-lane influence line ordinates are calculated as follows:

1. *Ordinates at* 0 *and* (0').

$$\frac{F_0}{P_3} = 1.0 - 2.1961(\tfrac{1}{12})^2 + 1.1961(\tfrac{1}{12})^3 = 0.9854$$

$$\frac{F_0}{P_2} = [-0.8038(\tfrac{1}{12}) + 1.3923(\tfrac{1}{12})^2 - 0.5885(\tfrac{1}{12})^3](-0.268)^0 = -0.0577$$

2. *Ordinates at* 1 *and* (1').

$$\frac{F_1}{P_3} = 1.0 - 2.1961(\tfrac{11}{12})^2 + 1.1961(\tfrac{11}{12})^3 = 0.0760$$

$$\frac{F_1}{P_2} = \frac{F_0}{P_2}(-0.268) = (-0.0577)(-0.268) = 0.01546$$

3. *Ordinates at* 2 *and* (2').

$$\frac{F_2}{P_3} = [-0.8038(\tfrac{11}{12}) + 1.3923(\tfrac{11}{12})^2 - 0.5885(\tfrac{11}{12})^3](-0.268)^0 = -0.0202$$

$$\frac{F_2}{P_2} = \frac{F_0}{P_2}(-0.268)^2 = (-0.0577)(-0.268)^2 = -0.0041$$

4. *Ordinates at 3 and (3').*

$$\frac{F_3}{P_3} = \frac{F_2}{P_3}(-0.268) = (-0.0202)(-0.268) = 0.0054$$

$$\frac{F_3}{P_2} = \frac{F_0}{P_2}(-0.268)^3 = (-0.0577)(-0.268)^3 = 0.0011$$

5. *Ordinates at 4 and (4').*

$$\frac{F_4}{P_3} = \frac{F_2}{P_3}(-0.268)^2 = (-0.0202)(-0.268)^2 = -0.0014$$

$$\frac{F_4}{P_2} = \frac{F_0}{P_2}(-0.268)^4 = (-0.0577)(-0.268)^4 = -0.0003$$

A summary of these influence line ordinates $(F_0/P)$ or $(F_m/P)$ at the various floor beams is given in Table 6.6. The coefficients $\eta_m/l$ are evaluated from Table 6.3 for $\gamma = 0.50$ at the various support points. The sum effect of $(F_m/P)(\eta_m/l)$ gives

$$\left(\frac{F_m}{P}\right)\left(\frac{\eta_m}{l}\right) = 0.206 \qquad \text{adjoining loaded lanes}$$

$$\left(\frac{F_m}{P}\right)\left(\frac{\eta_m}{l}\right) = 0.0605 \qquad \text{critical loaded lane}$$

The general equation can now be applied:

$$\Delta M_c = Q_0 la \frac{Q_{ix}}{Q_0} \sum \frac{F_m}{P} \frac{\eta_m}{l} \tag{6.20}$$

First the computation of the terms $Q_{ix}/Q_0$, is required. For truck in critical lane (Fig. 6.29):

$$\frac{Q_{ix}}{Q_0} = \frac{8}{\pi} \sin \frac{\pi g}{b} \cos \frac{\pi e}{b} \sin \frac{\pi d_2}{b} \sin \frac{\pi x}{b}$$

where $g = 1.0$
$\quad\quad 3 = 3.0$
$\quad\quad d_2 = 35$
$\quad\quad x = 32$
$\quad\quad b = 60$ ft

Then

$$\frac{Q_{ix}}{Q_0} = \frac{8}{\pi} \sin \frac{\pi}{60} \cos \frac{\pi}{60} \sin \frac{35\pi}{60} \sin \frac{32\pi}{60}$$

$$\frac{Q_{ix}}{Q_0} = \frac{8}{3.1416}[(0.0523)(0.9877)(0.9658)(0.9944)] = 0.1263 \quad \text{for} \quad d_2$$

**Table 6.6  Evaluation of Expression $\sum F_m/P \cdot \eta_m/l$**

$\gamma = 0.50$

| Loading Case | Values Computed | $m = 4'$ | $3'$ | $2'$ | $1'$ | $0'$ | $0$ | $1$ | $2$ | $3$ | $4$ | $\sum \dfrac{F_m}{P} \cdot \dfrac{\eta_m}{l}$ |
|---|---|---|---|---|---|---|---|---|---|---|---|---|
| Adjoining lane loaded | $\dfrac{\eta_m}{l}$ | −0.00196 | −0.01935 | −0.04932 | −0.04858 | 0.11450 | 0.11450 | −0.04858 | −0.04932 | −0.01935 | −0.00196 | |
| | $\dfrac{F_m}{P}\ P_2$ | −0.001 | 0.005 | 0.020 | 0.076 | 0.985 | −0.058 | 0.015 | −0.004 | 0.001 | −0.000 | |
| | $P_3$ | −0.000 | 0.001 | −0.004 | 0.015 | −0.058 | 0.985 | 0.076 | −0.020 | 0.005 | −0.001 | |
| | Total | −0.001 | 0.006 | −0.024 | 0.091 | 0.927 | 0.927 | 0.091 | −0.024 | 0.006 | −0.001 | |
| | $\dfrac{F_m}{P} \cdot \dfrac{\eta_m}{l}\ P_2$ | — | −0.00012 | 0.00118 | −0.00442 | 0.10614 | 0.10614 | −0.00442 | 0.00118 | −0.00012 | — | 0.206 |
| Critical lane loaded | $P_2$ | 0.002 | −0.009 | 0.034 | −0.127 | 0.601 | 0.601 | −0.127 | 0.034 | −0.009 | 0.002 | |
| | $\dfrac{F_m}{P}\ P_3^a$ | 0.000 | 0.001 | 0.003 | −0.010 | 0.036 | −0.135 | 0.800 | 0.378 | −0.091 | 0.024 | |
| | $P_4^a$ | 0.000 | 0.000 | 0.002 | −0.006 | 0.024 | −0.091 | 0.378 | 0.800 | −0.135 | 0.036 | |
| | Total | 0.002 | −0.008 | 0.037 | −0.138 | 0.641 | 0.450 | 0.659 | 0.820 | −0.160 | 0.042 | |
| | $\dfrac{F_m}{P} \cdot \dfrac{\eta_m}{l}$ | — | 0.00015 | −0.00182 | 0.00670 | 0.07339 | 0.05152 | −0.03201 | −0.04044 | 0.00310 | −0.00008 | 0.0605 |

$^a$ Two thirds of $P_3$ and $P_4$ is used since they are 8-k rather than 12-k loads.

For truck in adjoining lane; $d_1 = 25$ ft (Fig. 6.29):

$$\frac{Q_{ix}}{Q_0} = \frac{8}{\pi}\left[\sin\frac{\pi}{60}\cos\frac{3\pi}{60}\sin\frac{25\pi}{60}\sin\frac{32\pi}{60}\right]$$

$$\frac{Q_{ix}}{Q_0} = \frac{8}{3.1416}[(0.0523)(0.9877)(0.9658)(0.9944)] = 0.1263 \qquad \text{for } d_1$$

For truck in other adjoining lane; $d_0 = 13$ ft, $d_3 = 47$ ft:

$$\frac{Q_{ix}}{Q_0} = \frac{8}{\pi}\left[\sin\frac{\pi}{60}\cos\frac{3\pi}{60}\sin\frac{13\pi}{60}\sin\frac{32\pi}{60}\right]$$

$$\frac{Q_{ix}}{Q_0} = \frac{8}{3.1416}[(0.0523)(0.9877)(0.6293)(0.9944)] = 0.0823 \qquad \text{for } d_0 \text{ and } d_3$$

Equation 6.20 gives the following calculations for critical lane loaded:

$$\Delta M_c = Q_0 la\frac{Q_{ix}}{Q_0}\sum\frac{F_m}{P}\frac{\eta_m}{l}$$

$$Q_0 \text{ for 12-k wheel} = \frac{P}{2c} = \frac{1.3\times12}{2\times12} = 0.65 \text{ k/in.}$$

$$\Delta M_c = 0.65(12)(18)(0.1263)(0.0605) = 1.072 \text{ k-ft per rib}$$

For adjoining lane loaded:

$$Q_0 \text{ for 16-k wheel} = \frac{1.3\times16}{2\times22} = 0.867 \text{ k/in.}$$

$$\Delta M_c = 0.867(12)(18)(0.1263)(0.206) = 4.873 \text{ k-ft per rib}$$

For other lane loaded:

$$\Delta M_c = 0.867(12)(18)(0.0823)(0.206) = 2.175 \text{ k-ft per rib}$$

Note $d_3 = 47.0$.

$$\Delta M_c = 3.175 \text{ k-ft per rib}$$

Resulting moment: the total induced rib moment, due to floor beam flexibility, and including the four-lane AASHTO reduction factor (0.75) gives;

$$\text{total } \Delta M_c = [1.072 + 4.873 + 3.175 + 3.175][0.75] = 9.22 \text{ k-ft per rib}$$

The floor-beam-moment modification, due to flexibility of the floor beams, is found from equation 6.21:

$$\Delta M_F = Q_0\left(\frac{l}{\pi}\right)^2\frac{Q_{ix}}{Q_0}\left[\frac{F_0}{P} - \sum\frac{F_0}{P}\gamma_m\right] \qquad (6.21)$$

where $(F_0/P)$ = induced reactions on rigid supports (equation 6.18)
$\gamma_m$ = flexible-floor-beam influence line reaction coefficients (Table 6.5)

The rigid-floor-beam reactions, due to the truck positioned as shown in Fig. 6.31, are computed from equations 6.18a and 6.18b. Application of the equations gives the induced reaction at support 0, where the coefficients for equation 6.18a are

$$\text{span } 0\text{--}1, \ 0\text{--}1' : \qquad y = 7 \text{ ft}, \qquad P = 16 \text{ k}$$

and the coefficients for equations 6.18b are

$$\text{span } 1\text{--}2 : \qquad y = 9 \text{ ft}, \qquad P = 4 \text{ k}, \qquad m = 1$$

Application of these equations and the respective coefficients gives

$$R_0 - [1 - 2.1962(\tfrac{7}{12})^2 + 1.1962(\tfrac{7}{12})^3]2$$
$$+ \tfrac{1}{4}[-0.8038(\tfrac{9}{12}) + 1.3923(\tfrac{9}{12})^2 - 0.5885(\tfrac{9}{12})^3] = 0.962$$

For the induced reaction at support 1, where the coefficients for equation 6.18a are

$$\text{span } 0\text{--}1 : \qquad y = 5 \text{ ft}, \qquad P = 16 \text{ k}$$
$$\text{span } 1\text{--}2 : \qquad y = 9 \text{ ft}, \qquad P = 4 \text{ k}$$

and the coefficients for equation 6.18b are

$$\text{span } 1'\text{--}0 : \qquad y = 7 \text{ ft}, \qquad P = 16 \text{ k}, \qquad m = 1$$

application of the equations and the respective coefficients gives

$$R_1 = 1 - 2.1962(\tfrac{5}{12})^2 + 1.1962(\tfrac{5}{12})^3 + [-0.8038(\tfrac{7}{12})$$
$$+ 1.3923(\tfrac{7}{12})^2 - 0.5885(\tfrac{7}{12})^3] + \tfrac{1}{4}[1 - 2.1962(\tfrac{9}{12})^2$$
$$+ 1.1962(\tfrac{9}{12})^3] = 0.651$$

For the induced reaction at support 2, the coefficients for equation 6.18a are

$$\text{span } 1\text{--}2 : \qquad y = 3 \text{ ft}, \qquad P = 4 \text{ k}$$

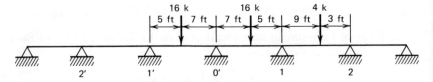

**Figure 6.31** Floor-beam truck loading—maximum $R_0$.

and the coefficients for equation 6.18b are

$$\text{span } 0\text{--}1: \qquad y = 5 \text{ ft}, \qquad P = 16 \text{ k}, \qquad m = 1$$

$$\text{span } 0\text{--}1': \qquad y = 7 \text{ ft}, \qquad P = 16 \text{ k}, \qquad m = 2$$

From the equations and these coefficients we get

$$R_2 = [-0.8038(\tfrac{7}{12}) + 1.3923(\tfrac{7}{12})^2 - 0.5885(\tfrac{7}{12})^3][-0.2679]$$
$$+ [-0.8038(\tfrac{5}{12}) + 1.3923(\tfrac{5}{12})^2 - 0.5885(\tfrac{5}{12})^3]$$
$$+ \tfrac{1}{4}[1 - 2.1962(\tfrac{3}{12})^2 + 1.1962(\tfrac{3}{12})^3] = 0.114$$

For the induced reaction at support $1'$, where the coefficients for equation 6.18a are

$$\text{span } 1'\text{--}0: \qquad y = 5 \text{ ft}, \qquad P = 16 \text{ k}$$

and the coefficients for equation 6.18b are

$$\text{span } 0\text{--}1: \qquad y = 7 \text{ ft}, \qquad P = 16 \text{ k}, \qquad m = 1$$

$$\text{span } 1\text{--}2: \qquad y = 9 \text{ ft}, \qquad P = 4 \text{ k}, \qquad m = 2$$

from the equations and these coefficients we get

$$R'_1 = 1 - 2.1962(\tfrac{5}{12})^2 + 1.1962(\tfrac{5}{12})^3 + [-0.8038(\tfrac{7}{12})$$
$$+ 1.3923(\tfrac{7}{12})^2 - 0.5885(\tfrac{7}{12})^3] + \tfrac{1}{4}[-0.8038(\tfrac{9}{12})$$
$$+ 1.3923(\tfrac{9}{12})^2 - 0.5885(\tfrac{9}{12})^3][-0.2679] = 0.598$$

For the induced reaction at support $2'$, where the coefficients for equation 6.18b are

$$\text{span } 1'\text{--}0: \qquad y = 5 \text{ ft}, \qquad P = 16 \text{ k}, \qquad m = 1$$

$$\text{span } 0\text{--}1: \qquad y = 7 \text{ ft}, \qquad P = 16 \text{ k}, \qquad m = 2$$

$$\text{span } 1\text{--}2: \qquad y = 9 \text{ ft}, \qquad P = 4 \text{ k}, \qquad m = 3$$

substitution of the coefficients into the referenced equations gives

$$R'_2 = [-0.8038(\tfrac{5}{12}) + 1.3923(\tfrac{5}{12})^2 - 0.5885(\tfrac{5}{12})^3]$$
$$+ [-0.8038(\tfrac{7}{12}) + 1.3923(\tfrac{7}{12})^2 - 0.5885(\tfrac{7}{12})^3][-0.2679]$$
$$+ \tfrac{1}{4}[-0.8038(\tfrac{9}{12}) + 1.3923(\tfrac{9}{12})^2 - 0.5885(\tfrac{9}{12})^3][-0.2679]^2$$
$$= -0.107$$

The coefficient $\gamma_m$ is found from Table 6.5 with $\gamma = 0.50$. These values and the $F_0/P$ coefficients are given in Table 6.7. Multiplying $(F_0/P)$ times the respective coefficient $\gamma_m$ and summing these values gives $\sum F_m/P\gamma_m = 0.724$, and the final term $[F_0/P - \sum F_0/P \gamma_m]$ is

$$\text{Reaction } 0: \qquad [0.962 - 0.724] = 0.238$$

**Table 6.7    Moment Relief in Floor Beams; Computation of $(F_0/P)-(\bar{F}_0/P)$**

| Values Computed | Reference | $m = 2'$ | $1'$ | 0 | 1 | 2 | $\bar{F}_0/P$ | $(F_0/P)-(\bar{F}_0/P)$ |
|---|---|---|---|---|---|---|---|---|
| $\dfrac{F_m}{P}$ | Computed | −0.107 | 0.598 | 0.962 | 0.651 | 0.114 | | |
| $\gamma_{0m}$ | Table 6.5 | 0.069 | 0.256 | 0.4184 | 0.256 | 0.069 | | |
| $\dfrac{F_m}{P}\ \gamma_m$ | — | −0.007 | 0.153 | 0.403 | 0.170 | 0.008 | 0.724 | 0.238 |

Since

$$Q_0 = \frac{P}{2g} = \frac{16}{24}(1+I)$$

and

$$(1+I) = \left(1 + \frac{50}{125+60}\right) = 1.207$$

Therefore

$$Q_0 = \tfrac{2}{3}(1.207) \times 12 = 10.16 \text{ k/ft}$$

The term $Q_{ix}/Q_0$ is as computed for the rib moment relief for three trucks or

$$\frac{Q_{ix}}{Q_0} = 0.1263 + 0.1263 + 0.0823 = 0.3349$$

Considering the reduction in load intensity according to AASHTO specifications, when three trucks are on the bridge (90%), $\Delta M_F$ is

$$\Delta M_F = Q_0 \left(\frac{l}{\pi}\right)^2 \frac{Q_{ix}}{Q_0}\left[\frac{F_0}{P} - \sum \frac{F_0}{P} \nu_m\right]$$

$$\Delta M_F = 10.16\left(\frac{60}{\pi}\right)^2 (0.3349)[0.238](0.90)$$

$$M_F = 267 \text{ k-ft}$$

TOTAL STRESSES

The forces and stresses in the main longitudinal girders are found by conventional techniques by loading the three-span continuous girders with the AASHTO loading and the dead load. The maximum induced positive and negative moments are summarized in Table 6.8, as are the rib and floor-beam moments. The respective section moduli, as computed previously, for these various elements are summarized in Table 6.9. Using these section properties and the moments given in Table 6.8, the induced bending stresses are then computed giving the stresses shown in Table 6.10.

A summary of the critical stress locations and values are shown in Fig. 6.32 and are given below.

The total floor-beam flange stress is

$$7.06 + 23.66 - 5.59 = \underline{25.13 \text{ ksi}} < f_b = 27 \text{ ksi} \qquad \text{Use A572 grade 50}$$

The total stress in the bottom of the rib at point A is given by

$$20 \times 1.25 = 25 \text{ ksi}$$

$$+0.20 + 9.60 + 3.67 + 15.02 = \underline{28.52 \text{ ksi}} > 20 \times 1.25 \qquad \text{Say no good:}$$

A572 Grade 42 is tried.

$$23 \times 1.25 = 28.75 \text{ ksi}$$

**Table 6.8   Summary of Moments**

| Element | Rigid-System Dead Load | Rigid-System Live Load | Elastic-System Live Load |
|---|---|---|---|
| Rib at midspan | 0.49 k-ft | 23.94 k-ft | 0.22 k-ft |
| Rib at floor beam | −0.98 k-ft | −23.90 k-ft | — |
| Floor beam, 2 ft 0 in., right of center | 337 k-ft | 1,130 k-ft | −267 k-ft |
| Main girder position | | | |
| Spans 1 and 3 | 37,811 k-ft | 28,257 k-ft | — |
| Span 2 | 36,607 k-ft | 32,485 k-ft | — |
| Main girder supports | 70,260 k-ft | 41,141 k-ft | — |

**Table 6.9   Summary of Section Moduli (in.³)**

| Element | Rigid System Deck Plate | Rigid System Bottom of Element | Elastic System Deck Plate | Elastic System Bottom of Element |
|---|---|---|---|---|
| Rib at midspan | 101.7 | 29.92 | 77.40 | 30.10 |
| Rib at floor beam | 101.7 | 29.92 | — | — |
| Floor beam 2 ft 0 in. right of center | 2,917 | 573 | 2917 | 573 |
| Main girder positive spans 1, 2, and 3 | 58,722 | 31,615 | — | — |
| Main girder at supports | 85,580 | 53,430 | — | — |

**Table 6.10  Summary of Stresses (ksi)**

| | Rigid System | | Elastic System | |
|---|---|---|---|---|
| Element | Deck Plate | Bottom of Element | Deck Plate | Bottom of Element |
| Rib at midspan | DL, −0.057 | 0.20 | −1.43 | 3.67 |
| | LI −2.82 | 9.60 | | |
| Rib at floor beam | DL, +0.114 | −0.40 | — | — |
| | LL, −2.82 | −9.60 | | |
| Floor beam 2 ft 0 in. | DL, −1.39 | 7.06 | 1.09 | −5.59 |
| right of center | LL, −4.65 | 23.66 | | |
| | Spans 1 and 3, −13.50 | | | |
| Main girder positive | Span 2, −14.11 | 25.05 | — | — |
| spans 1, 2, and 3 | | 26.23 | | |
| Main girder supports | +15.62 | −25.02 | — | — |
| Deck plate | Transverse, ±21.1 | — | — | — |
| | Longitudinal, ±6.3 | | | |

At point B:

$$-0.40 - 9.60 - 14.11 = \underline{-24.11 \text{ ksi}}$$
$$< 23.0 \times 1.25 \qquad \text{Use A572 Grade 42 Ribs}$$

The total stress in the deck plate is calculated. Point C longitudinal stress is given by

$$-0.057 - 2.82 - 1.43 - 14.11 - 6.3 = \underline{-24.72 \text{ ksi}}$$

Point C transverse stress is

$$-1.39 + 1.09 - 4.65 - 21.1 = \underline{-26.05 \text{ ksi}}$$

At point D the longitudinal stress is given by

$$+0.114 + 2.82 + 15.62 + 6.3 = \underline{24.85 \text{ ksi}}$$

and the transverse stress is 21.1 ksi.

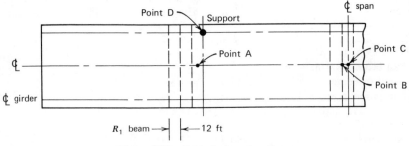

**Figure 6.32**  Critical stress locations.

## 6.5   ESTIMATION OF FLOOR-BEAM FORCES AND DEFORMATION

Application of a computer-oriented technique (1) that evaluates the response of continuous orthotropic plates on flexible supports has led to the development of an expression that estimates the live-load response of transverse floor beams (15). Use of this equation eliminates the need for consideration of rigid- and elastic-support effects, as the entire bridge is considered as an interacting flexible system.

A study of the effects various parameters have on the response of the loaded system, that is, floor-beam stiffness, deck stiffness, floor-beam spacing, and number of traffic lanes, led to the following equation:

$$\log\left(\frac{f}{f^*}\right) = -0.11N - 0.86(1 - \log S) + (-0.0485N + 0.046)\log\left(\frac{D_y}{EI}\right)$$

$$(6.26)$$

where  $f/f^* = (\delta/\delta^*),\ (M/M^*),\ (V/V^*)$
$\delta$ = girder-system deflection
$M$ = girder-system moment
$V$ = girder-system shear
$\delta^*$ = simple-beam deflection
$M^*$ = simple-beam moment
$V^*$ = simple-beam moment
$D_y$ = primary plate bending stiffness/width
$EI$ = floor-beam stiffness

The girder-system values $(f)$ represent the response of the floor beams when they are interacting as part of an orthotropic bridge. The $f^*$ values are the functions, as computed when the floor beams are isolated as a simple beam and are subjected to a set of single HS20 axles, as determined by the number of lanes. This equation is similar to the live-load distribution factor $(S/5.5)$, as used in composite-beam design.

In Design Example 6.1, the induced moments in the floor beam required extensive computations to account for deck and floor beam flexibility. As an alternate to such a study, an approximate analysis can be performed using equation 6.26.

The position of the trucks across the bridge width (see Fig. 6.33) is now considered. Although the span width is technically six lanes, the wheel loads overhanging the supports tend to reduce the moments caused by the interior loads. The span may therefore be considered to have five lanes for preliminary design. Equation 6.26 may be used to determine $M/M^*$

$$\log_{10}\left(\frac{M}{M^*}\right) = -0.11N - 0.86(1 - \log_{10} S) + (-0.0485N + 0.046)\log_{10}\left(\frac{D_y}{EI}\right)$$

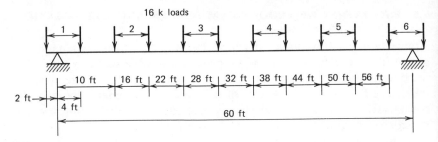

**Figure 6.33** Truck loading for approximate analysis.

where  $N = 5$  lanes
$S = 12$  ft
$D_y = 2.76 \times 10^4$  k-ft$^2$/ft
$EI = 4.73 \times 10^6$  k-ft$^2$

Substituting these values into the above equation gives;

$$\frac{M}{M^*} = 0.906$$

The induced simple beam moment due to the loads shown in Fig. 6.33 is $M^* = 1192$ k-ft. The maximum bending moment can now be determined.

$$M = \frac{M}{M^*} \times M^* = 0.906 \times 1192$$

$$M = 1080 \text{ k-ft}$$

This moment compares with the total computed moment of $M = 1091$ k-ft.

## 6.6  PLATE–GIRDER SYSTEM BEHAVIOR

### 6.6.1  Orthotropic Plate

As an alternative to the approximate Pelikan-Esslinger method, it is possible to examine the behavior of the orthotropic plate on flexible supports directly (1, 4, 15). This direct solution, however, necessitates the use of a computer.

A plate that has a specified grid or nodes, as shown in Fig. 6.1, is considered. It is assumed that the plate is subjected to some load $q$ and thus deforms some amount $w_0$, $w_a$, $w_b$, $w_l$, and $w_r$ at the respective nodes 0, $a$, $b$, $l$, and $r$, as shown in Fig. 6.34. Assuming now that the deflected surface $w(x, y)$ is given by a paraboloid,

$$\bar{w} = Ax^2 + Bx + C + Dy + Ey^2 \tag{6.27}$$

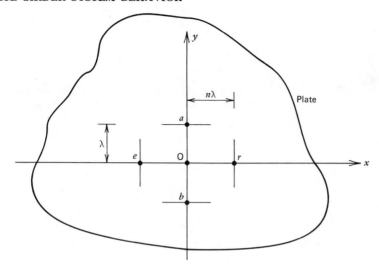

**Figure 6.34**  Plate difference mesh.

such that $\bar{w}$ is identical to $w$ at the five given points $(l, r, 0, a, b)$. Applying the conditions at the various points gives the equations shown below. At

$$x = y = 0 \qquad \bar{w} = w_0$$
$$\bar{w} = w_0 = C \tag{a}$$

At

$$x = -n\lambda, \qquad y = 0 \qquad \bar{w} = w_l$$

$$\bar{w} = w_1 = A\overline{n\lambda}^2 - Bn\lambda + C + D(0) + E(0)$$

$$w_1 = A\overline{n\lambda}^2 + Bn\lambda + C \tag{b}$$

At

$$x = +n\lambda, \qquad y = 0 \qquad \bar{w} = w_r$$
$$w_r = A\overline{n\lambda}^2 + Bn\lambda + C \tag{c}$$

At

$$x = 0, \qquad y = -\lambda \qquad \bar{w} = w_b$$
$$w_b = C - D\lambda + E\lambda^2 \tag{d}$$

$$x = 0, \qquad y = \lambda \qquad \bar{w} = w_a$$
$$w_a = C + D\lambda + E\lambda^2 \tag{e}$$

Solving for constants $A, B, C, D, E$ gives

$$A = \frac{1}{2n\lambda^2}(w_r - 2w_0 + w_1)$$

$$B = \frac{1}{2n\lambda}(w_r - w_1)$$

$$C = w_0 \tag{f}$$

$$D = \frac{1}{2\lambda}(w_a - w_b)$$

$$E = \frac{1}{2\lambda^2}(w_a - 2w_0 + w_b)$$

Substituting equations f into equation 6.27 gives

$$\bar{w} = \frac{1}{2\lambda^2}\left[(w_r - 2w_0 + w_1)\frac{x^2}{n^2} + (w_r - w_1)\frac{x\lambda}{n} + w_0 2\lambda^2 \right.$$
$$\left. + (w_a - w_b)y\lambda + (w_a - 2w_0 + w_b)y^2 \right] \tag{6.28}$$

Equation 6.28 is used, in part, to evaluate the biharmonic equation

$$D_x\frac{\partial^4 w}{\partial x^4} + 2H\frac{\partial^4 w}{\partial x^2 \partial y^2} + D_y\frac{\partial^4 w}{\partial y^4} = q \tag{6.12}$$

First the mesh pattern is extended to more points, as shown in Fig. 6.35, because all terms would vanish if the $\partial^4 w/\partial x^4$ and so on were taken of equation 6.28.

The partial derivatives of equation 6.28 are now taken:

$$\frac{\partial^2 \bar{w}}{\partial x^2} = \frac{1}{n\lambda^2}(w_r - 2w_0 + w_i)$$

$$\frac{\partial^2 \bar{w}}{\partial y^2} = \frac{1}{\lambda^2}(w_a - 2w_0 + w_b)$$

The fourth-order differential can be written as

$$\frac{\partial^4 \bar{w}}{\partial x^4} = \frac{\partial^2}{\partial x^2}\left(\frac{\partial^2 w}{\partial x^2}\right) = \frac{1}{n\lambda^2}\left[\frac{\partial^2 \bar{w}}{\partial x^2}\bigg|_r - 2\frac{\partial^2 \bar{w}}{\partial x^2}\bigg|_0 + \frac{\partial^2 \bar{w}}{\partial x^2}\bigg|_1\right] \tag{g}$$

where expansion is made with respect to nodes $r$, $0$, and $l$. These expansions are

$$\frac{\partial^2 \bar{w}}{\partial x^2}\bigg|_r = \frac{1}{n\lambda^2}(w_{rr} - 2w_r + w_0)$$

$$\frac{\partial^2 \bar{w}}{\partial x^2}\bigg|_0 = \frac{1}{n\lambda^2}(w_r - 2w_0 + w_l)$$

$$\frac{\partial^2 \bar{w}}{\partial x^2}\bigg|_1 = \frac{1}{n\lambda^2}(w_0 - 2w_1 + w_{ll})$$

in accordance with the mesh pattern given in Fig. 6.35 substituting these relationships into expression g gives

$$\frac{\partial^4 \bar{w}}{\partial x^4} = \frac{1}{n\lambda^4}(w_{rr} - 4w_r + 6w_0 - 4w_1 + w_{11}) \tag{h}$$

Similarly,

$$\frac{\partial^4 \bar{w}}{\partial y^4} = \frac{1}{\lambda^4}(w_{aa} - 4w_a + 6w_0 - 4w_b + w_{bb}) \tag{i}$$

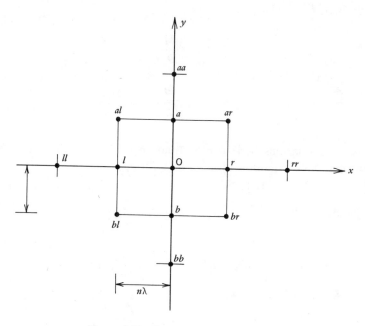

**Figure 6.35** General difference mesh.

The mixed partial $(\partial^4 w / \partial x^2 \partial y^2)$ is found by expanding $(\partial^2 w / \partial y^2)$ about nodes $r$, $0$, and $l$ in accordance with

$$\frac{\partial^4 \bar{w}}{\partial x^2 \partial y^2} = \frac{\partial^2}{\partial y^2}\left(\frac{\partial^2 w}{\partial x^2}\right) = \frac{\partial^2}{\partial y^2}\left(\frac{w_r - 2w_0 + w_l}{n^2 \lambda^2}\right) = \frac{1}{n^2 \lambda^2}\left[\frac{\partial^2 w}{\partial y^2}\bigg|_r - 2\frac{\partial^2 w}{\partial y^2}\bigg|_0 + \frac{\partial^2 w}{\partial y^2}\bigg|_1\right]$$

However,

$$\frac{\partial^2 w}{\partial y^2}\bigg|_r = \frac{w_{ar} - 2w_r + w_{br}}{\lambda^2}$$

$$\frac{\partial^2 w}{\partial y^2}\bigg|_0 = \frac{w_a - 2w_0 + w_b}{\lambda^2}$$

$$\frac{\partial^2 w}{\partial y^2}\bigg|_1 = \frac{w_{al} - 2w_l + w_{bl}}{\lambda^2}$$

Therefore

$$\frac{\partial^4 w}{\partial x^2 \partial y^2} = \frac{1}{n^2 \lambda^4}\left[w_{ar} - 2w_r + w_{br} - 2w_a + 4w_0 - 2w_b + w_{al} - 2w_l + w_{bl}\right] \quad \text{(j)}$$

If

$$\beta = \frac{H}{D_y}$$

and

$$\alpha = \frac{H}{\sqrt{D_x D_y}}$$

substituting equations h, i, j into equation 6.12 gives the final general orthotropic plate equation in difference form:

$$\left[w_0\left(6n^4 + 8n^2\beta + 6\left(\frac{\beta}{\alpha}\right)^2\right) + 4(w_r + w_l)\left(\left(\frac{\beta}{\alpha}\right)^2 + n^2\beta\right)\right.$$

$$-4(w_a + w_b)(n^2\beta + n^4) + (w_{rr} + w_{ll})\left(\frac{\beta}{\alpha}\right)^2 + (w_{aa} + w_{bb})n^4$$

$$\left. + (w_{ar} + w_{br} + w_{al} + w_{bl}) \cdot 2n^2\beta\right] = \frac{qn^4\lambda^4}{D_y} \quad (6.29)$$

The moments and reactions can also be developed in difference form (1, 4):

$$M_x = -\frac{D_x}{n\lambda^2}(w_l - 2w_0 + w_r) \tag{6.30}$$

$$M_y = -\frac{D_y}{\lambda^2}(w_a - 2w_0 + w_b) \tag{6.31}$$

$$R_x = -D_x\left[\frac{1}{2n\lambda^3}(w_{rr} - 2w_r + 2w_l - w_{ll})\right.$$

$$\left. + \frac{2\xi}{2n\lambda^3}(-2_{al} + 2w_l - w_{bl} + w_{ar} - 2w_r + w_{br})\right] \tag{6.32}$$

$$R_y = -D_y\left[\frac{1}{2\lambda^3}(w_{aa} - 2w_a + 2w_b - w_{bb})\right.$$

$$\left. + \frac{2\beta}{2n^2\lambda^3}(w_{al} - 2w_a + w_{ar} - w_{bl} + 2w_b - w_{br})\right] \tag{6.33}$$

where

$$\xi = \frac{H}{D_x}$$

## 6.6.2 Orthotropic Plate with Interacting Girders

It is now assumed that the orthotropic plate is supported on interacting girders, as shown in Fig. 6.36. Each girder has a stiffness $EI_x$ and $EI_y$ and the plate has stiffness $D_x$, $D_y$, and $H$. Equilibrium of the intersecting plate and girders is given by

$$\sum F = 0$$

$$q_T = q_{PL} + q_{Bx} + q_{By} \tag{6.34}$$

where $q_T$ = externally applied load

$q_{PL}$ = load resisted by the plate

$q_{Bx}$ = load resisted by the beam in the $x$ direction

$q_{By}$ = load resisted by the beam in the $y$ direction

From beam theory,

$$\frac{d^4w}{dy^4} = \frac{P_y}{EI_y}$$

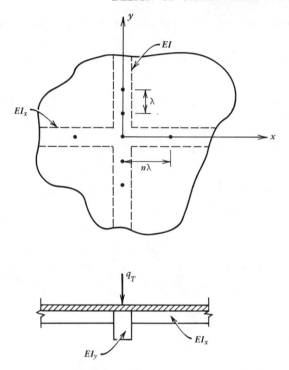

**Figure 6.36**  Intersecting girders and plate.

or in difference form,

$$\frac{P_y\lambda^4}{EI_y} = (w_{aa} - 4w_a + 6w_0 + 4w_b + w_{bb}) \tag{6.35}$$

Similarly,

$$\frac{d^4w}{dx^4} = \frac{P_x}{EI_x}$$

$$\frac{P_x n\lambda^4}{EI_x} = (w_{ll} - 4w_1 + 6w_0 - 4w_r + w_{rr}) \tag{6.36}$$

The forces ($P_y$ and $P_x$) given in these equations are per unit length of beam. The equilibrium of forces in the plate and girder equation 6.34 are forces per unit area. Therefore, converting $P_x$, $P_y$ in force per area, or

$$q_{Bx} = \frac{P_x}{\lambda} \quad \text{and} \quad q_{By} = \frac{P_y}{n\lambda}$$

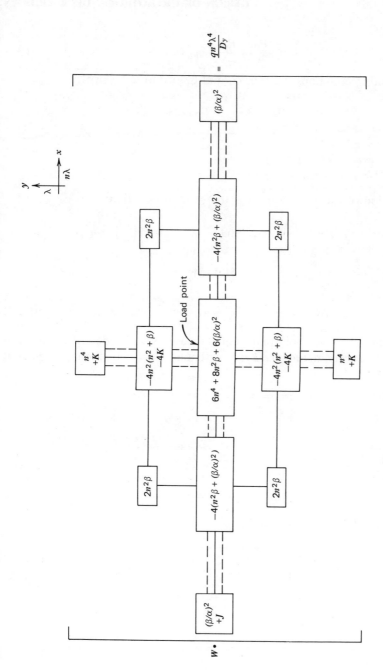

**Figure 6.37** General difference pattern.

Therefore

$$\frac{P_{Bx}}{\lambda} + \frac{P_{By}}{n\lambda} + q_{PL} = q_T \tag{6.37}$$

Substituting equations 6.29, 6.35, and 6.36 into 6.37 gives

$$Jw_{Bx}[\ \ ] + Kw_{By}[\ \ ] + w_{PL}[\ \ ] = \frac{qn^4\lambda^4}{D_y}$$

where $J = EI_x/D_y\lambda$
$\quad\quad K = (EI_y/\lambda D_y)n^3$
$\quad\quad [\ \ ] = $ mesh point parameters

Expansion of this equation gives the general equation:

$$w_0\left[6n^4 + 8n^2\beta + 6\left(\frac{\beta}{\alpha}\right)^2 + 6K + 6J\right] - 4(w_r + w_l)\left[\left(\frac{\beta}{\alpha}\right)^2 + n^2\beta - 4J\right]$$

$$-4(w_a + w_b)(n^2\beta + n^4 - 4K) + (w_{rr} + w_{ll})\left[\left(\frac{\beta}{\alpha}\right)^2 + J\right]$$

$$+(w_{aa} + w_{bb})(n^4 + K + (w_{ar} + w_{br} + w_{al} + w_{bl})2n^2\beta = \frac{qn^4\lambda^4}{Dy} \tag{6.37}$$

This general orthotropic plate equation can also be described in mesh form, as shown in Fig. 6.37. Note that if $K = J = 0$, then this general

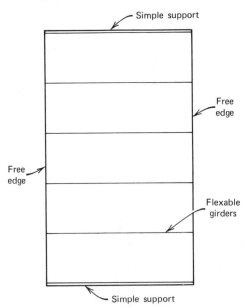

**Figure 6.38**  Bridge boundaries.

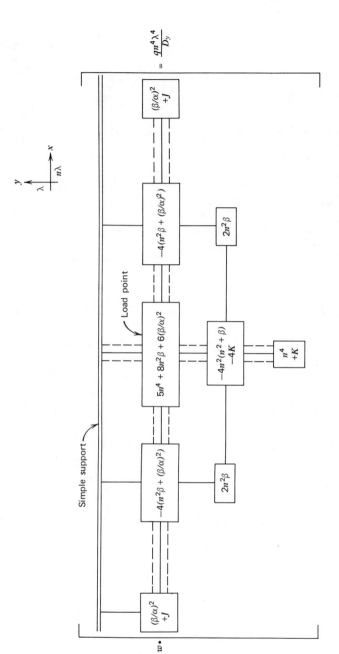

**Figure 6.39** Mesh pattern—load adjacent to simple support.

$$= \frac{qn^4\lambda^4}{D_y}$$

221

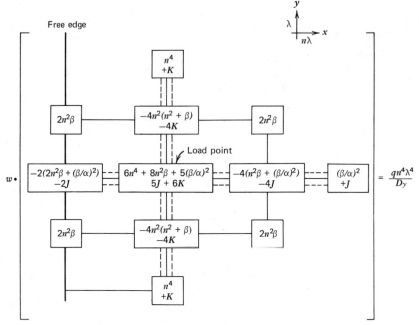

**Figure 6.40** Mesh pattern—load adjacent to free edge.

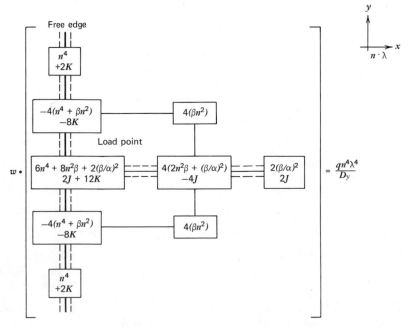

**Figure 6.41** Mesh pattern—load on free edge.

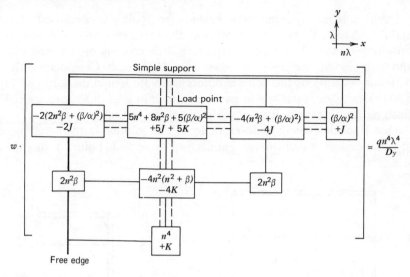

**Figure 6.42** Mesh pattern—load adjacent to simple support and adjacent to free edge.

equation 6.37 reduced to equation 6.29, representing an orthotropic plate without girders. If the plate is isotropic, that is, $\alpha = \beta = 1$ and has no girders $(K = J = 0)$ and a square mesh $(n = 1)$, then equation 6.37 becomes

$$w_0[20]+[w_r + w_l][-8]+[w_a + w_b][-8]+[w_{rr} + w_{ll}][1]$$

$$+[w_{aa} + w_{bb}][1]+[w_{ar} + w_{br} + w_{al} + w_{bl}][2] = \frac{q\lambda^4}{D} \quad (6.38)$$

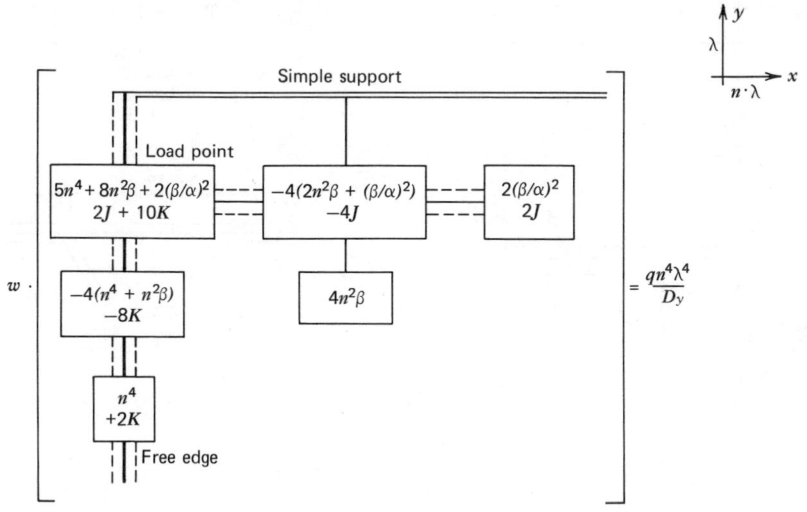

**Figure 6.43** Mesh pattern-load on free edge and adjacent to simple support.

In the case of an orthotropic bridge, the system is composed of two main longitudinal girders, transverse floor beams, and a continuous orthotropic deck. This type of structure then consists of two free edges and two simple supports, as shown in Fig. 6.38. Considering these boundary conditions the general difference plate (equation 6.29 and Fig. 6.37) is modified, resulting in patterns shown in Figs. 6.39 through 6.43. Each pattern is then written at the respective point on the structure, after the mesh spacing configuration is defined. The solution of the resulting set of equations gives deformations at each of these node points. Using these

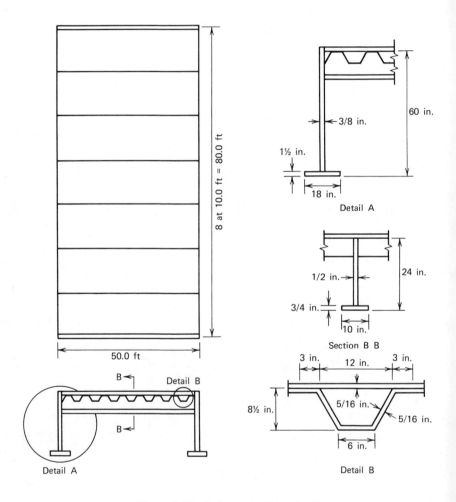

**Figure 6.44**   Orthotropic bridge details.

deflections the induced moments are determined from equations 6.30 and 6.31:

$$M_{x,y} = -\frac{EI}{(\text{spacing})^2}(W_a^l - 2W_0 + W_b^r)$$

A typical bridge and the general finite difference grid pattern are shown in Figs. 6.44 and 6.45, respectively.

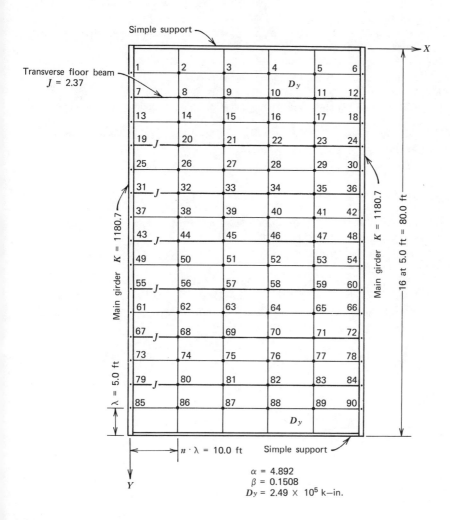

**Figure 6.45** Mesh pattern of bridge—example of orthotropic bridge plan slope deflection method.

## REFERENCES

1 C. P. Heins and C. T. G. Looney, "Bridge Analysis Using Orthotropic Plate Theory," *J. Struct. Div. Am. Soc. Civil Eng.*, Vol. 93, No. ST2, February 1968.

2 W. Pelikan and M. Esslinger, "Die Stahlfahrbahn Berechnung und Konstruktion," M.A.N., *Forschungsheft*, No. 7, 1957.

3 P. C. Loveys, "Orthotropic Steel Plate Deck Bridges," Canadian Institute of Steel Construction, November 1963.

4 C. P. Heins and C. T. G. Looney, "Bridge Tests Predicted by Finite Difference Plate Theory," *Struct. Div. Am. Soc. Civil Eng.*, Vol. 95, No. ST2, February 1969.

5 G. H. Powell and D. W. Ogden, "Analysis of Orthotropic Steel Plate Bridge Decks," *J. Struct. Div. Am. Soc. Civil Eng.*, Vol. 95, ST5, May 1969.

6 D. R. Cusens and R. P. Pama, *Bridge Deck Analysis*, Wiley, 1975.

7 E. J. Shields, "Artistic Orthotropic Design Also Saves Money," *Eng. News Rec.*, November 21, 1963, pp. 26–30.

8 E. J. Shields, "Popular Street Bridge. Design and Fabrication," *Civil Eng.*, February 1966, pp. 52–55.

9 *Eng. News—Rec.*, "California's Orthotropic Bridge Set to Go," October 29, 1964, pp. 22–25.

10 *Eng. News—Rec.*, "The San Mateo-Hayward Bridge: A Fabrication Dream," June 23, 1966, pp. 22–28.

11 G. Hardenberg, "Design of the Superstructure of the Port Mann Bridge," *Eng. Jl*, July 1961.

12 *Orthotropic Bridges—Theory and Design*, Lincoln Arc Welding Foundation, Cleveland Ohio, 1967.

13 *Design Manual for Orthotropic Plate Deck Bridges*, American Institute of Steel Construction, New York, 1963.

14 D. H. Hall, "Orthotropic Bridges—PD-2032," Bethlehem Steel Co., Bethlehem, Pa., March 1971.

15 P. G. Perry and C. P. Heins, "Rapid Design of Orthotropic Bridge Floor Beams," *J. Struct. Div. Am. Soc. Civil Eng.*, Vol. 98, ST11, November 1972.

# 7

# Design of Straight Steel Box Girders

## 7.1 GENERAL

Box girder bridges have received wide attention in the design office because of their desirable characteristics. In particular, this type of bridge is very effective in resisting bending because of the wide bottom flanges. Because of the enclosed shape of the box girder, it offers substantial rigidity in resisting torsion of the bridge. In addition, the box can resist corrosion more readily, as half of the steel surface is contained within the section. Along with these basic advantages, the box-shape girder is an aesthetically pleasing structure.

Generally, there are two basic types of box structures. One type consists of multiple box units with composite concrete decks with spans of 120 ft or less and is used for general highway interchanges and overpasses. The second type box girders used is for major long-span structures, such as those girders used in orthotropic bridges, where the spans are 200 ft or more and are used for major river crossings.

The design criteria that are discussed herein pertain to the smaller box girder bridges, the specifications for which are given in the AASHTO Code (1). The code is described and then it is applied in a design example.

## 7.2  DESIGN CODE—BOX BEAMS

The design guides given pertain to composite box girders if they meet the following definition.

### 7.2.1  AASHTO 1.7.102: Definition

A simple and continuous span steel-concrete multi-box girder bridge of moderate length may be of single cell and of width (C.L. to C.L. of top flanges) equal to spacing between boxes. The overhang of the deck slab (including curbs and parapets) shall be limited to 60 percent of the distance between boxes, but in no case greater than 6 feet.

### 7.2.2  AASHTO 1.7.103: Lateral Load Distribution

The live load bending moment for each box girder shall be determined by applying to the girder the fraction $w_L$ of a wheel load (front and rear) according to the following:

$$w_L = 0.1 + 1.7R + \frac{0.85}{N_w}$$

where      $R = N_w$/number of box girders, $R$ shall not be less than 0.5 nor greater than 1.5
$N_w = W_c/12$ (reduced to nearest whole number)
$W_c$ = roadway width between curbs (in feet)

### 7.2.3  AASHTO 1.7.104: Web Plates

A. *Vertical Shear.* The design shear $V_w$ for a web shall be calculated using:

$$v_w = \frac{v_v}{\cos \theta}$$

where      $V_v$ = vertical shear
$\theta$ = angle of inclination of the web plate to the vertical

B. *Secondary Bending.* If web plate inclination is 1 to 4 or less and the width of the bottom flange is no greater than 20 percent of the span, secondary bending can be neglected. If these conditions are exceeded, appropriate analysis is needed.

### 7.2.4.  AASHTO 1.7.105: Bottom Flange Plates

A. *Tension Flanges.* For simply supported spans, the bottom flange ($W_f$) is completely effective if $W_f \le L/5$. If $W_f$ exceeds $L/5$, the effective width equal to

$L/5$ shall be considered effective ($L$ = span length). For continuous spans, the above applies where $L$ = span length between points of contraflexure.

B. *Compression Flanges Unstiffened.* If $b/t \leq 6140/\sqrt{F_y}$, where $b$ = flange width between webs in inches and $t$ = flange thickness in inches, then the allowable stress $f_b = 0.55F_y$.

If $b/t$ exceeds this value, but is less than 60, the allowable stress is governed by:

$$f_b = 0.55F_y - 0.224F_y\left\{1 - \sin\left(\frac{\pi}{2}\right)\left(\frac{13,300\sqrt{K} - b\sqrt{F_y/t}}{7160}\right)\right\}$$

If $b/t$ exceeds $13,300/\sqrt{F_y}$, the allowable stress is governed by:

$$f_b = 57.6\left(\frac{t}{b}\right)^2 \times 10^6$$

If $b/t > 45$, longitudinal stiffeners should be considered.

C. *Compression Flanges Stiffened Longitudinally.* Longitudinal stiffeners shall be equally spaced across the flange and shall have a moment of inertia (each stiffener) about an axis parallel to the flange at the base of the stiffener at least equal to:

$$I_s \geq \phi t^3 w$$

where $\quad \phi = 0.07K^3 n^4$ for $n > 1$
$\qquad \phi = 0.125K^3$ for $n = 1$
$\qquad W$ = width of flange between stiffeners or between web and nearest stiffener
$\qquad n$ = number of stiffeners
$\qquad K$ = buckling coefficient (assume values between 2 and 4)

For the flange with stiffeners, if

$$\frac{w}{t} \leq 3070\frac{\sqrt{K}}{\sqrt{F_y}}$$

then

$$f_b = 0.55F_y$$

For greater values of $w/t$ but not exceeding $6650\sqrt{K}/\sqrt{F_y}$ or 60, whichever is less, the allowable stress is:

$$f_b = 0.55F_y - 0.224F_y\left[1 - \sin\left(\frac{\pi}{2}\right)\left(\frac{6650\sqrt{K} - w/t\sqrt{F_y}}{3580\sqrt{K}}\right)\right]$$

or values of $w/t > 6650\sqrt{K}/\sqrt{F_y}$ but less than 60, the flange stress shall not exceed

$$f_b = 14.4K\left(\frac{t}{w}\right)^2 \times 10^6$$

When longitudinal stiffeners are used, it is preferable to have at least one transverse stiffener placed near the point of dead-load contraflexure. The size of this stiffener should be equal to the longitudinal stiffener.

If longitudinal stiffeners are placed at their maximum $w/T$ ratio for the allowable of $0.55F_y$ and the longitudinal stiffeners exceed 2, then transverse stiffeners should be considered.

D. *Compression Flanges Stiffened Longitudinally and Transversely.* Longitudinal stiffeners shall have equal spacings across the flange width with each stiffener having a moment of inertia (about an axis parallel to the flange and at the base of stiffener) equal to:

$$I_s \geq 8t^3 w$$

The transverse stiffeners shall be proportioned such that the moment of inertia (about an axis through the centroid) shall be equal to

$$I_t \geq 0.10(n+1)^3 w^3 \frac{f_s}{E} \frac{A_f}{a}$$

where    $A_f$ = area of bottom flange including longitudinal stiffeners
         $a$ = spacing of transverse stiffeners
         $f_s$ = maximum longitudinal bending stress in the flange of the panels
                on either side of the transverse stiffener
         $E$ = modulus of elasticity of steel

For the flange including stiffeners, the $w/t$ ratio for the longitudinal stiffeners shall not exceed

$$\frac{w}{t} \leq \frac{3070\sqrt{K_1}}{\sqrt{F_y}}$$

where

$$K_1 = \frac{[1+(a/b)^2]^2 + 87.3}{(n+1)^2(a/b)^2[1+0.1(n+1)]}$$

for the allowable stress of $f_b = 0.55F_y$. For greater values of $w/t$, but not exceeding $6650\sqrt{K_1}/\sqrt{F_y}$ or 60, whichever is less, the flange stress including stiffeners shall not exceed

$$f_b = 0.55F_y - 0.224F_y\left[1 - \sin\left(\frac{\pi}{2}\right)\left(\frac{6650\sqrt{K_1} - w\sqrt{F_y}/t}{3580\sqrt{K_1}}\right)\right]$$

For values of $w/t$ exceeding $6650\sqrt{K_1}/\sqrt{F_y}$ but less than 60, the allowable flange stress is:

$$f_b = 14.4K_1\left(\frac{t}{w}\right)^2 \times 10^6$$

The maximum $K_1 = 4$, when $K_1 = 4$, the transverse stiffeners shall have a spacing "$a$" equal to or less than $4w$. If $a/b > 3$, transverse stiffeners are not necessary.

The transverse stiffeners need not be connected to the flange plate, but to the webs of the box and to each longitudinal stiffener. The connection to the web shall be designed to resist the vertical force determined by the formula

$$R_w = \frac{F_y S_s}{2b}$$

where $S_s$ = section modulus of the transverse stiffener.

The connection to each longitudinal stiffener shall be designed to resist the vertical force determined by:

$$R_s = \frac{F_y S_s}{nb}$$

E. *Compression Flange Stiffeners*. The $b'/t'$ ratio of any outstanding element of the flange stiffeners shall not exceed

$$\left(\frac{b}{t}\right)' \leq \frac{2600}{\sqrt{F_y}}$$

where    $b'$ = width of any outstanding stiffener element
         $t'$ = thickness of outstanding stiffener element

## Design Example

STRAIGHT BOX BEAM BRIDGE (COMPOSITE)

A two-span continuous composite box beam bridge of 120 ft–120 ft spans, as shown in Fig. 7.1 is designed here. *Given:*

1. Two equal spans of 120 ft 0 in.
2. Slab: $7\frac{1}{2}$ in
3. Spacing between boxes: 7 ft–10 in.
4. Clear roadway: 44 ft–0 in.
5. Beam steel: A441 at interior supports, A36 elsewhere
6. Future wearing surface: 25 psf
7. A haunch: 1 in. × 3 in. over the beam flanges
8. Concrete: $f'_c = 3000$ psi
9. Loading: HS20-44
10. Truck loading: $5 \times 10^5$ cycles

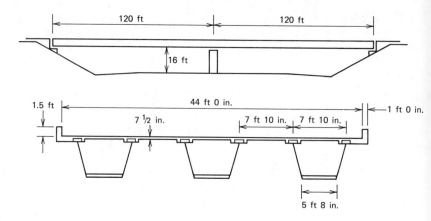

**Figure 7.1** Plan and elevation of continuous box girder bridge.

Dead load A (no composite action)

Slab: $\dfrac{7.5}{12} \times 15.67 \times 0.150$          1.468 k/ft

Haunch: $4 \times \dfrac{3}{12} \times \dfrac{1}{12} \times 0.150$     0.012

Steel (estimate):                    $\underline{0.420}$
                                     1.900 k/ft

Dead load B (composite action); equally distributed to all girders)

Parapet: $1.0 \times 1.5 \times 0.150 \times \frac{2}{3}$          0.150 k-ft
Railing: $(1516/\text{ft}) \times \frac{2}{3}$                     0.010
Asphalt paving: $0.025\,\text{ft} \times 15.67$          $\underline{0.360}$
                                                     0.520 k/ft

DEAD LOAD A—BARE STEEL GIRDER

The induced dead-load moment at 0.20 points along the girder can be determined by use of influence lines (2). The total positive and negative areas under the influence line diagrams for moment at various points are given in Table 7.1. Using these values, the moments are tabulated using the general expression

$$M = A \times w \times L^2 \tag{7.1}$$

| Location | $A \times L^2$ | $M$ (k-ft) |
|----------|----------------|------------|
| At 0.2 | $0.055 \times \overline{120}^2 = 792 \text{ ft}^2$ | $792 \times 1.9 = 1504$ |
| At 0.4 | $0.070 \times \overline{120}^2 = 1008$ | $1008 \times 1.9 = 1915$ |
| At 0.6 | $0.045 \times \overline{120}^2 = 648$ | $648 \times 1.9 = 1231$ |
| At 0.8 | $-0.020 \times \overline{120}^2 = -288$ | $-288 \times 1.9 = -547$ |
| At $B$ | $-0.125 \times \overline{120}^2 = -1800$ | $-1800 \times 1.9 = -3420$ |

The dead-load shears are also found from the influence line data using the general expression

$$R \quad \text{or} \quad V = A \times w \times L \tag{7.2}$$

where $A =$ area under the diagram that is loaded. Applying this equation and using the data from Table 7.1 gives

| Location | $A \times L$ | $V$ (k) |
|----------|--------------|---------|
| $R_A = V_{ab}$ | $0.375 \times 120 = 45$ | $45 \times 1.9 = 85.5 \text{ k}$ |
| $V_{BA}$ | $0.625 \times 120 = 75$ | $75 \times 1.9 = 143 \text{ k}$ |

DEAD LOAD B—COMPOSITE SECTION

The superimposed dead load B, as previously given, is 0.520 k/ft. Using the same procedure as given for dead load A and the same influence line ordinates gives the following induced moments.

| Locations | $A \times L^2$ | $M$(k-ft) |
|-----------|----------------|-----------|
| At 0.2 | 792 | $792 \times 0.520 = 412$ |
| At 0.4 | 1008 | $1008 \times 0.520 = 524$ |
| At 0.6 | 648 | $648 \times 0.520 = 337$ |
| At 0.8 | -288 | $-288 \times 0.520 = -150$ |
| At $B$ | -1800 | $-1800 \times 0.520 = -936$ |

**Table 7.1  Influence Line Ordinates for Moments and Shears**

| Unit Load | At 0.2 | At 0.4 | At 0.6 | At 0.8 | At $B$ |
|---|---|---|---|---|---|
| | Moments/$PL^a$ | | | | |
| A | 0 | 0 | 0 | 0 | 0 |
| 0.2 | 0.1504 | 0.1008 | 0.0512 | 0.0016 | −0.0432 |
| 0.4 | 0.1032 | 0.2064 | 0.1090 | 0.0128 | −0.0840 |
| 0.6 | 0.0608 | 0.1216 | 0.1824 | 0.0432 | −0.0960 |
| 0.8 | 0.0256 | 0.0512 | 0.0708 | 0.1024 | −0.0720 |
| B | 0 | 0 | 0 | 0 | 0 |
| 0.2 | −0.0144 | −0.0288 | −0.0432 | −0.0576 | −0.0720 |
| 0.4 | −0.0192 | −0.0384 | −0.0576 | −0.0768 | −0.0960 |
| 0.6 | −0.0168 | −0.0336 | −0.0504 | −0.0672 | −0.0840 |
| 0.8 | −0.0096 | −0.0192 | −0.0288 | −0.0384 | −0.0480 |
| C | 0 | 0 | 0 | 0 | 0 |
| + | 0.0675 | 0.0950 | 0.0825 | 0.0300 | 0 |
| Area − | −0.0125 | −0.0250 | −0.0375 | −0.0500 | −0.1250 |

| Unit Load | $V_{AB} = R_A$ | $V_{BA}$ | $V_{BC}$ | $R_B$ |
|---|---|---|---|---|
| | Shears/P | | | |
| A | 1.0 | 0 | 0 | 0 |
| 0.2 | 0.752 | −0.248 | 0.048 | 0.296 |
| 0.4 | 0.516 | −0.484 | 0.084 | 0.568 |
| 0.6 | 0.304 | −0.696 | 0.096 | 0.792 |
| 0.8 | 0.128 | −0.872 | 0.072 | 0.944 |
| B | 0 | −1/0 | 0/1.0 | 1.0 |
| 0.2 | −0.072 | −0.072 | 0.872 | 0.944 |
| 0.4 | −0.096 | −0.096 | 0.696 | 0.792 |
| 0.6 | −0.084 | −0.084 | 0.484 | 0.568 |
| 0.8 | −0.048 | −0.048 | 0.248 | 0.296 |
| C | 0 | 0 | 0 | 0 |
| + | 0.4375 | | 0.625 | 1.250 |
| − | −0.0625 | −0.6250 | 0 | 0 |

$^a$ $P$ indicates load, $L$ indicates span length.

The computations for the dead-load B shears are similar to those for dead-load A, They are

| Location | $A \times L$ | $V$ |
|----------|-----------|-----|
| $R_A = V_{ab}$ | $0.375 \times 120 = 45$ | $0.520 \times 45 = 23.4$ |
| $V_{BA}$ | $0.625 \times 120 = 75$ | $0.520 \times 75 = 39.0$ |

LIVE LOAD

The distribution of live loads to each respective box section is determined by the following equation (Section 1.6.103 of AASHTO):

$$w_L = 0.1 + 1.7R + \frac{0.85}{N_w}$$

where
$R = N_w/$number of box girders
$N_w = w_c/12$, reduced to nearest whole number
$w_c =$ roadway width between curbs (ft)
$R$ should not be less than 0.5 nor greater than 1.5
$W_L$ is the fraction of a front and rear wheel to be applied to the box girder

For this box girder design

$N_w = \frac{44}{12} = 3.67$     Use 4.0

$R = \frac{4}{3} = 1.33$

$w_L = 0.1 + 1.7 \times 1.33 + 0.85/4$

$w_L = 2.58$ wheels to each box or 1.29 lanes to each box

$$\text{Impact} = \frac{50}{125 + 120} = 0.204$$

As given in Section 1.2.5 of AASHTO, the live loading can consist of a truck on lane loading plus a concentrated load. These loads are multiplied by the impact factor and the distribution facor to give the design loading and are given below.

MOMENTS

$$\text{Truck: } P = 16.0(2.58)(1.204) = 49.7 \text{ k}$$

$$\text{Lane: uniform } W = 0.64(1.29)(1.204) = 0.994 \text{ k/ft}$$

$$\text{concentrated } P = 18(1.29)(1.204) = 28.0 \text{ k}$$

These loads can now be used in conjunction with the influence line ordinates given in Table 7.1 to determine the maximum positive and negative live-load moments. The influence line ordinates times the load and the span length gives the induced moment.

For truck loading, the positive moment is given by:

$$\text{At } 0.2: \ 49.2 \times 120 \left( 0.1504 + 0.125 + \frac{0.098}{4} \right) = 1789 \text{ k-ft}$$

$$\text{At } 0.4: \ 49.2 \times 120 \left( 0.2064 + 0.160 + \frac{0.150}{4} \right) = 2403 \text{ k-ft}$$

$$\text{At } 0.6: \ 49.2 \times 120 \left( 0.1824 + 0.145 + \frac{0.120}{4} \right) = 2132 \text{ k-ft}$$

$$\text{At } 0.8: \ 49.2 \times 120 \left( 0.1024 + 0.068 + \frac{0.043}{4} \right) = 1082 \text{ k-ft}$$

$$\text{At } 0.9: \ 49.2 \times 120 \left( 0.0515 + 0.011 - \frac{0.007}{4} \right) = 359 \text{ k-ft}$$

The negative moment is generally controlled by lane loading, but should be checked at a point close to the support for truck loading. The negative moment for truck loading is

$$\text{At } 0.8: \ 49.7 \times 120 \left( -0.0768 - 0.065 - \frac{0.065}{4} \right) = -941 \text{ k-ft}$$

The total negative moments for Lane loading is the sum of several moments.

$$\text{At B lane: } 994 \text{ k/ft} \times \overline{120}^2 (-0.1250) = 1789 \text{ k-ft}$$

$$\text{Concentrated load: } 28.0 \times 120(-0.0960)2 = \underline{\ 645}$$

$$-2434 \text{ k-ft}$$

$$\text{At } 0.8: \ 0.99 \text{ k/ft} \times \overline{120}^2 (-0.0500) = \ 716 \text{ k-ft}$$

$$28.0 \text{ k} \times 120(-0.0768) = \underline{\ 258}$$

$$-974 \text{ k-ft}$$

The results of these live-load moments and dead-load moments are given in Fig. 7.2, which shows a "moment envelope."

SHEARS

The live-load shears are also found by using influence line ordinates, as given in Table 7.1. These ordinates are multiplied by the load for concentrated effects.

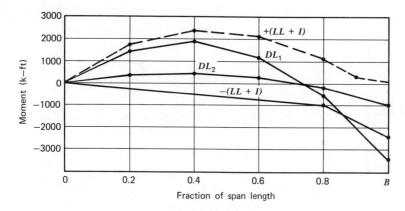

**Figure 7.2** Moment envelope.

For truck loading, $R_a = V_{ab}$:

$$\text{At } A\colon 49.7\left(1 + 0.86 + \frac{0.72}{4}\right) = 101 \text{ k}$$

$$\text{At } 0.2\colon 49.7\left(0.75 + 0.62 + \frac{0.49}{4}\right) = 74 \text{ k}$$

$$\text{At } 0.4\colon 49.7\left(0.52 + 0.40 + \frac{0.28}{4}\right) = 49 \text{ k}$$

$$\text{At } 0.6\colon 49.7\left(0.30 + 0.20 + \frac{0.10}{4}\right) = 26 \text{ k}$$

For truck loading, $V_{ba}$:

$$\text{At } 0.4\colon 49.7\left(0.48 + 0.34 + \frac{0.20}{4}\right) = 43 \text{ k}$$

$$\text{At } 0.6\colon 49.7\left(0.70 + 0.56 + \frac{0.44}{4}\right) = 68 \text{ k}$$

$$\text{At } 0.8\colon 49.7\left(0.87 + 0.77 + \frac{0.65}{4}\right) = 90 \text{ k}$$

$$\text{At } B\colon 49.7\left(1 + 0.93 + \frac{0.84}{4}\right) = 106 \text{ k}$$

For lane loading, $R_a = V_{ab}$:

$$\text{At } A\colon 1.204 \times 0.826 \times 120(0.4375) = 52 \text{ k}$$
$$1.204 \times 33.5(1) \qquad\qquad = 40 \text{ k}$$
$$\overline{\qquad\qquad\qquad\qquad\qquad\quad 92 \text{ k}}$$

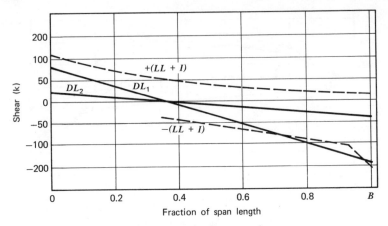

**Figure 7.3** Shear envelope.

For lane loading, $R_b$:

$$\text{At } B: 1.137 \times (0.826 \times 120(1.250) = 141 \text{ k}$$
$$1.137 \times 33.5 \times (1) = \underline{\phantom{0}38 \text{ k}}$$
$$179 \text{ k}$$

It should be noted that the impact factor for the lane loading is modified in that the span length $2L$ is used in the impact equation as recommended by the AASHTO code for shear.

A summary of the dead-load and live-load shears is given in Fig. 7.3, representing the shear envelope of the bridge.

SECTION PROPERTIES

To establish the geometry of the box beams, a survey was conducted to establish the characteristics of typical box beams (3). The results of this survey are given in Table 7.2. In addition to these data, a study on typical box girder bridges (4) has resulted in characteristics of box girders as a function of number of lanes, number of girders, and span length. These results are given in Tables 7.3 and 7.4 for bare steel sections and composite sections and are showing Figs. 7.4 and 7.5. The resulting properties of these sections are given in Tables 7.5 and 7.6. All of this information can be used in establishing preliminary design properties and cost estimation.

After several preliminary trials, the box sections, as shown in Figs. 7.6a through 7.6c, are assumed for three locations along the girders, as shown in Fig. 7.7. The properties for these sections are computed below.

For the section at bent $(0.9L-B)$

**Table 7.2  Prototype Box-Girder Geometry[a]**

| Sample Box Girder | $d/L$ | $d/w$ | $d/t_w$ | Slab Overhang/$w$ | $b'/t$ | Compact? |
|---|---|---|---|---|---|---|
| 1 | — | 0.69 | 90 | 0.50 | — | No |
| 2 | 0.036 | 0.721 | 142 | 0.50 | 7.5 | No |
| 3 | 0.035 | 0.422 | 149 | 0.42 | 9.4 | No |
| 4 | 0.039 | 0.458 | 96 | 0.44 | 7.7 | No |
| 5 | 0.033 | 0.661 | 120 | — | 8.7 | No |
| Average | 0.036 | 0.590 | 119 | 0.47 | 8.3 | No |
| AASHTO Compact | ≥0.033 | — | ≤70 | ≤0.60 | ≤8.4 | Yes |
| Noncompact | ≥0.033 | | ≤150 | ≤0.60 | ≤11.6 | No |

[a] $d$ = total depth of composite box girder; $L$ = span length; $w$ = distance between webs at top of box girder; $t_w$ = web thickness; $b'$ = width of projecting flange element from web; $t$ = flange thickness.

**Table 7.3  Dimensions of Braced Sections (in.)**

| Bridge | Type | Span (ft) | A | B | C | D | $T_1$ | $T_2$ | $T_3$ | $T_4$ |
|---|---|---|---|---|---|---|---|---|---|---|
| 2L | $2G^a$ | | 108 | 96 | 12 | 19.44 | 0.625 | 0.375 | 0.5 | 0.0878 |
| 3L | 3G | 50 | 98 | 86 | 12 | 19.47 | 0.5625 | 0.375 | 0.5 | 0.0980 |
| 4L | 4G | | 92 | 80 | 12 | 19.50 | 0.5 | 0.375 | 0.5 | 0.105 |
| 2L | 2G | | 110 | 96 | 14 | 48.125 | 1.0 | 0.5 | 0.75 | 0.0878 |
| 3L | 3G | 100 | 99 | 86 | 13 | 48.125 | 1.0 | 0.4375 | 0.75 | 0.0980 |
| 4L | 4G | | 92 | 80 | 12 | 48.125 | 1.0 | 0.4375 | 0.75 | 0.105 |
| 2L | 2G | | 110 | 96 | 14 | 74.75 | 1.5 | 0.625 | 1.0 | 0.0878 |
| 3L | 3G | 150 | 99 | 86 | 13 | 74.75 | 1.5 | 0.5625 | 1.0 | 0.0980 |
| 4L | 4G | | 92 | 80 | 12 | 74.75 | 1.5 | 0.5625 | 1.0 | 0.105 |

[a] $L$ indicates lanes, $G$ indicates girders, i.e., 2L 2G indicates two lanes—two girders.

**Table 7.4  Dimensions of Composite Sections (in.)**

| Bridge | Type | Span (ft) | A' | B' | C' | D' | E' | F' | T1' | T2' | T3' | T4' |
|---|---|---|---|---|---|---|---|---|---|---|---|---|
| 2L | $2G^a$ | | 192 | 96 | 12 | 19.44 | 25.75 | 42.0 | 0.625 | 0.375 | 0.5 | 0.0878 |
| 3L | 3G | 50 | 172 | 86 | 12 | 19.47 | 25.75 | 37.0 | 0.5625 | 0.375 | 0.5 | 0.0980 |
| 4L | 4G | | 160 | 80 | 12 | 19.50 | 25.75 | 34.0 | 0.5 | 0.375 | 0.5 | 0.105 |
| 2L | 2G | | 192 | 96 | 14 | 48.125 | 54.625 | 41.0 | 1.0 | 0.5 | 0.75 | 0.0878 |
| 3L | 3G | 100 | 172 | 86 | 13 | 48.125 | 54.625 | 36.5 | 1.0 | 0.4375 | 0.75 | 0.0980 |
| 4L | 4G | | 160 | 80 | 12 | 48.125 | 54.625 | 34.0 | 1.0 | 0.4375 | 0.75 | 0.105 |
| 2L | 2G | | 192 | 96 | 14 | 74.75 | 81.5 | 41.0 | 1.5 | 0.625 | 1.0 | 0.0878 |
| 3L | 3G | 150 | 172 | 86 | 13 | 74.75 | 81.5 | 36.5 | 1.5 | 0.5625 | 1.0 | 0.0980 |
| 4L | 4G | | 160 | 80 | 12 | 74.75 | 81.5 | 34.0 | 1.5 | 0.5625 | 1.0 | 0.105 |

[a] $L$ indicates lanes, $G$ indicates girders.

**Table 7.5    Section Properties of Braced Sections**

| Bridge | Type | Span (ft.) | $Y$ (in.) | $I_x$ (in.$^4$) | $K_T$ (in.$^4$) | $I_w$ (in.$^6$) |
|--------|------|-----------|-----------|-----------------|-----------------|-----------------|
| 2L | $2G^a$ |     | 6.94 | 6,550 | 18,900 | 3,220,000 |
| 3L | $3G$   | 50  | 7.16 | 6,090 | 17,000 | 2,200,000 |
| 4L | $4G$   |     | 7.21 | 5,720 | 15,600 | 1,710,000 |
| 2L | $2G$   |     | 18.6 | 67,400 | 140,000 | 9,450,000 |
| 3L | $3G$   | 100 | 18.9 | 61,700 | 118,000 | 2,940,000 |
| 4L | $4G$   |     | 19.1 | 58,300 | 107,000 | 1,720,000 |
| 2L | $2G$   |     | 30.3 | 236,000 | 378,000 | 3,710,000 |
| 3L | $3G$   | 150 | 30.7 | 216,000 | 311,000 | 216,000 |
| 4L | $4G$   |     | 31.0 | 205,000 | 280,000 | 528,000 |

$^a$ $L$ indicates lanes, $G$ indicates girders.

**Table 7.6    Section Properties of Composite Sections**

| Bridge | Type | Span (ft.) | $Y$ (in.) | $I_x$ (in.$^4$) | $K_T$ (in.$^4$) | $I_w$ (in.$^6$) |
|--------|------|-----------|-----------|-----------------|-----------------|-----------------|
| 2L | $2G^a$ |     | 19.8 | 27,200 | 56,800 | 3,640,000 |
| 3L | $3G$   | 50  | 19.8 | 24,700 | 49,100 | 2,170,000 |
| 4L | $4G$   |     | 19.7 | 23,100 | 44,600 | 1,540,000 |
| 2L | $2G$   |     | 37.8 | 178,000 | 250,000 | 15,800,000 |
| 3L | $3G$   | 100 | 37.9 | 159,000 | 197,000 | 16,400,000 |
| 4L | $4G$   |     | 37.8 | 150,000 | 176,000 | 15,400,000 |
| 2L | $2G$   |     | 52.1 | 511,000 | 551,000 | 98,700,000 |
| 3L | $3G$   | 150 | 52.3 | 460,000 | 433,000 | 101,000,000 |
| 4L | $4G$   |     | 52.1 | 433,000 | 385,000 | 92,400,000 |

$^a$ $L$ indicates lanes, $G$ indicates girders.

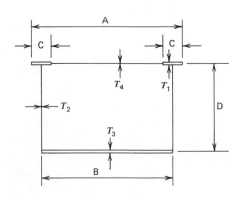

**Figure 7.4**  Braced steel box girder.

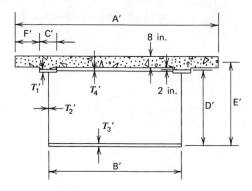

**Figure 7.5** Composite box girder.

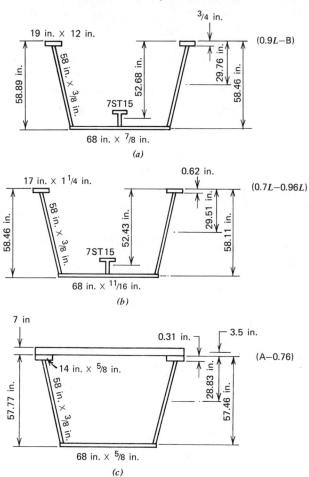

**Figure 7.6** Box girder dimensions.

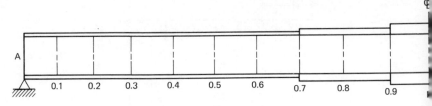

**Figure 7.7** Girder section locations.

1. *Centroid Location (reference to top flange).*

| Element | $A$ | $y$ | $Ay$ |
|---------|-----|-----|------|
| Top flanges | 57.0 | 0.75 | 43.0 |
| Webs | 43.5 | 29.76 | 1295.0 |
| $T$ | 4.41 | 52.68 | 232.0 |
| Bottom flange | 59.5 | 58.46 | 3478.0 |
| | 164.41 in.$^2$ | | 5048.0 in$^3$. |

The above data give:

$$\bar{y} = \frac{5048}{164.41} = 30.70 \text{ in.}$$

2. *Section Modulus.*

| Element | $I_0$ | $A$ | $y$ | $y^2$ | $Ay^2$ |
|---------|-------|-----|-----|-------|--------|
| Top flanges | 10.6 | 57 | 29.95 | 897.0 | 51,129.0 |
| Webs | 11,582.0 | 43.5 | 0.94 | 0.88 | 38.3 |
| $T$ | 19.0 | 4.4 | 21.98 | 483.1 | 2,125.6 |
| Bottom flange | 4 | 59.5 | 27.76 | 770.6 | 45,850.7 |
| | 11,615.6 in.$^4$ | | | | 99,143.6 |

Using the above data gives:

$$I_T = 11,615.6 + 99,143.6 = 110,759.2 \text{ in.}^4$$

$$S_T = \frac{110,759.2}{30.70} = 3607.8 \text{ in.}^3 \qquad S_B = \frac{110,759.2}{28.19} = 3929.0 \text{ in.}^3$$

For the section at $0.7L$ to $0.9L$:
1. *Centroid Location* (*reference to top flange*).

| Element | A | y | Ay |
|---|---|---|---|
| Top flanges | 42.5 | 0.62 | 26 |
| Webs | 43.5 | 29.51 | 1284 |
| T | 4.4 | 52.43 | 231 |
| Bottom flange | 46.75 | 58.11 | 2717 |
| | 137.16 in.$^2$ | | 4258 in.$^3$ |

Thus

$$\bar{y} = \frac{4258}{137.16} = 31.04 \text{ in.}$$

2. *Section Modulus.*

| Element | $I_0$ | A | y | $y^2$ | $Ay^2$ |
|---|---|---|---|---|---|
| Top flanges | 6.0 | 42.5 | 30.42 | 925.4 | 39,329.5 |
| Webs | 11,582.0 | 43.5 | 1.53 | 2.3 | 100.1 |
| T | 19.0 | 4.4 | 21.39 | 457.5 | 2,013.0 |
| Bottom flange | 2.0 | 46.75 | 27.07 | 732.8 | 34,258.4 |
| | 11,609.0 in.$^4$. | | | | 75,700.9 in.$^4$ |

Using the above data gives:

$$I_T = 75,700.9 + 11,609.0 = 87,309.9 \text{ in.}^4$$

$$S_T = \frac{87,309.9}{31.04} = 2812.8 \text{ in.}^3, \qquad S_B = \frac{87,309.9}{27.42} = 3184.2 \text{ in.}^2$$

For the section at $A$ to $0.7L$—steel section:

1. *Centroid Location.*

| Element | A | y | Ay |
|---|---|---|---|
| Top flanges | 17.5 | 0.31 | 5.0 |
| Webs | 43.5 | 28.83 | 1256.0 |
| Bottom flange | 42.5 | 57.46 | 2442.0 |
| | 103.5 in.$^2$ | | 3703.0 in.$^3$ |

Thus

$$\bar{y} = \frac{3703.0}{103.5} = 35.78 \text{ in.}$$

2. *Section Modulus.*

| Element | $I_0$ | $A$ | $y$ | $y^2$ | $Ay^2$ |
|---|---|---|---|---|---|
| Top flanges | 2.0 | 17.5 | 35.47 | 1258.1 | 22,017.1 |
| Webs | 11,582.0 | 43.5 | 6.90 | 47.6 | 2071.0 |
| Bottom flange | 1.0 | 42.5 | 21.68 | 470.0 | 19,976.0 |
| | 11,585.0 in.$^4$ | | | | 44,064.1 in.$^4$ |

From the above data:

$$I_T = 11,585.0 + 44,064.1$$
$$I_T = 55,649.1 \text{ in.}^4$$

$$S_T = \frac{55,649.1}{35.78} = 1555.3 \text{ in.}^3 \qquad S_B = \frac{55,649.1}{21.99} = 2530.7 \text{ in.}^3$$

For the section at $A$ to $0.7L$ (composite, $n = 30$) the effective flange width of box girders is in accordance with AASHTO Section 1.7.98 for I girders, except the computed width is doubled for box girders.

$$120 \times \tfrac{1}{4} = 30 \text{ ft}$$
$$\text{CC beams} = 15.67 \text{ ft}$$
$$12 \times \frac{7}{12} = 7 \text{ ft governs} \qquad b_e = 2 \times 7 = 14 \text{ ft}$$

Note that 7.0 in. is used, as 0.5 in. is for wearing surface and is not considered effective. The transformed width of flange is

$$b_t = \frac{14 \times 12}{30} = 5.6 \text{ in.}$$

1. *Centroid Location* (reference at center of gravity of slab).

| Element | $A$ | $y$ | $Ay$ |
|---|---|---|---|
| Slab | $5.6 \times 7 = 39.2$ | — | — |
| Steel | 103.5 | 39.28 | 4065.5 |
| | 142.7 in.$^2$ | | 4065.5 in.$^3$ |

Thus

$$\bar{y} = \frac{4065.5}{142.7} = 28.5 \text{ in.}$$

## 2. Section Modulus.

| Element | $I_0$ | $A$ | $y$ | $y^2$ | $Ay^2$ |
|---|---|---|---|---|---|
| | 160.0 | 39.2 | 28.5 | 812.3 | 31,840.0 |
| | 55,649.1 | 103.5 | 10.78 | 116.2 | 12,027.6 |
| | 55,809.1 in.$^4$ | | | | 43,867.6 in.$^4$ |

From the above data:

$$I_T = 55,809.1 + 43,867.6$$
$$I_T = 99,676.7 \text{ in.}^4$$

Steel: $S_T = \dfrac{99,676.7}{25} = 3987.0 \text{ in.}^3$ $\quad$ $S_B = \dfrac{99,676.7}{32.77} = 3041.7 \text{ in.}^3$

Concrete: $S_T = \dfrac{99,676.7}{32} \times 30 = 93446.9 \text{ in.}^3$

For the section at $A$ to $0.7L$ (composite, $n = 10$) the transformed width of flange is

$$b_t = \frac{14 \times 12}{10} = 16.8 \text{ in.}$$

## 1. Centroid Location (reference at center of gravity of slab).

| Element | $A$ | $y$ | $Ay$ |
|---|---|---|---|
| Slab | $16.8 \times 7 = 117.6$ | — | — |
| Steel | 103.5 | 39.28 | 4065.5 |
| | 221.1 in.$^2$ | | 4065.5 in.$^3$ |

Thus

$$\bar{y} = \frac{4065.5}{221.1} = 18.4 \text{ in.}$$

2. *Section Modulus.*

| Element | $I_0$ | $A$ | $y$ | $y^2$ | $Ay^2$ |
|---------|-------|-----|-----|-------|--------|
| Slab    | 480.2    | 117.6 | 18.4  | 338.6 | 39,814.7 |
| Steel   | 55,649.1 | 103.5 | 20.88 | 436.0 | 45,123.4 |
|         | 56,129.3 |       |       |       | 84,938.1 |

Using the above data gives:

$$I_T = 56,129.3 + 84,938.1$$
$$I_T = 141,067.4 \text{ in.}^4$$

Steel: $S_T = \dfrac{141,067.4}{14.9} = 9467.6 \text{ in.}^3$    $S_B = \dfrac{141,067.4}{42.87} = 3290.6 \text{ in.}^3$

Concrete: $S_T = \dfrac{141,067.4}{21.9} \times 10 = 64,414.3 \text{ in.}^3$

CHECK OF COMPRESSION FLANGES ($0.55 F_y$ is used)

$$\frac{b}{t} = \frac{6140}{\sqrt{F_y}}.$$

For $F_y = 46$ ksi

$$\frac{b}{t} = 28.6$$

For $F_y = 36$ ksi

$$\frac{b}{t} = 32.4$$

At $0.7L$ to $0.9L$:

$$\frac{b}{t} = \frac{17}{1.25} = 13.6 < 32.4 \qquad \text{O.K.}$$

At $A$ to $0.7L$:

$$\frac{b}{t} = \frac{14}{0.625} = 22.4 < 32.4 \qquad \text{O.K.}$$

BENDING STRESSES

By utilizing the moments given in Fig. 7.2 and as given in Table 7.7, the maximum bending stresses can now be determined.

*Stresses Support B.* The allowable stress for A441 steel as given in AASHTO Section 1.7.1 is, for thickness greater than $\frac{3}{4}$ in. to a maximum of $1\frac{1}{2}$ in., $f_{all} = 25$ ksi. For load A:

For load A:

$$M_{DL} = -3420 \text{ k-ft}$$

For load B:

$$M_{SDL} = -936 \text{ k-ft}$$
$$M_{LL+I} = \underline{-2434 \text{ k-ft}}$$
$$-6790 \text{ k-ft}$$

$$f_{top} = \frac{6790 \times 12}{3607.8} = +22.6 \text{ ksi}, \qquad f_b = \frac{6790 \times 12}{3929.0} = -20.7 \text{ ksi}$$

Fatigue is now checked according to AASHTO Section 1.7.3. The roadway is considered to be moderately traveled or $ADT < 2500$, which corresponds to a truck loading of 500,000 stress cycles. The governing structural detail classes on categories consist of butt weld flanges (category B), fillet welded webs to flanges (category B), stiffeners to webs (category C), or shear studs to flanges (category C). In this instance, category C governs, which allows a live-load stress range $F_{sr} = 19.0$ ksi. The actual stress range is

$$F_{LL+I} = \frac{2434 \times 12}{3602.8} = 8.1 \text{ ksi}$$

which is less than the allowable 19.0 ksi; therefore the section is satisfactory.

Allowable compressive stress is checked in accordance with AASHTO Section 1.7.105. If $w/t \le 6140/\sqrt{F_y}$, then $F_a = 0.55F_y = 25$ ksi.

$$\frac{w}{t} = \frac{6140}{\sqrt{46,000)}} = 28.6$$

$$\left(\frac{w}{t}\right)_{act} = \frac{68}{(\frac{7}{8})} = 77.7 > 28.6$$

Therefore a longitudinal stiffener is needed. As shown in Fig. 7.6a, a structural tee $7ST15$ is assumed and is welded at midwidth giving $w = 34$ in. The required stiffness is given by

$$I_s = \phi t^3 w$$

where $\phi = 0.125 \text{ k}^3$

$$k = 2 \quad \text{to} \quad 4$$

It is assumed that $k = 3.0$ and with $w = 34$ in. and $t = \frac{7}{8}$ in. $I_s$ is computed as

$$I_s = (0.125)(27)(\tfrac{7}{8})^3 \times 34$$
$$I_s = 76.9 \text{ in.}^4$$

For the 7 *ST* 15 Tee, the stiffness about the base is

$$I_{\text{base}} = 19.0 + 4.42(5.35)^2$$
$$= 145.5 \text{ in.}^4 > 76.9 \text{ in.}^4 \qquad \text{O.K.}$$

The effective width is now equal to $(w/t)_{\text{act}}( = 34.0)$, the permissible $w/t$, if $F_a = 25$ ksi is governed by

$$\frac{w}{t} = \frac{3070\sqrt{K}}{\sqrt{F_y}}$$

$$-\frac{3070\sqrt{3}}{\sqrt{46{,}000}} = \frac{5317.4}{214.5} = 24.8 < 34.0$$

Therefore the allowable stress is governed by

$$F_b = 0.55F_y - 0.224F_y\left[1 - \sin\frac{\pi}{2}\left(\frac{6650\sqrt{K} - (w/t)\sqrt{F_y}}{3580\sqrt{K}}\right)\right]$$

$$F_b = 25 - 10.3\left[1 - \sin\frac{\pi}{2}\left(\frac{11{,}518 - 34 \times 214.5}{6200.7}\right)\right]$$

$$F_b = 25 - 10.3\left[1 - \sin\frac{\pi}{2}(0.68)\right]$$

$$F_b = 25 - 10.3[1 - \sin(1.07)]$$

$$F_b = 25 - 10.3[1 - \sin 61°]$$

$$F_b = 25 - 1.29$$

$$F_b = 23.71 \text{ ksi}$$

The actual stress $F_{\text{act}} = 20.7$ ksi; therefore flange is adequate.

*Stresses at 0.9L (12 ft from Bent) (A36 steel $f_{\text{all}} = 20$ ksi).* Selecting the moments from Fig. 7.2 at 0.9L gives for load A:

$$M_{DL} = -2000 \text{ k-ft} \qquad\qquad M_{DL} = -2000 \text{ k-ft}$$

For load B:

$$M_{SDL} = -500 \qquad\qquad M_{SDL} = -500$$

$$\underline{M_{LL+I} = -1700} \qquad\qquad \underline{M_{LL+I} = +359}$$

$$-4200 \text{ k-ft (max)} \qquad\qquad -2141 \text{ k-ft (min)}$$

$$f_{top} = \frac{4200 \times 12}{2812.8} = 17.9 \text{ ksi}, \qquad f_b = \frac{4200 \times 12}{3184.2} = 15.8 \text{ ksi}$$

Fatigue is checked according to AASHTO Section 1.7.3, 500,000 cycles.

$$F_{sr} = \frac{1700 \times 12}{2812.8} = 7.3 \text{ ksi} < 19 \text{ ksi} \qquad \text{O.K.}$$

*Stresses at 0.4L Composite Section (A36 steel $f_{all}$ = 20 ksi).* For load A:

$$M_{DL} = +1915 \text{ k-ft}$$

For load B:

$$M_{SDL} = +524$$

$$M_{LL+I} = +2403 \text{ k-ft}$$

The following are the equations for steel:

$$f_{DL} = \frac{1915 \times 12}{2531} = 9.1 \text{ ksi}$$

$$f_{SDL} = \frac{524 \times 12}{3042} = 2.1$$

$$f_{LL+I} = \frac{2403 \times 12}{3291} = \underline{8.8}$$

$$20.0 \text{ ksi}$$

The equations for concrete are

$$f_{SDL} = \frac{524 \times 12}{93,446.9} = 0.067$$

and

$$f_{LL+I} = \frac{2403 \times 12}{64,414} = \underline{0.448}$$

$$0.515 \text{ ksi} < 1.2 \text{ ksi (allowable)}$$

Fatigue is checked according to AASHTO Section 1.7.3.

$$F_{sr} = 8.8 < 19.0 \text{ ksi} \qquad \text{O.K.}$$

WEB STRESSES AND STIFFENERS

Web to depth limitation is determined from AASHTO Section 1.7.10:

Steel section:
$$\frac{d}{L} \geq \frac{1}{25}$$

$\left\{\begin{array}{l}\text{Concrete and steel} \\ \text{Steel}\end{array}\right\}$ composite section:
$$\frac{d'}{L} \geq \frac{1}{25}$$
$$\frac{d}{L} \geq \frac{1}{30}$$

At support B:

$$\frac{d}{L} = \frac{58.89/12}{120} = \frac{1}{24.5} \qquad \text{O.K.}$$

At 0.4L:

$$\frac{d'}{L} = \frac{64.77/12}{120} = \frac{1}{22.2} \qquad \text{O.K.}$$

$$\frac{d}{L} = \frac{57.77/12}{120} = \frac{1}{24.9} \qquad \text{O.K.}$$

Web thickness is determined according to AASHTO Section 1.7.70.

$$t \geq \frac{D\sqrt{f_b}}{23000}$$

without web stiffeners but in no case should $t \leq D/170$

where $\quad D =$ depth of web

$\qquad f_b =$ calculated compression stress in flange

The maximum compressive stress occurs at the bent $B$,

where $\quad f_b = -20.7$ ksi and $D = 56.51$; therefore

$$f_b = 56.6 \frac{\sqrt{20,700}}{23,000} = 0.353 \text{ in.}$$

Thus a $\frac{3}{8}$ in. (0.375 in.) web plate is satisfactory without longitudinal stiffeners for all sections.

Transverse intermediate stiffeners are determined from AASHTO Section 1.7.71.

The webs of plate girders should be stiffened at intervals not to exceed

$$d \leq 11,000 \frac{t}{\sqrt{f_v}}$$

where $t$ = thickness of web plate, which is $\frac{3}{8}$ in.

$f_v$ = average shearing stress in psi at point considered

nor be greater than the girder depth.

The design shear for the web $V_w$ is calculated using the following, as given in AASHTO Section 1.7.104(A).

$$V_w = \frac{V_v}{\cos \theta}$$

where $V_v$ = vertical shear

$\theta$ = angle of inclination of web plate w.r. to vertical

The total vertical shear at each point is found from Fig. 7.3 and is given in Table 7.7. The total web shear $V_w$, web stress $f_v$, and required stiffener are computed as follows:

| Section | $v_v$ | $\cos \theta$ | $V_w = V_v/\cos \theta$ | $A_t = 2A_w$ | $f_v = (V_w/A_t)$ | $\sqrt{f_v}$ | $d_{max}$ (in.) |
|---|---|---|---|---|---|---|---|
| End | | | | | | | |
| Support A | 209.9 | 0.973 | 215.7 | $2 \times 58 \times \frac{3}{8} = 43.5$ | 4.96 | 70.4 | 58.6 |
| 0.8L | 215.0 | 0.973 | 221.0 | $2 \times 58 \times \frac{3}{8} = 43.5$ | 5.10 | 71.4 | 57.8 |
| Center | | | | | | | |
| Support B | 288.0 | 0.973 | 296.0 | $2 \times 58 \times \frac{3}{8} = 43.5$ | 6.80 | 82.5 | 50.0 |

The following sequence of stiffener spacing, from Support A is used (Fig. 7.8):

$$2 \text{ at } 2 \text{ ft } 0 \text{ in.} = \quad 4 \text{ ft } 0 \text{ in.}$$

$$18 \text{ at } 4 \text{ ft } 8 \text{ in.} = \quad 84 \text{ ft } 0 \text{ in.}$$

$$\underline{8 \text{ at } 4 \text{ ft } 0 \text{ in.} = \quad 32 \text{ ft } 0 \text{ in.}}$$

$$120 \text{ ft } 0 \text{ in.}$$

where the maximum spacing is less than $d_{max}$ given in the above table. The first two stiffeners are placed such that they do not exceed one-half of the maximum value of 58.0 in, as recommended in AASHTO Section 1.7.71.

The moment of inertia provided by each stiffener, according to Section 1.7.71, must satisfy

$$I = \frac{d_0 t^3 J}{10.92}$$

where $J = 25(D^2/d^2) - 20$, but not less than 5.0

$D$ = depth of web

$d$ = required spacing between stiffeners

$t$ = web plate thickness

$d_0$ = actual spacing between stiffeners

**Table 7.7   Summary of Design Moments (k-ft)**

| | | | Combined Moment | | | |
|---|---|---|---|---|---|---|
| Location | $DL_1$ | $DL_2$ | $LL+I$ $(+)$ | $LL+I$ $(-)$ | $M+$ | $M-$ |
| 0.2 | 1504 | 400 | 1789 | — | 3693 | — |
| 0.4 | 1915 | 509 | 2403 | — | 4827 | — |
| 0.6 | 1231 | 327 | 2132 | — | 3690 | — |
| 0.8 | −547 | −145 | 1082 | −974 | 1082 | 1666 |
| B | −3420 | −909 | — | −2419 | — | 6748 |

| | Shears (k) | | | |
|---|---|---|---|---|
| | $DL_1$ | $DL_2$ | $LL+I$ | Reaction (k) |
| A | 85.5 | 23.4 | 101 | 209.9 |
| B | 143 | 39 | 106 | 288 |
| 0.8L | 100 | 25 | 90 | 215 |

The required stiffness at support B is examined:

$$J = 25\left(\frac{58}{50}\right)^2 - 20$$

$$J = 13.64$$

$$I = 48 \times \left(\frac{3}{8}\right)^3 \times \frac{13.64}{10.92}$$

$$I = 3.15 \text{ in.}^4$$

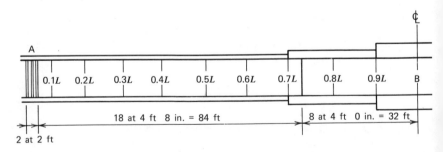

**Figure 7.8**   Girder stiffener locations.

Assuming a 4 in. $\times \frac{5}{16}$ in. plate,

$$I = \tfrac{1}{3}(\tfrac{5}{16})(4)^3 = 6.7 \text{ in.}^4 > 3.15 \text{ in.}^4$$

The following provisions must also be considered:

$$\text{Width plate} \geq 2 \text{ in.} + \frac{d}{30} \text{ nor} \leq \frac{b_f}{4}$$

Thickness plate $> \frac{1}{16}$ (plate width)

Width plate (min.) $= 2 + \frac{58}{30} = 3.9$ in. $< 4.0$ in.     O.K.

Thickness (min.) $= \frac{4}{16} < \frac{5}{16}$ in.     O.K.

Use 4 in. $\times \frac{5}{16}$ in. plate stiffeners welded to webs.

Bearing stiffeners are now determined. At support B:

$$R_B = 2(143) + 2(39) + 179 = 543 \text{ k per box}$$
$$= 271 \text{ k per bearing stiffeners}$$

As given in AASHTO Section 1.7.73, bearing stiffeners should be designed as columns and the thickness should not be less than

$$t \geq \frac{b'}{12} \sqrt{\frac{F_y}{33,000}}$$

where $b' =$ width of stiffeners. When two plates are used, the participating portion of the web is equal to $18 \times t_{\text{web}}$ or $18 \times t_{\text{bearing plate}}$. Assuming a $\frac{7}{16}$ in. bearing plate,

$$18 \times \tfrac{7}{16} \text{ in.} = 7.9 \text{ in.}$$

Two sets of stiffeners $4\frac{1}{2}$ in $\times \frac{3}{4}$ in. are used, as shown in Fig. 7.8, and a spacing of 6.0 in. is assumed, as shown in Fig. 7.9. The stresses in these stiffeners are induced by assuming an equivalent column. The area of the equivalent column is

$$A = 4(4.5)(\tfrac{3}{4}) + [2(3.9) + 6](0.4375) = 19.5 \text{ in.}^2$$

Then

$$I_x = \tfrac{1}{3}(\tfrac{3}{4})(4.5)^3 \times 4 = 91 \text{ in.}^4$$

$$r = \sqrt{\frac{I}{A}} = \sqrt{\frac{91}{19.5}} = 2.2 \text{ in.}$$

$$\frac{L}{r} = \left(\frac{56.51}{2.2}\right) = 25.7$$

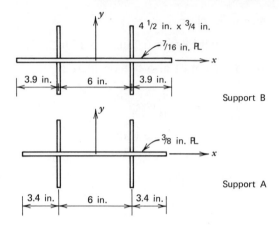

**Figure 7.9**  Bearing stiffeners.

The allowable compressive stress is in accordance with AASHTO Section 1.7.1 or

$$F_a = 16,000 - 0.38\left(\frac{L}{r}\right)^2$$

$$F_a = 16,000 - 0.38(25.7)^2$$

$$F_a = 15.75 \text{ ksi (allowable)}$$

$$F_{\text{act}} = \frac{271}{19.5} = 13.9 \text{ ksi} < 15.75 \qquad \text{O.K.}$$

The required thickness of stiffener is checked.

$$t \geq \frac{4.5}{12}\sqrt{\frac{36,000}{33,000}}$$

$t \geq 0.39$ in. which is less than $\frac{3}{4}$ in. is assumed.
At support A:

$$R_A = 85.5 + 23.4 + 101 = 209.9 \text{ k per box}$$

$$R_A = 105 \text{ k per bearing stiffener}$$

Stiffeners of 4 in. $\times \frac{1}{2}$ in. as shown in Fig. 7.9, are assumed. The column areas of these stiffeners are

$$A = 4(4)(\tfrac{1}{2}) + [2(3.4) + 6](\tfrac{3}{8}) = 12.8 \text{ in.}^2$$

Therefore

$$I_x = \frac{1}{3}(\frac{1}{2})(4)^3 \times 4 = 42.7 \text{ in.}^4$$

$$r = \left(\frac{42.7}{12.8}\right)^{1/2} = 3.3 \text{ in.}$$

$$\left(\frac{L}{r}\right) = \left(\frac{56.52}{3.3}\right) = 17.1$$

$$F_a = 16,000 - 0.38(17.1)^2 = 15.89 \text{ ksi}$$

$$F_{act} = \frac{105}{12.8} = 8.2 \text{ ksi} < 15.89 \quad \text{O.K.}$$

SHEAR CONNECTORS

To ensure interaction between the slab and girder, shear connectors must be provided as governed by AASHTO Section 1.7.100.

$$S_r = \frac{V_r Q}{I}$$

where $S_r$ = range in horizontal per inch. For this structure

$$V_r = 101 \text{ k (at } R_A)$$

$$I_{(n=10)} = 141,067.4 \text{ in.}^4$$

$$Q = A_{slab}\, x\bar{y} = 117.6(14.88 + 3.50) = 2161 \text{ in.}^3$$

$$S_r = \frac{101 \times 2161}{141,067.4} = 1.54 \text{ k/in.}$$

Now $4\frac{3}{4}$ in. $\times 4$ in. studs at 500,000 cycles across each flange are tried. The allowable range of horizontal shear $Z_r$, in pounds, of an individual connector is

$$Z_r = \alpha d^2$$

where $\quad d$ = diameter of stud = 0.75 in.

$\qquad \alpha = 10,600$

Therefore

$$Z_r = 10,600 \times (0.75)^2$$

$$Z_r = 5962.5 \text{ lb. per connector if necessary}$$

$$Z_r = 4 \times 5962.5$$

$$Z_r = 23.85 \text{ k per flange}$$

$$\text{spacing of connectors} = \left(\frac{23.85}{1.54}\right)^2 = 30.98 \text{ in.}$$

Use maximum spacing of 24.0 in.

The ultimate strength is now checked. The required number of connectors between supports and maximum positive moment is

$$N = \frac{P}{\phi S_u}$$

where    $S_u = 0.4 d^2 (f'_c E_c)^{1/2}$

$S_u = 4 (\tfrac{3}{4})^2 (3000 \times 3 \times 10^6)^{1/2}$

$S_u = 21,300$ lb per stud

$S_u = 21.3 \times 4 = 85.2$ k per flange

$\phi = 0.85$

$P = A_s F_y$ or $0.85 f'_c bc$, whichever is smaller

$P = 103.5 \times \tfrac{36}{2} = 1863$ k per flange

$P = 0.85 \times 3.0 \times 7 \times (14 \times 12)/2 = 1499$ k per flange ← governs

Then

$$N = \frac{P}{\phi S_u} = \frac{1499}{0.85 \times 85.2} = 20.7,$$

$N = 20.7$ connectors per flange between abutment and $0.4L$ (maximum positive moment) or a distance of $0.4 \times 120 = 48$ ft.

Maximum spacing selected is 2.0 ft, or $\tfrac{48}{2} = 24$ rows, which is greater than 20.7; therefore conditions are met.

## REFERENCES

1  American Association of State Highway and Transportation Officials *Standard Specification for Highway Bridges*, AASHTO, 11th ed. 1973, Washington, D.C.

2  American Institute of Steel Construction *Moments, Shears and Reaction for Continuous Highway Bridges*, New York, 1966.

3  R. S. Humphreys, "Interaction Effects of Bending and Torsion on Box Girders," M. S. Thesis, Civil Engineering Department, University of Maryland, College Park, Maryland, December 1975.

4  J. C. Olenik and C. P. Heins, "Diaphragm Spacing Requirements for Curved Girder Bridges," Civil Engineering Report No. 58, University of Maryland, College Park, Maryland, August 1974.

CHAPTER **8**

# Design of Curved Steel Box Girders

## 8.1  GENERAL

The design of curved box girders involves consideration of the interaction between bending and torsional forces, as illustrated in Chapter 5 for I girder bridges. However, in the design of box girders the interaction between adjacent boxes need not be considered because of the high torsional rigidity of an individual box element. Thus external diaphragms attached between these boxes need not be utilized. However, because of the flexibility of the slender webs, it is necessary to move internal diaphragms to reduce cross-sectional deformations, which can induce excessive normal stresses. In addition to these internal cross diaphragms, top lateral bracing is generally recommended during fabrication and erection of the box. Such bracing can be utilized if properly attached in the dead-load design process, as is illustrated in Design Example 8.1.

Examples of the type of box-girder bridges that are presently being constructed for highway interchanges (1) are shown in Figs. 8.1 through 8.4. Figure 8.2 shows a typical detail of the cross-diaphragms and top lateral bracing. Figure 8.3 shows the erected box beams with stay-in-place forms prior to placement of the concrete. Figure 8.4 shows the completed

**Figure 8.1** Steel box girder.

**Figure 8.2** Top lateral bracing and cross diaphragms.

**Figure 8.3**  Erection of stay in place forms.

**Figure 8.4**  Completed structure.

structure. As can be seen from this photograph, the box girder offers a very pleasing structural element.

## 8.2  DESIGN DATA

### 8.2.1  Lateral Bracing

The top lateral bracing is needed primarily to stiffen the box girder during erection. However, the bracing can be used as part of the basic box in considering structural performance. As shown in Figs. 8.2 and 8.5, the lateral bracing is generally attached along or near the top flange. The configuration of the bracing can be as shown in Figs. 8.6 through 8.8. This bracing can be reduced into an equivalent uniform thick plate (2, 3) by applying the equations 8.1 to 8.3 for systems I through III, respectively.

$$t_{eq} = \frac{E}{G} \frac{Sw}{\dfrac{d^3}{A_d} + \dfrac{2}{3}\dfrac{S^3}{A_f}} \tag{8.1}$$

$$t_{eq} = \frac{E}{G} \frac{Sw}{\dfrac{2d^3}{A_d} + \dfrac{w^3}{4A_v} + \dfrac{S^3}{6}\dfrac{1}{A_f}} \tag{8.2}$$

$$t_{eq} = \frac{E}{G} \frac{Sw}{\dfrac{d^3}{2A_d} + \dfrac{S^3}{6}\dfrac{1}{A_f}} \tag{8.3}$$

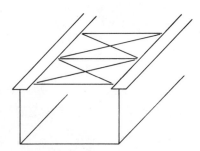

**Figure 8.5**  Lateral bracing of box section.

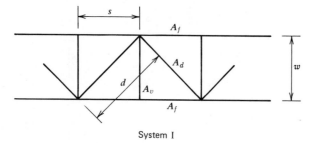

System I

**Figure 8.6** Bracing configuration I.

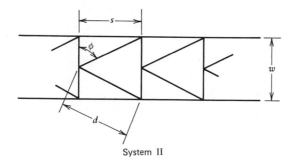

System II

**Figure 8.7** Bracing configuration II.

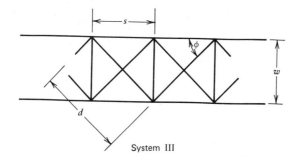

System III

**Figure 8.8** Bracing configuration III.

The box girder in cross section then has the general configuration shown in Fig. 8.9. Thus the torsional constant $K_T$ is computed as

$$K_T = \frac{4A_0^2}{\oint ds/t} \tag{8.4a}$$

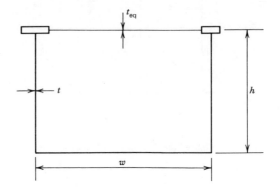

**Figure 8.9** Equivalent plate thickness of braced section.

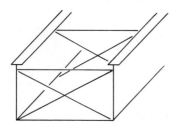

**Figure 8.10** Transverse bracing of box section.

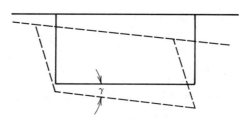

**Figure 8.11** Distorted box section.

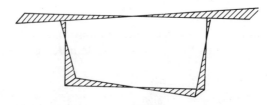

**Figure 8.12** Induced distortional normal stresses.

or

$$K_T = \frac{4[hw]^2}{[(2h+w)/t + w/t_{eq}]} \qquad (8.4b)$$

rather than the stiffness of an open cross section.

## 8.2.2 Diaphragms

Without the aid of internal cross-diaphragms, as shown in Fig. 8.10, the cross section does not retain its shape, resulting in the configuration shown in Fig. 8.11. Such a distortion further induces normal stresses, as shown in Fig. 8.12. The evaluation of these normal stresses has been presented by Dabrowski (4) and application of this theory by Oleinik and Heins (5) has resulted in the following design equations when curved girders are subjected to dead and live loads.

For dead load:

$$\frac{\sigma_f}{\sigma_b} = \frac{\bar{A}(s)^2}{(WA^*)^{1/4}R} \qquad (8.5a)$$

where $\sigma_f$ = maximum induced distortional normal stress in the bottom flange of the box beam under study (ksi)

$\sigma_b$ = maximum induced normal bending stress in the bottom flange of the box beam under study (ksi)

$\bar{A}$ = coefficient = $0.220L^3 + 30.8L^2 - 697.0L$

$s$ = diaphragm spacing as a fraction of span length

$R$ = radius of curvature (ft)

$L$ = span length (ft)

$WA^*$ = distortional stiffness (k-in.$^4$), which is a function of the section; may be approximated as $WA^* = 30 \times 10^3(24.5 \times 10^2 L^2 - 5 \times 10^6)$, which is applicable for the range in parameters $50\,\text{ft} \leq L \leq 150\,\text{ft}$, $1.5\,\text{ft} \leq d \leq 6.0\,\text{ft}$, $6.5\,\text{ft} \leq W \leq 8.0\,\text{ft}$

$d$ = depth of box section (ft)

$W$ = width of bottom flange of box section (ft)

A more approximate equation can be used (6):

$$\frac{\sigma_f}{\sigma_b} = (10L - 350)\frac{s^2}{R} \qquad (8.5b)$$

where the parameters are as defined previously.

For live load:

$$\frac{\sigma_f}{\sigma_b} = \frac{(28.1)(L)^3(s)^2}{(WA^*)^{1/4}(R)^2} + \frac{(105)(L)^{1/2}(s)}{(WA^*)^{1/4}} \qquad (8.6a)$$

where $L$, $s$, $WA^*$, $R$ are as defined previously, and $WA^*$ may be approximated as $WA^* = 30 \times 10^3 (87 \times 10^2 \, L^2 - 16 \times 10^6)$ and is limited to those box beam ranges as given previously.

A more approximate equation can be used:

$$\frac{\sigma_f}{\sigma_b} = (-0.25 \times 10^{-2} \, L + 1.3)s \qquad (8.6b)$$

where the parameters are as defined previously.

Equations 8.5 and 8.6 give the magnitudes of the induced distortional normal stress ($\sigma_f$) if geometry of the box girder and the bending stress ($\sigma_b$) are known. The evaluation of the bending stress, however, warrants an analysis of a single-span *curved* box beam. To eliminate this difficulty, several additional equations have been developed (5, 6) that relate the bending response of identical boxes when the beam is straight and curved. For dead load:

$$KD = 1 + \frac{1}{10(R/L)^2} \qquad (8.7)$$

$$\sigma_b = \sigma_{b_s} \times KD \qquad (8.8)$$

where  $KD$ = dead-load modification factor
$\sigma_b$ = induced dead-load bending stress in curved box beam
$\sigma_{b_s}$ = induced dead-load bending stress in straight box beam
$R/L > 1.0$

For live load:

$$KL = 1 + \frac{1}{10(R/L)} \qquad (8.9)$$

$$\sigma_b = \sigma_{b_s} \times KL \qquad (8.10)$$

where  $KL$ = live-load modification factor
$\sigma_b$ = induced live-load bending stress in curved box beam subjected to one full AASHTO truck (72 k)
$\sigma_{b_s}$ = induced live-load bending stress in straight box beam subjected to one full AASHTO truck (72 k)
$R/L > 1.0$

Thus an equivalent straight box beam is examined and the resulting stress is determined. These stresses are then multiplied by equations 8.7 and 8.9. They are then substituted into equations 8.5 and 8.6 to evaluate the distortional stress ($\sigma_f$) directly. The application of these equations is now illustrated in part.

## Design Example 8.1

A single box girder ramp structure, as shown in Fig. 8.13, is considered here. The girder has a span length of 80.0 ft and a radius of curvature of 500.0 ft, as shown in Fig. 8.14. *Given:*

1. Span: $L = 80$ ft.
2. $R = 500$ ft 0 in.
3. Clear roadway: 14 ft.
4. Beam steel: A36.
5. Future wearing surface: 25 psf.
6. Haunch: 1 in. $\times$ 3 in. over beam flanges.
7. Concrete: $f'_c = 3000$ psi.
8. Loading: HS20-44.

Dead load A (no composite action)

| | |
|---|---|
| Slab: $7.5/12 \times 14.0 \times 0.150$ | 1.313 |
| Haunch: $4 \times \frac{3}{12} \times \frac{1}{12} \times 0.150$ | 0.012 |
| Steel (estimate) | 0.580 |
| | 1.905 |

Dead load B (composite action)

| | |
|---|---|
| Parapet: $1 \times 1.5 \times 0.150 \times 2$ | 0.450 |
| Railing: $(15\ 16/\text{ft}) \times 2$ | 0.030 |
| Asphalt paving: $0.025 \times 14$ | 0.350 |
| | 0.830 k/ft |

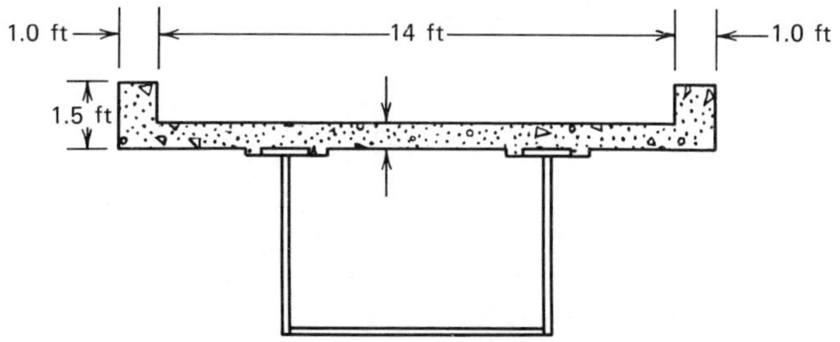

**Figure 8.13** General box girder bridge.

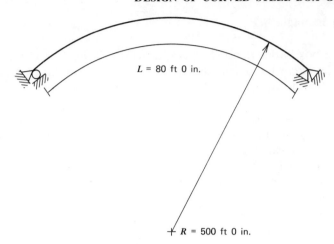

**Figure 8.14** Plan view of girder.

DEAD-LOAD A; BARE STEEL GIRDER

The induced dead-load moment is first computed assuming a straight girder of $L = 80.0$ ft:

$$M_D = \frac{wL^2}{8} = 1.91 \times \frac{(80)^2}{8} = 1528.0 \text{ k-ft}$$

The magnification of this moment as a result of curvature is found from equation 8.7, or

$$M_{D_{\text{curved}}} = K_D M_{D_{\text{straight}}}$$

when

$$K_D = 1 + \frac{1}{10(R/L)^2}$$

Therefore

$$K_D = 1 + \frac{1}{10(500/80)^2}$$

$$K_D = 1.0026$$

$$M_{D_{\text{curved}}} = 1.003 \times 1528 = 1533 \text{ k-ft}$$

The induced torsion moment due to a uniform load is found by applying the equations given in Chapter 5. As shown in Fig. 8.15, the average moment arm of the uniform load to the chord is $0.75e$, or

$$T_{DL} = w(0.75e)L$$

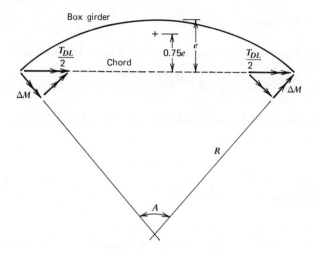

**Figure 8.15** Dead load torque.

The central angle and eccentricity can be computed by the following equations:

$$A = 57.3 \frac{L}{R} \text{ (degrees)}$$

$$e = 2R \sin^2 \left(\frac{A}{4}\right)$$

For the girder geometry given in Figs. 8.14 and 8.15,

$$A = 57.3 \left(\frac{80}{500}\right) = 9.167°$$

$$e = 2(500) \sin^2 \left(\frac{9.167}{4}\right)° = 1.60 \text{ ft}$$

Therefore

$$T_{DL} = 1.91(0.75 \times 1.60)80$$
$$T_{DL} = 183.36 \text{ k-ft}$$

The maximum dead-load reaction or shear is

$$R = 76.4k$$

DEAD LOAD B; COMPOSITE $n = 30$

The induced moment and torque due to the superimposed dead load are

$$M_{SDL} = \left(\frac{wL^2}{8}\right)KD = \frac{0.83 \times (80)^2}{8} \times 1.003 = 666.0 \text{ k-ft}$$

$$T_{SDL} = 0.830(0.75 \times 1.60)80 = 79.7 \text{ k-ft}$$

The induced maximum reaction or shear is

$$R = 33.2k$$

LIVE LOAD; COMPOSITE $n = 10$

A single HS20 vehicle is positioned along the single box girder, as shown in Fig. 8.16. This loading gives a maximum straight girder bending moment of

$$M_{LL} = 1165.0 \text{ k-ft}$$

The effects of curvature on this induced moment are determined from equation 8.9:

$$K_L = 1 + \frac{1}{10(R/L)}$$

$$K_L = 1 + \frac{1}{10(500/80)}$$

$$K_L = 1.016$$

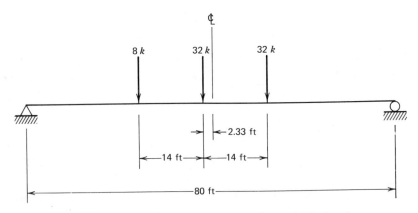

**Figure 8.16** AASHTO live load. Critical truck loading for bending.

Therefore the live-load moment is

$$M_{LL} = 1.016 \times 1165.0$$
$$M_{LL} = 1183.6 \text{ k-ft}$$

The live-load torsion moment is found from the equations given in Chapter 5, where

$$T_{LL} = (e-c)8 + e(32) + (e-c)32$$

as shown in Fig. 8.17, where the central axle is positioned at midspan. The term $(e-c)$ can be approximated by

$$(e-c) = e\left[1 - \left(\frac{x}{0.5L}\right)^2\right]$$

As computed previously for dead load, $e = 1.51$ ft; therefore

$$(e-c) = 1.60\left[1 - \left(\frac{14}{0.5 \times 80}\right)^2\right] = 1.41 \text{ ft}$$

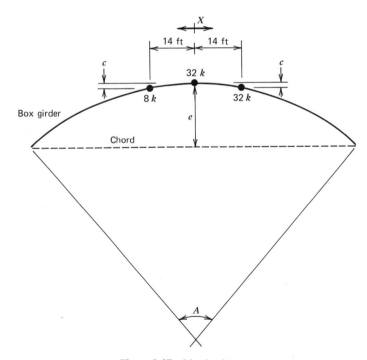

**Figure 8.17**  Live load torque.

and

$$T_{LL} = (1.41)8 + 1.60(32) + 1.41(32)$$
$$T_{LL} = 107.6 \text{ k-ft}$$

The impact, as given by AASHTO, is

$$I = \frac{50}{L + 125} = \frac{50}{80 + 125} = 0.244$$

and therefore

$$M_{LL+I} = 1.244 \times 1183.6 = 1472.4 \text{ k-ft}$$
$$T_{LL+I} = 1.244 \times 107.6 = 133.9 \text{ k-ft}$$

The maximum induced reaction is computed as

$$R = 63.6 \text{ k}$$

SECTION PROPERTIES

The basic girder section geometry is that shown in Fig. 8.18, as obtained from the study of preliminary data given in Chapter 7. The section properties of this box are computed below.

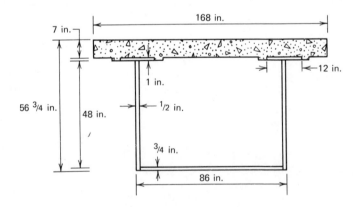

**Figure 8.18** Isolated box girder.

*Bare Steel Section*

1. *Centroid Location* (*reference top flanges*).

| Element | $A$ | $y$ | $Ay$ |
|---|---|---|---|
| Top flanges | 28 | 0.50 | 14.0 |
| Webs | 48 | 25.0 | 1200.0 |
| Bottom flange | 64.50 | 49.375 | 3184.7 |
| | 140.50 in.$^2$ | | 4398.7 in.$^3$ |

The above data give

$$y = \frac{4398.7}{140.5} = 31.3 \text{ in.}$$

2. *Section Modulus.*

| Element | $I_0$ | $A$ | $y$ | $y^2$ | $Ay^2$ |
|---|---|---|---|---|---|
| Top flanges | 2.3 | 28 | 30.8 | 948.7 | 26,563.6 |
| Webs | 9216.0 | 48.0 | 6.3 | 39.7 | 1,905.1 |
| Bottom flange | 3.0 | 64.5 | 18.1 | 327.6 | 21,130.8 |
| | 9221.3 in.$^4$ | | | | 49,599.5 in.$^4$ |

$$I_T = 49,599.5 + 9221.3$$
$$I_T = 58,820.8 \text{ in.}^4$$

$$S_T = \frac{58,820.8}{31.30} = 1879.3 \text{ in.}^3 \qquad S_B = \frac{58,820.8}{18.45} = 3188.1 \text{ in.}^3$$

*Composite Steel* $n = 30$. The effective flange width, as per AASHTO Section 1.7.98, is

$$12 \times \tfrac{7}{12} = 7 \text{ ft} \qquad be = 2 \times 7 \text{ ft} = 14 \text{ ft}$$

The transformed width of flange is

$$b_t = \frac{14 \times 12}{30} = 5.6 \text{ in.}$$

1. *Centroid Location (reference at CG of slab).*

| Element | A | y | Ay |
|---------|---|---|-----|
| Slab | $5.6 \times 7 = 39.2$ | — | — |
| Steel | 140.5 | 34.80 | 4889.4 |
| | 179.7 in.$^2$ | | 4889.4 in.$^3$ |

From the above data:

$$\bar{y} = \frac{4889.4}{179.7} = 27.2 \text{ in.}$$

2. *Section Modulus.*

| Element | $I_0$ | A | y | $y^2$ | $Ay^2$ |
|---------|-------|---|---|-------|--------|
| Slab | $\frac{1}{12}39.2(7)^2 = 160.0$ | 39.2 | 27.2 | 739.8 | 29,001.7 |
| Steel | 5882.8 | 140.5 | 7.6 | 57.8 | 8,115.3 |
| | 58,980.8 in.$^4$ | | | | 37,117.0 in.$^4$ |

Thus

$$I_T = 58,980.8 + 37,113$$

$$I_T = 96,093.5 \text{ in.}^4$$

$$\text{Steel: } S_T = \frac{96,093.8}{23.7} = 4054.6 \text{ in.}^3$$

$$S_B = \frac{96,093.8}{26.05} = 3688.8 \text{ in.}^3$$

$$\text{Concrete: } S_T = \frac{96,093.8}{30.7} \times 30 = 93,902.7 \text{ in.}^3$$

*Composite* $n = 10$. The transformed width of the concrete flange is

$$b_t = \frac{14 \times 12}{10} = 16.8 \text{ in.}$$

1. *Centroid Location (reference at center of gravity of slab).*

| Element | $A$ | $y$ | $\overline{Ay}$ |
|---|---|---|---|
| Slab | $16.8 \times 7 = 117.6$ | — | — |
| Steel | 140.5 | 34.80 | 4889.4 |
| | 258.1 in.$^2$ | | 4889.4 in.$^3$ |

From these data:

$$\bar{y} = \frac{4889.4}{258.1} = 18.9 \text{ in.}$$

2. *Section Modulus.*

| Element | $I_0$ | $A$ | $y$ | $y^2$ | $Ay^2$ |
|---|---|---|---|---|---|
| Slab | 480.2 | 117.6 | 18.9 | 357.2 | 42,007.9 |
| Steel | 58,820.8 | 140.5 | 15.9 | 252.8 | 35,519.8 |
| | 59,301.0 in.$^4$ | | | | 77,527.7 in.$^4$ |

These data give

$$I_T = 59,301.0 + 77,527.7$$
$$I_T = 136,828.0 \text{ in.}^4$$

For steel:  $S_T = \dfrac{136,828}{15.4} = 8884.9 \text{ in.}^3$   $S_B = \dfrac{136,828}{34.35} = 3983.4 \text{ in.}^3$

For concrete:  $S_T = \dfrac{136,828}{22.4} \times 30 = 183,251.8 \text{ in.}^3$

BENDING STRESSES

For load A:

$$M_D = 1533 \text{ k-ft}$$

For load B:

$$M_{SDL} = 666.0 \text{ k-ft}$$
$$M_{LL+I} = 1472.4 \text{ k-ft}$$

For steel, bottom:

$$f_{DL} = \frac{1533 \times 12}{3188} = 5.77$$

$$f_{SDL} = \frac{666.0 \times 12}{3688.8} = 2.16 \quad (n = 30)$$

$$f_{LL+I} = \frac{1472.4 \times 12}{3983.4} = \frac{4.44}{12.36} \text{ ksi} < 20 \text{ ksi} \quad (n = 10)$$

For steel, top:

$$f_{DL} = \frac{1533 \times 12}{1879} = 9.79$$

$$f_{SDL} = \frac{664 \times 12}{4054.6} = 1.9$$

$$f_{LL+I} = \frac{1472.4 \times 12}{8885} = \underline{1.99}$$

$$13.68 < 20 \text{ ksi}$$

TRANSVERSE DIAPHRAGM REQUIREMENTS

Diaphragm spacing at 16.0 ft or $16/80 = 0.20$ of the span is assumed. The induced normal stress $\sigma_f$ is given below.

For dead load:

$$\frac{\sigma_f}{\sigma_b} = (10L - 350)\frac{s^2}{R}$$

$$\frac{\sigma_f}{\sigma_b} = (10 \times 80 - 350)\frac{(0.2)^2}{500}$$

$$\frac{\sigma_f}{\sigma_b} = 450 \times \frac{(0.2)^2}{500} = 0.036$$

However, $\sigma_{b_{max}} = 7.9$ ksi (assuming the bare-steel section supports *DL* plus *SDL*). Therefore $\sigma_f = 7.9 \times 0.036 = 0.28$ ksi. Hence total dead-load bottom flange stress including distortion is

$$f_{DL} = 7.9 + 0.28$$

$$f_{DL} = 8.18 \text{ ksi}$$

For live load, using equation 8.6b, the distortion stress is

$$\frac{\sigma_f}{\sigma_b} = (-0.25 \times 10^{-2} L + 1.3)S$$

$$\frac{\sigma_f}{\sigma_b} = (-0.25 \times 10^{-2} \times 80 + 1.3)(0.2)$$

$$= (-0.2 + 1.3)0.20$$

$$= 0.22$$

$$\sigma_f = 0.22\sigma_b$$

$$\sigma_f = 4.44 \times 0.22 = 0.98 \text{ ksi}$$

$$f_{LL} = 0.98 + 4.44 = 5.42 \text{ ksi}$$

The total stress is given by

$$f_T = 8.18 + 5.42$$

$$f_T = 13.6 \text{ ksi} < 20 \text{ ksi}$$

Diaphragms are spaced at 16.0 ft internally.

Fatigue is checked according to AASHTO Section 1.7.3. The roadway used here is considered to be heavily traveled and thus truck loading is over $2 \times 10^6$. The box girder plates are fillet welded together and thus fall into category B. The allowable stress range is $F_{sr} = 18$ ksi, where the stress range is defined as the algebraic difference between the maximum line stress and the minimum line stress.

For this structure, $F_{sr} = 4.4$ ksi, which is less than 18 ksi, therefore fatigue does not control.

The compressive stress is checked according to AASHTO Section 1.7.105:

$$\frac{b}{t} \leq \frac{6140}{\sqrt{F_y}} = 28.6 \quad \text{for} \quad F_y = 46 \text{ ksi}$$

$$\left(\frac{b}{t}\right)_{actual} = \frac{14}{1} = 14 < 28.6 \qquad \text{Use 25 ksi}$$

Web-to-depth limitations are checked according to AASHTO Section 1.7.10:

$$\text{composite:} \quad \frac{d'}{L} \geq \frac{1}{25}$$

$$\text{steel:} \quad \frac{d}{L} \geq \frac{1}{30}$$

$$\text{steel:} \quad \frac{d}{L} = \frac{49.75/12}{80} = \frac{4.15}{80} = \frac{1}{19.2} > \frac{1}{30} \quad \text{O.K.}$$

$$\text{composite:} \quad \frac{d'}{L} = \frac{56.75/12}{80} = \frac{4.7}{80} = \frac{1}{17} > \frac{1}{25} \quad \text{O.K.}$$

Web thickness is determined according to AASHTO Section 1.7.70.

$$t \geq \frac{D\sqrt{f_b}}{23,000}$$

where $D$ = web depth = 48 in.
$f_b$ = 13.6 ksi

$$t \geq 48 \times \frac{\sqrt{13,600}}{23,000} = 0.25 \text{ in.}$$

A $\frac{1}{2}$-in. web plate is used, and thus longitudinal stiffeners are not required.

Transverse intermediate stiffeners are covered in AASHTO Section 1.7.71.

The web should be stiffened at intervals not exceeding the girder depth nor

$$d \leq \frac{11,000t}{\sqrt{f_v}}$$

$$R_{max} = R_{DL} + R_{SDL} + R_{LL+I}$$
$$= 76.4 + 33.2 + 63.6$$
$$= 173.2 \text{ k}$$

$$A_{webs} = w \times 48 \times \tfrac{1}{2} = 48 \text{ in.}^2$$

$$f_v = \frac{173.2}{48} = 3.6 \text{ ksi}$$

$$d \leq 11,000 \times \tfrac{1}{2}/\sqrt{3600}$$

$$d \leq 91.7 \quad \text{Use 48-in. spacing}$$

Two stiffeners are spaced at 2.0-ft intervals and the remaining at 4.0-ft intervals. The size of the stiffeners must be in accordance with AASHTO Section 1.7.71:

$$\text{plate width} \geq 2 + \frac{d}{30}$$

$$= 2 + \tfrac{48}{30} = 3.6 \text{ in.} \qquad \text{Use 4-in. plate.}$$

$$\text{plate thickness} \geq \tfrac{1}{16} \text{ plate} = \tfrac{4}{16} = \tfrac{1}{4} \text{ in.} \qquad \text{Use } \tfrac{3}{8}\text{-in.}$$

A 4 in. $\times \frac{3}{8}$ in. plate is tried.

$$I_{req} = \frac{d_0 t^3 J}{10.92}$$

where $J = 25 \dfrac{D^2}{d^2} - 20$

$$J = 25 \left( \frac{48}{48} \right)^2 - 20$$

$$J = 5$$

Then

$$I_{req} = \frac{48(\tfrac{1}{2})^3 \times 5}{10.92} = 2.75 \text{ in.}^4$$

$$I_{act} = \tfrac{1}{3}(\tfrac{3}{8})(4)^3 = 8 \text{ in.}^4 > 2.75 \text{ in.}^4$$

Therefore a 4 in. $\times \frac{3}{8}$ in. stiffener welded to web should be used.
Pairs of 4-in. $\times \frac{3}{8}$-in. bearing stiffeners should be used.

TORSIONAL EFFECTS

The induced shearing stress $f_v$ is the combined effect of bending and torsion and is given by

$$f_v = \frac{VQ}{I_t} + \frac{T}{2A_0 t}$$

where $Q = 117.6 \times 18.9 + 2 \times 1 \times 14 \times 14.9 + 2 \times \tfrac{1}{2} \times 14.4 \times 7.2 = 2743.5 \text{ in.}^3$
$I = 108,359 \text{ in.}^4$
$t = 2 \times \tfrac{1}{2} = 1 \text{ in.}$
$A_0 = 85.5 \times 48.875 = 4178.8 \text{ in.}^2$

At center line:

$$V_{LL+I} = 27.6 \times 1.244 = 34.3 \text{ k}$$
$$T_{LL+I} = 126.9 \text{ k-ft}$$
$$T_{DL} = 173.8$$
$$T_{SDL} = \underline{\phantom{0}75.5} \text{ k-ft}$$
$$T_T = 376.2 \text{ k-ft}$$
$$f_v = \frac{34.3 \times 2743.5}{108,359 \times 1} + \frac{376.2 \times 12}{2 \times 4178.8 \times 1}$$
$$f_v = 0.868 + 0.54$$
$$f_v = 1.408 \text{ ksi} < 3.68 \text{ ksi at bearing,} \qquad \text{does not govern}$$

SHEAR CONNECTORS—AASHTO SECTION 1.7.100

$$S_r = \frac{V_r Q}{I}$$
$$V_r = 63.6 \text{ k}$$
$$Q = 2743.5 \text{ in.}^3$$
$$I = 108,359 \text{ in.}^4$$
$$S_r = \frac{63.6 \times 2743.5}{108,359}$$
$$= 1.61 \text{ k/in.}$$

Studs of $4\frac{3}{4}$-in. $\times$ 4-in. at $2 \times 10^6$ cycles across each flange are tried. The allowable range of horizontal shear $Z_r$, in pounds, of an individual connector is

$$Z_r = \alpha d^2 \qquad d = 0.75 \text{ in.} \qquad \alpha = 7850$$
$$Z_r = 7850 \times (0.75)^2$$
$$Z_r = 4416 \text{ lb per connector}$$

or

$$Z_r = 4 \times 4416$$
$$Z_r = 17.664 \text{ k per flange}$$

$$\text{spacing of connectors} = \left(\frac{17.66}{1.61}\right) 2 = 21.93 \text{ in.}$$

A maximum spacing of 18 in. should be used.

The ultimate strength is checked where the required number of connectors is given by

$$N_1 = \frac{P}{\phi Su}$$

$P = A_s F_y$   or   $0.85 f'_c bc$, whichever is smaller

$P = 140.5 \times 46 = 6463$ k

$P = 0.85 \times 3 \times 7 \times (14 \times 12)/2 = 1499$ k per flange governing

$Su = 0.4 d^2 (f'_c E_c)^{1/2}$

$Su = 0.4(\frac{3}{4})^2 (3000 \times 3 \times 10^6)^{1/2}$

$Su = 0.4 \times \frac{9}{16} (9 \times 10^9)^{1/2}$

$Su = \frac{3.6}{16} \times 9.48 \times 10^4$

$Su = 21,330$ per stud

$Su = 21.3 \times 4 = 85.2$ k per flange

$$N = \frac{P}{\phi Su} = \frac{1499}{0.85} \times 85.2$$

$N = 20.7$ connectors per flange between support
and center of span.

The spacing selected is 1.5 ft or $40/1.5 = 26.7$ rows, which is greater than the 20.7 required.

## REFERENCES

1  J. Buchanan, C. Yoo, and C. P. Heins, "Field Study of a Curved Box Beam Bridge," Civil Engineering Report No. 59, University of Maryland, College Park, Maryland, December 1974.

2  C. Kollbrunner and K. Basler, *Torsion in Structures*, Spring Valley, New York, 1966.

3  C. P. Heins, *Bending and Torsional Design in Structural Members*, D. C. Heath Co., Lexington Books, Lexington, Massachusetts, 1975.

4  R. Dabrowski, *Curved Thin Walled Girders*, Translation No. 144, Cement and Concrete Association, London, 1968.

5  J. Oleinik and C. P. Heins, "Diaphragms for Curved Box Beam Bridges," *J. Struct. Div. Am. Soc. Civil Eng.*, Vol. 101, No. ST 10, October 1975.

6  J. C. Oleinik and C. P. Heins, "Diaphragm Spacing Requirements for Curved Girder Bridges," Civil Engineering Report No. 58, University of Maryland, College Park, Maryland, August 1974.

# CHAPTER 9

# Slant-Legged Rigid-Frame Highway Bridges

## 9.1 GENERAL

In recent years engineers have treated the entire super- and substructures as one unit by constructing a continuous steel rigid frame with supporting legs, as shown in Fig. 9.1. The bridge shown in this photograph is located in Charlottesville, Virginia on I-64 over Route 250 and was recently field tested (2). This type of construction eliminates the need for concrete piers and positions the supports away from the lower roadway, thus giving a safer structure. Because of the interaction of the inclined leg with the main longitudinal girders, the structure becomes indeterminate to the third degree. The response of such a system under a uniform dead load can be predicted easily by slope deflection or moment distribution. However, in the case of live load, which necessitates the development of influence lines, the solution would be quite tedious. A recent study (1) has

**Figure 9.1**  Slant-legged bridge, Charlottesville, Virginia. (Courtesy of W. T. McKeel, Head Structures Research Section, Virginia Highway and Transportation Research Council).

evaluated the response of various inclined-leg bridges, resulting in a series of influence line tables. Some of these results are utilized in the design example.

### 9.1.1  Girder Stiffness

The responses of these bridges are particularly sensitive to the stiffness of the girders (1). Thus it is important to obtain a reasonable value for the girder stiffness ($I$) and area ($A$). A study of typical girder properties, for both bare steel and composite girders, has resulted in the following expressions:

Bare steel: $I_x = \dfrac{(1.458L + 72.7)L^3}{1000}$

$$A = 0.01955 - 0.00028I_x + 0.0000011I_x^2$$

composite: $I_x = -174450. + 9125.3L - 149.7L^2$
$$+ 1.08L^3 - 0.0025L^4$$

where $L$ = end span in feet
$I_x$ = inertia (in.$^4$)
$A$ = area (in.$^2$)

Since the stiffness equations are functions of only the girder span $L$, for a given series of girder dimensions, the inertia can readily be computed. When $I_x$ has been computed, the corresponding area $A$ can then be

**Table 9.1   Influence Line—Bridge Variables**

| $L_1$ (ft) | $L_2$ (ft) | $H$ (ft) | $\phi(0)$ |
|---|---|---|---|
| 60 | 60 | 20 | 45 |
| 60 | 60 | 20 | 60 |
| 60 | 80 | 20 | 45 |
| 60 | 80 | 20 | 60 |
| 80 | 80 | 20 | 45 |
| 80 | 80 | 20 | 60 |
| 60 | 100 | 20 | 45 |
| 60 | 100 | 20 | 60 |
| 80 | 100 | 20 | 45 |
| 80 | 100 | 20 | 60 |
| 100 | 100 | 20 | 45 |
| 100 | 100 | 20 | 60 |
| 60 | 120 | 20 | 45 |
| 60 | 120 | 20 | 60 |
| 80 | 120 | 20 | 45 |
| 80 | 120 | 20 | 60 |
| 100 | 120 | 20 | 45 |
| 100 | 120 | 20 | 60 |
| 120 | 120 | 20 | 45 |
| 120 | 120 | 20 | 60 |
| 60 | 140 | 20 | 45 |
| 60 | 140 | 20 | 60 |
| 80 | 140 | 20 | 45 |
| 80 | 140 | 20 | 60 |
| 100 | 140 | 20 | 45 |
| 100 | 140 | 20 | 60 |
| 120 | 140 | 20 | 45 |
| 120 | 140 | 20 | 60 |
| 140 | 140 | 20 | 45 |
| 140 | 140 | 20 | 60 |

determined. The two properties determined reflect the dead-load response and live-load response. Thus the envelopes of dead- and live-load moments can readily be obtained. Two sets of influence lines for $M$ $V$, and $P$, for a series of spans, have been prepared and can aid in the design of these bridges (1).

In developing the set of influence line tables, as given in reference 1, a range of parameters had to be selected. The end span $L_1$ was set equal to intervals of 60, 80, 100, 120, and 140 ft. The interior span $L_2$ had corresponding lengths of 60 to 140 ft for each interval of $L_1$. The height of the bridge was set at a constant value of 20 ft, with the inclined leg angle $\phi$ set equal to 45 and 60° for all length combinations. Table 9.1 gives all these pertinent dimensions for those bridges studied (1).

In addition to the development of the influence line table, work was initiated to convert the general computer program to develop force envelopes (1), examples of which are given herein.

The influence line ordinate tables are arranged in accordance with a joint designation, as shown in Fig. 9.2. The dimensions shown in this figure pertain Design Example 9.1; however, the notations are constant for all tabulated results. As shown, spans $L_1$ and span $L_2$ are subdivided into 10 divisions. The spacing between each joint $ID$ is then $L_1/10$ or $L_2/10$. For the design example this spacing equals 4.5 and 10 ft, respectively. The influence line tables are generated in the standard format, that is, unit loads are applied at joints 2 through 10, B, 17 through 25, C, and 32 through 40. The resulting forces, at given joint locations are tabulated in Tables 9.2 through 9.9.

## Design Example 9.1

SLANT-LEGGED RIGID FRAME BRIDGE—COMPOSITE

As shown in Fig. 9.3, a five girder rigid-frame composite bridge of span 45, 100, 45 ft is designed here.

*Given*:
1. Three spans of length 45, 100, 45 ft.
2. Slab: 8 in.
3. Spacing between frames: 8 ft 0 in.
4. Clear roadway: 40 ft 0 in.
5. Beam steel: A36 and A441 (girder flanges near junction with leg).
6. Haunch: 1 in. $\times$ 3 in. over beam flanges.
7. Future wearing surface: 25 psf.
8. Concrete $f'_c = 3000$ psi.
9. Loading: HS20-44.

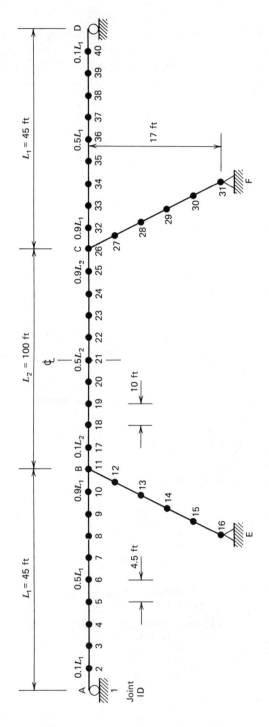

**Figure 9.2** Bridge joint identifications—matrix model of bridge structure.

## Table 9.2 Moment IL Table—Bare Steel

```
BARE STEEL          * MOMENT INFLUENCE LINE *          KIPS-IN./KIPS
                    *************************

        L1= 45.0 FT     L2= 100.0 FT     H= 17.0 FT     ANGLE= 60. DEG.

                              LOAD AT

JOINT
```

Table 9.3 Shear IL Table—Bare Steel

```
***********************************
BARE STEEL   * SHEAR  INFLUENCE  LINE *   KIPS/KIPS
             ***********************************

L1= 45.0 FT   L2= 100.0 FT   H= 17.0 FT   ANGLE= 60. DEG.

                        LOAD AT

JOINT
```

# Table 9.4 Thrust IL Table—Bare Steel

<pre>
*************************************
*     THRUST INFLUENCE LINE     *        KIPS/KIPS
*************************************
</pre>

BARE STEEL

L1= 45.0 FT    L2= 100.0 FT    H= 17.0 FT    ANGLE= 60. DEG.

| JOINT | .1 L1* | .2 L1* | .3 L1* | .4 L1* | .5 L1* | .6 L1* | .7 L1* | .8 L1* | .9 L1* | B | .1 L2* | .2 L2* | .3 L2* | .4 L2* | .5 L2* |
|---|---|---|---|---|---|---|---|---|---|---|---|---|---|---|---|
| BE | -.141 | -.278 | -.410 | -.533 | -.645 | -.742 | -.821 | -.880 | -.916 | -.925 | -.908 | -.877 | -.834 | -.779 | -.711 |
| 12 | -.141 | -.278 | -.410 | -.533 | -.645 | -.742 | -.821 | -.880 | -.916 | -.925 | -.908 | -.877 | -.834 | -.779 | -.711 |
| 13 | -.141 | -.278 | -.410 | -.533 | -.645 | -.742 | -.821 | -.880 | -.916 | -.925 | -.908 | -.877 | -.834 | -.779 | -.711 |
| 14 | -.141 | -.278 | -.410 | -.533 | -.645 | -.742 | -.821 | -.880 | -.916 | -.925 | -.908 | -.877 | -.834 | -.779 | -.711 |
| 15 | -.141 | -.278 | -.410 | -.533 | -.645 | -.742 | -.821 | -.880 | -.916 | -.925 | -.908 | -.877 | -.834 | -.779 | -.711 |
| 16 | -.141 | -.278 | -.410 | -.533 | -.645 | -.742 | -.821 | -.880 | -.916 | -.925 | -.908 | -.877 | -.834 | -.779 | -.711 |

| JOINT | .6 L2* | .7 L2* | .8 L2* | .9 L2* | C | | | | | | | | | |
|---|---|---|---|---|---|---|---|---|---|---|---|---|---|---|
| BE | -.633 | -.545 | -.447 | -.341 | -.227 | -.179 | -.138 | -.105 | -.078 | -.057 | -.040 | -.027 | -.016 | -.008 |
| 12 | -.633 | -.545 | -.447 | -.341 | -.227 | -.179 | -.138 | -.105 | -.078 | -.057 | -.040 | -.027 | -.016 | -.008 |
| 13 | -.633 | -.545 | -.447 | -.341 | -.227 | -.179 | -.138 | -.105 | -.078 | -.057 | -.040 | -.027 | -.016 | -.008 |
| 14 | -.633 | -.545 | -.447 | -.341 | -.227 | -.179 | -.138 | -.105 | -.078 | -.057 | -.040 | -.027 | -.016 | -.008 |
| 15 | -.633 | -.545 | -.447 | -.341 | -.227 | -.179 | -.138 | -.105 | -.078 | -.057 | -.040 | -.027 | -.016 | -.008 |
| 16 | -.633 | -.545 | -.447 | -.341 | -.227 | -.179 | -.138 | -.105 | -.078 | -.057 | -.040 | -.027 | -.016 | -.008 |

287

**Table 9.5  Deflection IL Table—Bare Steel**

```
BARE STEEL

************************************
* DEFLECTION INFLUENCE LINE *        IN./KIPS
************************************

L1= 45.0 FT    L2= 100.0 FT    H= 17.0 FT    ANGLE= 60. DEG.

                          LOAD AT
```

# Table 9.6 Moment IL Table—Composite

COMPOSITE

```
************************************
*   MOMENT  INFLUENCE  LINE   *          KIPS-IN./KIPS
************************************
```

L1= 45.0 FT    L2= 100.0 FT    H= 17.0 FT    ANGLE= 60. DEG.

*** LOAD AT ***

**Table 9.7  Shear IL Table—Composite**

COMPOSITE

```
**********************
*  SHEAR  INFLUENCE  LINE  *
**********************
```

L1= 45.0 FT    L2= 100.0 FT    H= 17.0 FT    ANGLE= 60. DEG.

LOAD AT

KIPS/KIPS

# Table 9.8 Thrust IL Table—Composite

```
                    ****************************
        COMPOSITE   *  THRUST  INFLUENCE  LINE  *        KIPS/KIPS
                    ****************************

        L1= 45.0 FT    L2= 100.0 FT    H= 17.0 FT    ANGLE= 60. DEG.

                              LOAD AT
 JOINT
```

| | .1 L1 | .2 L1 | .3 L1 | .4 L1 | .5 L1 | .6 L1 | .7 L1 | .8 L1 | .9 L1 | R .1 L2 | .2 L2 | .3 L2 | .4 L2 | .5 L2 |
|---|---|---|---|---|---|---|---|---|---|---|---|---|---|---|
| BE | -.110 | -.218 | -.322 | -.421 | -.512 | -.594 | -.665 | -.723 | -.766 | -.792 | -.820 | -.815 | -.786 | -.741 |
| 12 | -.110 | -.218 | -.322 | -.421 | -.512 | -.594 | -.665 | -.723 | -.766 | -.792 | -.820 | -.815 | -.786 | -.741 |
| 13 | -.110 | -.218 | -.322 | -.421 | -.512 | -.594 | -.665 | -.723 | -.766 | -.792 | -.820 | -.815 | -.786 | -.741 |
| 14 | -.110 | -.218 | -.322 | -.421 | -.512 | -.594 | -.665 | -.723 | -.766 | -.792 | -.820 | -.815 | -.786 | -.741 |
| 15 | -.110 | -.218 | -.322 | -.421 | -.512 | -.594 | -.665 | -.723 | -.766 | -.792 | -.820 | -.815 | -.786 | -.741 |
| 16 | -.110 | -.218 | -.322 | -.421 | -.512 | -.594 | -.665 | -.723 | -.766 | -.792 | -.820 | -.815 | -.786 | -.741 |

| | .6 L2 | .7 L2 | .8 L2 | .9 L2 | C | .9 L1 | .8 L1 | .7 L1 | .6 L1 | .5 L1 | .4 L1 | .3 L1 | .2 L1 | .1 L1 |
|---|---|---|---|---|---|---|---|---|---|---|---|---|---|---|
| BL | -.684 | -.614 | -.535 | -.448 | -.356 | -.314 | -.274 | -.236 | -.200 | -.164 | -.130 | -.097 | -.064 | -.032 |
| 12 | -.684 | -.614 | -.535 | -.448 | -.356 | -.314 | -.274 | -.236 | -.200 | -.164 | -.130 | -.097 | -.064 | -.032 |
| 13 | -.684 | -.614 | -.535 | -.448 | -.356 | -.314 | -.274 | -.236 | -.200 | -.164 | -.130 | -.097 | -.064 | -.032 |
| 14 | -.684 | -.614 | -.535 | -.448 | -.356 | -.314 | -.274 | -.236 | -.200 | -.164 | -.130 | -.097 | -.064 | -.032 |
| 15 | -.684 | -.614 | -.535 | -.448 | -.356 | -.314 | -.274 | -.236 | -.200 | -.164 | -.130 | -.097 | -.064 | -.032 |
| 16 | -.684 | -.614 | -.535 | -.448 | -.356 | -.314 | -.274 | -.236 | -.200 | -.164 | -.130 | -.097 | -.064 | -.032 |

# Table 9.9 Deflection IL Table—Composite

COMPOSITE

```
********************************
*  DEFLECTION INFLUENCE LINE  *        IN./KIPS
********************************

L1= 45.0 FT    L2= 100.0 FT    H= 17.0 FT    ANGLE= 60. DEG.

                        LOAD AT
```

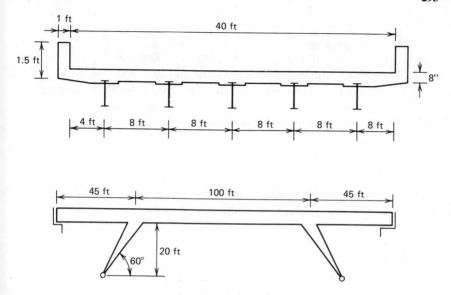

**Figure 9.3** Bridge details.

Dead load A (no composite action)

| | |
|---|---|
| Slab: $\frac{8}{12} \times 8.0 \times 0.150$ | 0.800 |
| Haunch: $2 \times \frac{3}{12} \times \frac{1}{12} \times 0.150$ | 0.006 |
| Steel: (estimate) | 0.238 |
| | 1.044 |

Dead load B (no composite action; equally distributed to all girders)

| | |
|---|---|
| Parapet: $1.0 \times 1.5 \times 0.150 \times \frac{2}{5}$ | 0.090 |
| Railing: $(15 \text{ lb/ft}) \times \frac{2}{5}$ | 0.012 |
| Asphalt paving: $0.25 \times 8.0$ | 0.200 |
| | 0.302 |

In this design example, it is assumed that dead load B is part of dead load A and thus is resisted by the bare-steel girder. This assumption is made because the influence lines have only been developed for the bare steel girder and composite section. After initial design, refinements in the load distribution can be made. The combined dead load is therefore

$$W_T = 1.044 + 0.302 = 1.346 \text{ k/ft per girder}$$

INFLUENCE LINES

For this structure with spans of 45, 100, and 45 ft; clear height $h = 17$ ft; and angle $\phi = 60°$, the influence lines for a steel and for a composite beam have been computed tables 9.2 through 9.4 and 9.6 through 9.8. The input and output verification for this design example is given in reference (1). These influence line ordinates that is, moments, shears, and thrust, have also been plotted, as shown in Fig. 9.4 through 9.8 for the *LL*-composite section only.

DEAD-LOAD MOMENTS

Using Table 9.2 and summing up the positive and negative ordinates gives the following dead-load moments. It should be noted that the spacing between ordinates in span AB and CD is 4.5 ft and in span BC it is 10 ft, as shown in Fig. 9.2. The resulting moments are computed below.
At 0.2 span BA:

$$(315.74 \times 4.5 \text{ ft} + 0.52 \times 10 \text{ ft}) = 1420.7 \text{(k-in./k)-ft}$$

$$(1.8 \times 4.5 \text{ ft} - 23.8 \times 10 \text{ ft}) = 10246.2 \text{ (k-in./k)-ft}$$
$$+ 1174.5 \text{ (k-in./k)-ft}$$

$$M_{DL} = \left(1174.5 \frac{\text{k-in.}}{\text{k}}\text{-ft}\right)\left(1.346 \frac{\text{k}}{\text{ft}}\right)\left(\frac{1}{12}\text{ ft.}\right)$$

$$M_{DL} = +132.3 \text{ k-ft}$$

Summation of Moment Ordinates

| Section | Positive | Negative | Load W | $M_{DL}$(k-ft) |
|---------|----------|----------|--------|----------------|
| 0.2 | 1,420.7 | −246.2 | 1.346 | 132.3 |
| 0.4 | 1,890.1 | −492.7 | 1.346 | 156.8 |
| 0.6 | 1377.6 | −738.7 | 1.346 | 71.7 |
| 0.8 | 29.2 | −1384.7 | 1.346 | −152.1 |
| BA | 51.8 | −3797.3 | 1.346 | −420.1 |
| BE | 2617.5 | −4374.4 | 1.346 | −197.1 |
| BC | 826.6 | −6028.7 | 1.346 | −589.1 |
| 0.1 | 3342.9 | −3495.4 | 1.346 | −17.1 |
| 0.2 | 6491.4 | −2444.0 | 1.346 | 454.0 |
| 0.3 | 9095.5 | −2048.2 | 1.346 | 790.8 |
| 0.4 | 10,785.8 | −1942.7 | 1.346 | 991.9 |
| 0.5 | 11,285.4 | −1838.4 | 1.346 | 1059.6 |

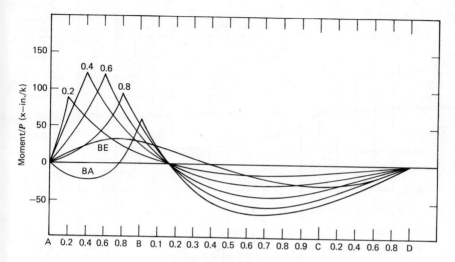

**Figure 9.4** Influence line for moment span AB (0.2, 0.4, 0.6, 0.8) joint *BA* leg, joint BE.

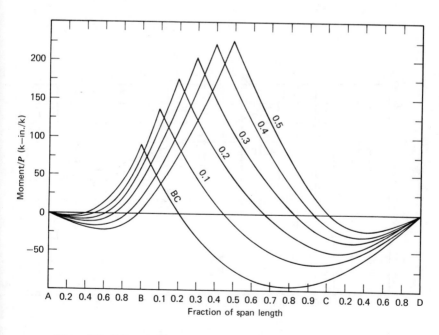

**Figure 9.5** Influence line for moment span BC (0.1, 0.2, 0.3, 0.4, 0.5).

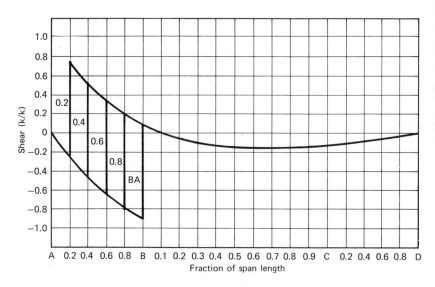

**Figure 9.6** Influence line for shear span AB (0.2, 0.4, 0.6, 0.8), joint BA.

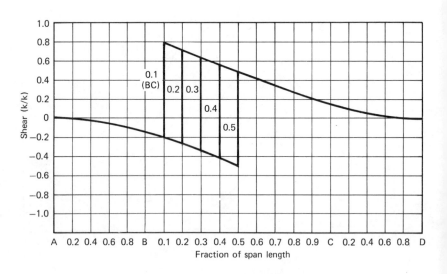

**Figure 9.7** Influence line for shear span BC [(0.1, 0.2, 0.3, 0.4, 0.5).

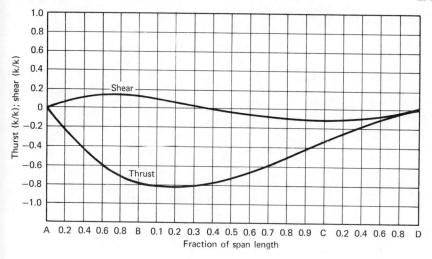

**Figure 9.8** Influence line for shear and thrust leg BE.

LIVE-LOAD MOMENTS

The distribution factor is given by

$$\frac{S}{5.5} = \frac{8}{5.5} = 1.45 \text{ wheels}$$

The impact factor is given by

$$I = \frac{50}{L + 125}$$

For end span $I$ is calculated according to

$$I = \frac{50}{45 + 125} = 0.294$$

For center span $I$ is obtained from

$$I = \frac{50}{100 + 125} = 0.222$$

For leg $I$ is determined from

$$I = \frac{50}{\dfrac{145}{2} + 125} = 0.253$$

For truck loading:

$$\text{Span AB: } 16.0 \times 1.45 \times 1.29 = 29.93 \text{ k}$$
$$\text{Span BC: } 16.0 \times 1.45 \times 1.22 = 28.30 \text{ k}$$
$$\text{Leg BE: } 16.0 \times 1.45 \times 1.25 = 29.07 \text{ k}$$

For lane loading:

$$\text{Span AB: } 0.64 \times 0.72 \times 1.29 = 0.594 \text{ k/ft}$$
$$18 \times 0.72 \times 1.29 = 16.72 \text{ k}$$
$$\text{Span BC: } 0.64 \times 0.72 \times 1.22 = 0.562 \text{ k/ft}$$
$$18 \times 0.72 \times 1.22 = 15.81 \text{ k}$$
$$\text{Leg BE: } 0.64 \times 0.72 \times 1.25 = 0.576 \text{ k/ft}$$
$$18 \times 0.72 \times 1.25 = 16.20 \text{ k}$$

Using Table 9.6 or the resulting plots shown in Fig. 9.4 and 9.5, the maximum live-load moments are computed as shown below.

The truck live-load moments for span AB are:

$$\text{At 0.2: } \left(83.27 + 48.05 + \frac{21.25}{4}\right) \times \frac{29.28}{12} = \;\;333.38 \text{ k-ft}$$

$$\left(-14.61 - 13.96 - \frac{13.51}{4}\right) \times \frac{29.28}{12} = \;-77.95 \text{ k-ft}$$

$$\text{At 0.4: } \left(119.81 + 50.01 + \frac{25.87}{4}\right) \times \frac{29.28}{12} = \;\;452.10 \text{ k-ft}$$

$$\left(-20.21 - 27.92 - \frac{27.05}{4}\right) \times \frac{29.28}{12} = -155.90 \text{ k-ft}$$

$$\text{At 0.6: } \left(117.75 + 49.89 + \frac{47.36}{4}\right) \times \frac{29.28}{12} = \;\;438.18 \text{ k-ft}$$

$$\left(-43.82 - 41.88 - \frac{40.57}{4}\right) \times \frac{29.28}{12} = -233.86 \text{ k-ft}$$

$$\text{At 0.8: } \left(90.66 + 33.40 + \frac{30.03}{4}\right) \times \frac{29.28}{12} = \;\;321.02 \text{ k-ft}$$

$$\left(-58.43 - 55.84 - \frac{54.10}{4}\right) \times \frac{29.28}{12} = -311.82 \text{ k-ft}$$

$$\text{At BA: } \left(57.54 - 4.88 - \frac{11.23}{4}\right) \times \frac{29.28}{12} = \;\;152.31 \text{ k-ft}$$

$$\left(-73.04 - 69.80 - \frac{67.63}{4}\right) \times \frac{29.28}{12} = -389.79 \text{ k-ft}$$

At BE: $\left(33.28 + 26.59 + \dfrac{26.17}{4}\right) \times \dfrac{29.28}{12} = \phantom{-}162.05$ k-ft

$\left(-29.24 - 27.09 - \dfrac{25.45}{4}\right) \times \dfrac{29.28}{12} = -152.97$ k-ft

The truck live-load moments for span BC are:

At BC: $\left(87.60 + 23.93 + \dfrac{21.03}{4}\right) \times \dfrac{28.30}{12} = \phantom{-}275.42$ k-ft

$\left(-95.52 - 91.55 - \dfrac{89.12}{4}\right) \times \dfrac{28.30}{12} = -493.72$ k-ft

At 0.1: $\left(134.61 + 70.22 + \dfrac{48.87}{4}\right) \times \dfrac{28.30}{12} = \phantom{-}511.87$ k-ft

$\left(-67.35 - 64.96 - \dfrac{57.07}{4}\right) \times \dfrac{28.30}{12} = -345.56$ k-ft

At 0.2: $\left(172.97 + 106.46 + \dfrac{83.91}{4}\right) \times \dfrac{28.30}{12} = \phantom{-}708.46$ k-ft

$\left(-51.70 - 44.26 - \dfrac{41.35}{4}\right) \times \dfrac{28.30}{12} = -250.68$ k-ft

At 0.3: $\left(201.33 + 131.71 + \dfrac{117.33}{4}\right) \times \dfrac{28.30}{12} = \phantom{-}854.60$ k-ft

$\left(-39.96 - 33.64 - \dfrac{24.75}{4}\right) \times \dfrac{28.30}{12} = -188.17$ k-ft

At 0.4: $\left(218.73 + 145.40 + \dfrac{137.81}{4}\right) \times \dfrac{28.30}{12} = \phantom{-}939.97$ k-ft

$\left(-31.04 - 24.58 - \dfrac{14.48}{4}\right) \times \dfrac{20.30}{12} = -139.71$ k-ft

At 0.5: $\left(224.59 + 147.36 + \dfrac{147.36}{4}\right) \times \dfrac{28.30}{12} = \phantom{-}964.06$ k-ft

$\left(-24.05 - 16.72 - \dfrac{8.50}{4}\right) \times \dfrac{28.30}{12} = -101.10$ k-ft

The lane live-load moments for span AB are:

At 0.2: $\left(2012.06 \times \dfrac{0.594}{12} + 83.27 \times \dfrac{16.72}{12}\right) = \phantom{-}215.68$ k-ft

$\left(-1080.56 \times \dfrac{0.594}{12} - 14.61 \times 16.12\right) = \phantom{-}-73.83$ k-ft

$$\text{At } 0.4: \left(3052.49 \times \frac{594}{12} + 119.81 \times \frac{16.72}{12}\right) = \quad 318.12 \text{ k-ft}$$

$$\left(-2508.0 \times \frac{594}{12} - 29.21 \times \frac{16.72}{12}\right) = -164.92 \text{ k-ft}$$

$$\text{At } 0.6: \left(3121.0 \times \frac{594}{12} + 117.75 \times \frac{16.72}{12}\right) = \quad 318.64 \text{ k-ft}$$

$$\left(-3763.72 \times \frac{594}{12} - 43.82 \times \frac{16.72}{12}\right) = -247.48 \text{ k-ft}$$

$$\text{At BA: } \left(899.12 \times \frac{594}{12} + 57.54 \times 16.72\right) = \quad 124.70 \text{ k-ft}$$

$$\left(-6830.43 \times \frac{594}{12} - 73.04 \times \frac{16.72}{12}\right) = -440.00 \text{ k-ft}$$

$$\text{At BE: } \left(1698.36 \times \frac{594}{12} + 33.28 \times \frac{16.72}{12}\right) = \quad 130.49 \text{ k-ft}$$

$$\left(-1951.58 \times \frac{594}{12} - 29.24 \times 16.72\right) = -137.41 \text{ k-ft}$$

The lane live-load moments for span BC are:

$$\text{At BC: } \left(1880.67 \times \frac{0.562}{12} + 87.6 \times \frac{15.81}{12}\right) = \quad 203.53 \text{ k-ft}$$

$$\left(-8177.27 \times \frac{0.562}{12} - 95.52 \times \frac{15.81}{12}\right) = -508.95 \text{ k-ft}$$

$$\text{At } 0.1: \left(3900.01 \times \frac{0.562}{12} + 134.61 \times \frac{15.81}{12}\right) = \quad 360.07 \text{ k-ft}$$

$$\left(-4796.72 \times \frac{0.562}{12} - 67.35 \times \frac{15.81}{12}\right) = -313.46 \text{ k-ft}$$

$$\text{At } 0.2: \left(6228.50 \times \frac{0.562}{12} + 172.97 \times \frac{15.81}{12}\right) = \quad 519.70 \text{ k-ft}$$

$$\left(-2925.17 \times \frac{0.562}{12} - 51.76 \times \frac{15.81}{12}\right) = -205.24 \text{ k-ft}$$

$$\text{At } 0.3: \left(8288.88 \times \frac{562}{12} + 201.33 \times \frac{15.81}{12}\right) = \quad 653.59 \text{ k-ft}$$

$$\left(-1985.93 \times \frac{0.562}{12} - 39.96 \times \frac{15.81}{12}\right) = -145.69 \text{ k-ft}$$

At 0.4: $\left(8499.75 \times \dfrac{0.562}{12} + 218.33 \times \dfrac{15.81}{12}\right) = 685.87$ k-ft

$\left(-1562.67 \times \dfrac{0.562}{12} - 31.04 \times \dfrac{15.81}{12}\right) = -114.11$ k-ft

At 0.5: $\left(10089.7 \times \dfrac{0.562}{12} + 224.59 \times \dfrac{15.81}{12}\right) = 768.61$ k-ft

$\left(-308.14 \times \dfrac{0.562}{12} - 24.05 \times \dfrac{15.81}{12}\right) = -96.65$ k-ft

Using these data, the live-load-moment envelope is plotted as shown in Fig. 9.9. This plot agrees with the computer data, as shown in Table 9.10. Also plotted is the shear envelope (Fig. 9.10). Live-load deflection envelope has also been developed, as given in Table 9.11.

SECTION PROPERTIES

Examination of the moment envelope (Fig. 9.9) indicates the following governing values for the designated spans, as shown in Figs. 9.11$a$ and 9.11$b$.

| Span | Section | $M_{DL}$ | $+(M_{LL} + I)$ | $-(M_{LL} + I)$ |
|---|---|---|---|---|
| A–0.5 | A | 150 | 450 | −200 |
| 0.5–0.85 | D | −200 | 300 | −380 |
| B | B | −590 | 270 | −510 |
| Midspan | C | 1050 | 950 | −100 |

Composite action is assumed at sections A and C and thus the estimation of a trial section is needed. As developed by Hacker (3), the estimated bottom flange is

$$A_{sb} = \frac{12}{F_b}\left(\frac{M_{DL}}{d_{cg}} + \frac{M_{SDL} + M_{LL} + M_I}{d_{cg} + t}\right)$$

where $M_{DL}$ = dead-load moment (k-ft)

$\quad M_{SDL}$ = superimposed dead load

$\quad M_{LL}$ = live-load moment (k-ft)

$\quad d_{cg}$ = distance between center of gravity of flanges of steel shape (in.)

$\quad\quad t$ = thickness of concrete slab (in.)

$\quad F_b$ = allowable steel bending stress (ksi)

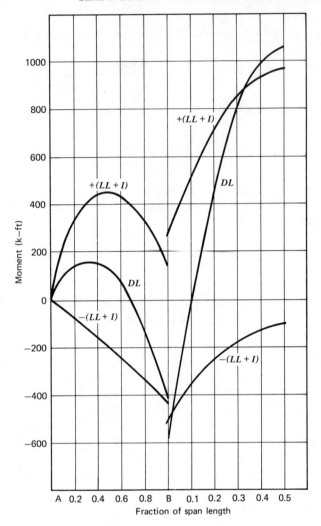

**Figure 9.9** Moment envelope.

At Section A (Fig. 9.12) for a span of $L = 45$ ft, the minimum depth is $d = L/25 = 45/25$ in. $\times 12 = 21.6$ in. $d_{cg} = 36$ in. is assumed. Applying the approximate equation gives

$$A_{sb} = \frac{12}{20} \left[ \frac{150}{36} + \frac{450}{42} \right] = 8.9 \text{ in.}^2$$

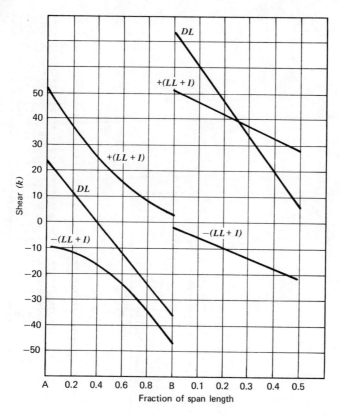

**Figure 9.10** Shear envelope.

A $12$ in. $\times \frac{1}{2}$ in. flange, as shown in Fig. 9.12, is assumed.
For steel:

$$
\begin{aligned}
\tfrac{1}{12}(\tfrac{3}{8}(36)^3 &= 1458.0 \\
2(\tfrac{1}{12})(12)(\tfrac{1}{2})^3 &= \phantom{000}0.25 \\
2(12 \times \tfrac{1}{2})(18.25)^2 &= \underline{3996.75} \\
I_T &= 5455.0 \ \text{in.}^4
\end{aligned}
$$

$$
S_T = S_B = \frac{5455}{18.5} = 294.9 \ \text{in.}^3
$$

**Table 9.10 Force and Deflection Envelopes**

```
                    ***********************
       COMPOSITE    *      ENVELOPE       *
                    ***********************

MOMENT      KIPS-FT    L1= 45.0 FT    L2= 100.0 FT    H= 17.0 FT    ANGLE= 60. DEG.
```

| JOINT | BE | 12 | 13 | 14 | 15 | F |
|---|---|---|---|---|---|---|
| POSITIVE | | .00 | .00 | .00 | .00 | .00 |
| NEGATIVE | | -78.51 | -78.51 | -78.51 | -78.51 | -78.51 |
| DEAD LOAD | | -133.67 | -133.67 | -133.67 | -133.67 | -133.67 |
| T(+)+DL | | -133.67 | -133.67 | -133.67 | -133.67 | -133.67 |
| T(-)+DL | | -212.18 | -212.18 | -212.18 | -212.18 | -212.18 |

MOMENT      KIPS-FT

| JOINT | A | 2 | 3 | 4 | 5 | BC | 7 | 8 | 9 | 10 | RA |
|---|---|---|---|---|---|---|---|---|---|---|---|
| POSITIVE | .00 | 202.57 | 342.93 | 426.94 | 465.04 | 276.82 | 450.70 | 406.86 | 330.22 | 237.02 | 156.64 |
| NEGATIVE | .00 | -41.79 | -83.59 | -126.39 | -167.80 | -515.07 | -250.79 | -292.60 | -334.40 | -378.17 | -445.99 |
| DEAD LOAD | .00 | 80.08 | 475.88 | 158.48 | -126.84 | -622.77 | -71.66 | -122.79 | -122.50 | -260.47 | -425.70 |
| POS.+DL | .00 | 282.65 | 475.84 | 585.43 | 621.84 | -345.95 | 522.36 | 305.07 | 207.71 | -223.45 | -269.05 |
| NEG.+DL | .00 | 58.29 | 49.32 | 33.10 | -11.39 | | -176.13 | -304.39 | -456.91 | -638.64 | -871.68 |

| JOINT | BE | 12 | 13 | 14 | E | BC | 17 | 18 | 19 | 20 | 21 |
|---|---|---|---|---|---|---|---|---|---|---|---|
| POSITIVE | 160.67 | -133.34 | -66.67 | .00 | .00 | 276.82 | -514.50 | -78.87 | -858.96 | 944.78 | 968.99 |
| NEGATIVE | -157.35 | -122.88 | -62.94 | -78.83 | .00 | -515.07 | -347.10 | -242.47 | -789.46 | -140.44 | -101.70 |
| DEAD LOAD | -197.07 | -157.66 | -78.83 | -12.16 | .00 | -622.77 | -453.98 | -749.41 | 992.35 | 1059.64 |
| POS.+DL | -30.40 | -24.32 | -12.16 | | | -345.95 | -497.39 | 1267.85 | 1601.74 | 1937.35 | 1851.30 | 957.94 |

SHEAR   KIPS

| JOINT | A | 2 | 3 | 4 | 5 | 6 | 7 | 8 | 9 | 10 | BA |
|---|---|---|---|---|---|---|---|---|---|---|---|
| POSITIVE | 45.02 | 45.02 | 38.10 | 31.63 | 24.98 | 20.33 | 15.59 | 11.61 | 8.42 | 5.64 | 3.48 |
| NEGATIVE | -10.31 | -10.46 | -13.22 | -16.70 | -20.35 | -24.13 | -29.02 | -34.70 | -41.40 | -46.80 | -51.45 |
| DEAD LOAD | 14.77 | 14.77 | 8.71 | 2.65 | -3.40 | -9.46 | -15.52 | -21.57 | -27.63 | -33.69 | -36.72 |
| POS + DL | 59.78 | 59.78 | 46.81 | 34.28 | 21.58 | 10.87 | 0.07 | -9.96 | -19.21 | -28.04 | -33.23 |
| NEG + DL | 4.46 | 4.31 | -4.51 | -14.04 | -23.75 | -33.59 | -44.54 | -56.28 | -69.03 | -80.48 | -88.17 |

| JOINT | BE | 12 | 14 | E | BC | 17 | 18 | 19 | 20 | 21 |
|---|---|---|---|---|---|---|---|---|---|---|
| POSITIVE | 8.49 | 8.49 | 8.49 | 8.49 | 46.33 | 46.33 | 41.75 | 37.06 | 32.32 | 27.58 |
| NEGATIVE | -8.30 | -8.30 | -8.30 | -8.30 | -6.36 | -9.78 | -13.87 | -18.28 | -22.88 | -27.58 |
| DEAD LOAD | -10.04 | -10.04 | -10.04 | -10.04 | 60.57 | 53.84 | 40.38 | 26.92 | 13.46 | 0.00 |
| POS + DL | -1.55 | -1.55 | -1.55 | -1.55 | 106.90 | 100.17 | 82.13 | 63.98 | 45.78 | 27.58 |
| NEG + DL | -18.34 | -18.34 | -18.34 | -18.34 | 54.21 | 44.05 | 26.50 | 8.63 | -9.42 | -27.58 |

**Table 9.10 Continued.**

```
                 COMPOSITE
                              ***********************************
                              *     DEFLECTION ENVELOPE   *      INCHES
                              ***********************************

                              L1= 45.0 FT    L2= 100.0 FT    H= 17.0 FT    ANGLE= 60. DEG.

**********************************************************************************************************
* JOINT  *     A        2        3        4        5        6        7        8        9       10        H  *
**********************************************************************************************************
* POSITIVE *  .0000    .0681    .1344    .1971    .2544    .3046    .3457    .3760    .3942    .4059    .3978 *
* NEGATIVE *  .0000   -.1090   -.2110   -.3023   -.3891   -.4264   -.4625   -.4723   -.4696   -.4611   -.4485 *
* DEAD LOAD*  .0000   -.3650   -.6630   -.8473   -.8950   -.8078   -.6113   -.3552   -.1137    .0153   -.0007 *
**********************************************************************************************************

* JOINT  *    12       13       14       15        E       17       18       19       20       21  *
**********************************************************************************************************
* POSITIVE *  .3621    .3238    .2336    .1221    .0000    .3374    .2851    .2183    .1500    .0800 *
* NEGATIVE * -.4225   -.3541   -.2547   -.1331    .0000   -.4268   -.4842   -.5519   -.5999   -.6110 *
* DEAD LOAD*  .0072    .0519    .0568    .0350    .0000   -.7432  -1.4054  -1.9613  -2.3280  -2.4658 *
**********************************************************************************************************
```

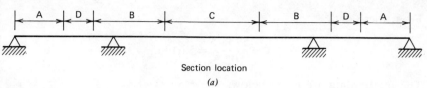

Section location

(a)

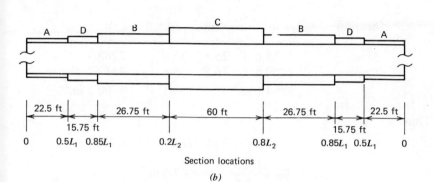

Section locations

(b)

**Figure 9.11**  Section length designation.

For a composite section:

1. *Centroid (reference about centroid of slab).*

| Section | $A$ | $\bar{y}$ | $A\bar{y}$ |
|---|---|---|---|
| Steel | 25.5 | 22.5 | 573.75 |
| Concrete | 76.8 | — | — |
| | 102.3 | | 573.75 |

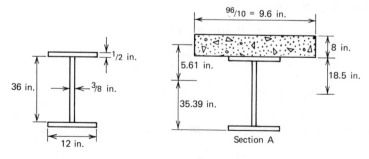

**Figure 9.12**  Section A girder dimensions.

From these data:

$$\bar{y} = \frac{573.75}{102.3} = 5.61 \text{ in.}$$

The inertia data are given below.

| Section | $I_0$ | $A$ | $d$ | $Ad^2$ |
|---------|-------|-----|-----|--------|
| Steel | 5455.0 | 25.5 | 16.9 | 7283.1 |
| Concrete | 409.6 | 76.8 | 5.61 | 2417.1 |
| | 5864.6 in.$^4$ | | | 9700.2 in.$^4$ |

Thus

$$I_T = 5864.6 + 9700.2 = 15564.8 \text{ in.}^4$$

$$\text{steel: } S_T = \frac{15564.8}{1.61} = 9667.6 \text{ in.}^3$$

$$S_B = \frac{15564.8}{35.4} = 439.7 \text{ in.}^3$$

$$\text{concrete: } S_T = \left(\frac{15564.8}{9.61}\right) \Big/ 10 = 161.96 \text{ in.}^3$$

For steel at section D (Fig. 9.13):

$$\tfrac{1}{12}(\tfrac{3}{8})(36)^3 = 1458.0$$
$$2(12 \times \tfrac{5}{8})(18.31)^2 = 5030.2$$
$$\overline{\phantom{2(12 \times \tfrac{5}{8})(18.31)^2 = } 6489.2 \text{ in.}^4}$$

$$S_T = S_B = \frac{6489.2}{18.63} = 348.4 \text{ in.}^3$$

For steel at section B (Fig. 9.14):

$$\tfrac{1}{12}(\tfrac{3}{8})(36)^3 = 1458.0$$
$$2(16 \times 1)(18.5)^2 = 10{,}952.0$$
$$\overline{\phantom{2(16 \times 1)(18.5)^2 = } 12{,}410.0 \text{ in.}^4}$$

$$S_T = S_B = \frac{12{,}410.0}{19} = 653.16 \text{ in.}^3$$

At section C (Fig. 9.15) applying the approximate equation gives

$$A_{sb} = \frac{12}{20}\left[\frac{1050}{36} + \frac{950}{42}\right] = 31 \text{ in.}^2$$

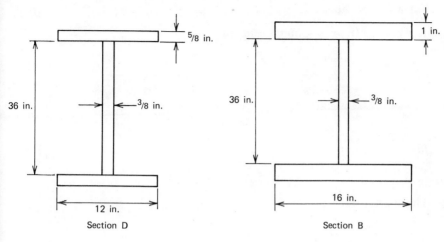

**Figure 9.13**  Section D girder dimensions.   **Figure 9.14**  Section B girder dimensions.

A 16 in. × 2 in. flange, as shown in Fig. 9.15, is assumed. For steel:

$$\frac{1}{12}(\tfrac{3}{8})(36)^3 = \quad 1458.0$$
$$2(\tfrac{1}{12})(16)(2)^3 = \quad\quad 21.3$$
$$2(16 \times 2)(19)^2 = \underline{23{,}104.0}$$
$$24{,}583.3 \text{ in.}^2$$

$$S_T = S_B = \frac{24{,}583.3}{20} = 1229.2 \text{ in.}^3$$

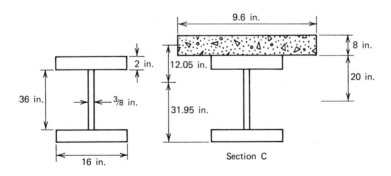

**Figure 9.15**  Section C girder dimensions.

For a composite section:

1. *Centroid (reference about centroid of slab).*

| Section | $A$ | $\bar{y}$ | $A\bar{y}$ |
|---------|-----|-----------|------------|
| Steel | 77.5 | 24 | 1860.0 |
| Concrete | 76.8 | — | — |
| | 154.3 in.$^2$ | | 1860.0 in.$^3$ |

From these data:

$$\bar{y} = \frac{1860}{154.3} = 12.05 \text{ in.}$$

The inertia data are given below.

| Section | $I_0$ | $A$ | $d$ | $Ad^2$ |
|---------|-------|-----|-----|--------|
| Steel | 23,583.3 | 77.5 | 11.95 | 11,067.2 |
| Concrete | 409.6 | 76.8 | 12.05 | 11,151.6 |
| | 24,992.9 in.$^4$ | | | 22,218.8 in.$^4$ |

These data give

$$I_T = 24{,}992.9 + 22{,}218.8 = 47{,}211.7 \text{ in.}^4$$

$$\text{steel: } S_T = \frac{47{,}211.7}{8.05} = 5864.8 \text{ in.}^3$$

$$S_B = \frac{47{,}211.7}{31.95} = 1477.7 \text{ in.}^3$$

$$\text{Concrete: } S_T = \left(\frac{47{,}211.7}{16.05}\right) \Big/ 10 = 294.154 \text{ in.}^3$$

Stresses (ksi) of Bottom Steel

| Section | Dead Load | | Live Load | | Total |
|---------|-----------|---|-----------|---|-------|
| A | $150 \times \dfrac{12}{294.9} =$ | 6.1 | $450 \times \dfrac{12}{439.7} = 12.3$ | | 18.4 |
| D | $200 \times \dfrac{12}{348.4} =$ | 6.9 | $380 \times \dfrac{12}{348.4} = 13.1$ | | 20.0 |
| B | $590 \times \dfrac{12}{653.3} = 10.8$ | | $510 \times \dfrac{12}{653.3} =$ | 9.4 | 20.0 |
| C | $1050 \times \dfrac{12}{1229.2} = 10.3$ | | $950 \times \dfrac{12}{1477.7} =$ | 7.7 | 18.0 |

These resulting stresses indicate a satisfactory section; however, some reduction in the section at locations A and C might be considered.

The remaining design of this bridge involves stiffeners and shear connectors. The other important element to be designed is the leg; since it has combined bending and axial load, the use of the AASHTO column equation is required.

LEG DESIGN

*Forces.* As shown in Figs. 9.4 and 9.8 the inclined support leg of the rigid frame is subjected to thrust, moment, and shears. The induced thrust, using the influence line diagram (Fig. 9.8), is computed as follows:

Dead load $(-7.27 \times 10 \, \text{ft} - 6.14 \times 4.5 \, \text{ft}) \times 1.346 \, \text{k/ft} = -133.67 \, \text{k}$

Live load

Truck: $16 \times 1.45 \times 1.258 = 29.28 \, \text{k}$
Lane: $0.64 \times 0.72 \times 1.258 = 0.586 \, \text{k/ft}$
$26.0 \times 0.72 \times 1.258 = 23.79 \, \text{k}$

Using the influence line diagram gives

Truck: $(-0.827 - 0.809 - 0.803/4)29.28 = 53.78 \, \text{k}$
Lane: $(-100.46 \times 0.586 - 0.827 \times 23.79) = 78.51 \, \text{k}$.

The total maximum thrust is therefore

$$133.67 + 78.51 = 212.1 \, \text{k}$$

The end moment, acting on the leg, as computed previously, and given in Table 9.10 is

$$M_{LL} = -152.97 \, \text{k-ft} \qquad (-157.35 \, \text{k-ft—computer})$$
$$M_{DL} = -197.10 \, \text{k-ft} \qquad (-197.07 \, \text{k-ft—computer})$$

Therefore the combined moment is $M_T = -350.07 \, \text{k-ft}(-354.42 \, \text{k-ft—}$ computer). The resultant shear is found by summing the moment about the support, which gives

$$V = \frac{350.07}{17/\sin 60°} = 17.9 \, \text{k}$$

The combined end actions are shown in Fig. 9.16. These forces are now used to design the leg for the combined action, utilizing the AASHTO equations (4) and the AISC equations (5). The AISC equations have recently been incorporated into the building specifications and have been developed for the design of tapered members. The following are the

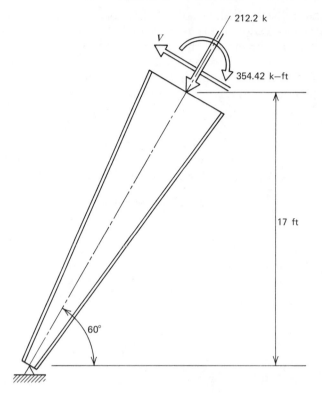

**Figure 9.16**  Leg BE forces.

design specifications as given by the AISC and modified appropriately using the AASHTO criteria.

1. *General Specifications.*  A tapered member must possess at least one axis of symmetry that is perpendicular to the plane of bending if moments are present. The flanges should be of equal and constant area. The depth should vary linearly as

$$d_0\left(1+\gamma\frac{z}{L}\right)$$

where    $L$ = unbraced length
         $d_0$ = depth at smaller end
         $d_L$ = depth at larger end
         $\gamma = (d_L - d_0)/d_0$ and must be less than the smaller of $0.268L/d_0$
            or 6.0
         $z$ = distance from the smaller end of a tapered member

2. *Allowable Compression Stress Specifications.* On the cross section of axially loaded tapered compression members,

$$F_a = \frac{F_y}{2.12}\left[1 - \frac{(KL/r)^2 F_y}{4\pi^2 E}\right] \quad \text{for} \quad \frac{KL}{r} \leq C_c$$

$$F_a = \frac{\pi^2 E}{2.12}\left[\frac{1}{(KL/r)^2}\right] \quad \text{for} \quad \frac{KL}{r} \geq C_c$$

where $C_c = (2\pi^2 E/F_y)^{1/2}$

For A36 steel, the above equations become

$$F_a = 16,980.0 - 0.53(KL/r)^2$$

$$F_a = \frac{135,000,740.0}{(KL/r)^2}$$

$$C_c = 126.1$$

The slenderness ratio $(KL/r)$ may be governed by

$$\frac{KL}{r_{oy}} \quad \text{for weak axis bending}$$

$$\frac{K_\gamma L}{r_{ox}} \quad \text{for strong axis bending}$$

where     $K$ = effective length factor for a prismatic member (0.80)
          $K_\gamma$ = effective length factor for a tapered member as determined by a rational analysis (see Figs. 9.17 through 9.24)
          $r_{ox}$ = strong axis radius of gyration at the smaller end of a tapered member
          $r_{oy}$ = weak axis radius of gyration at the smaller end of a tapered member

3. *Allowable Bending Stress Specifications.* Tension and compression on extreme fibers of tapered flexural members are given by

$$F_b = \frac{2}{3}\left[1.0 - \frac{F_y}{6B\sqrt{F_{s\gamma}^2 + F_{w\gamma}^2}}\right]F_y \leq 0.55\,F_y$$

unless $F_b \leq F_y/3$, in which case,

$$F_b = B\sqrt{F_{s\gamma}^2 + F_{w\gamma}^2}$$

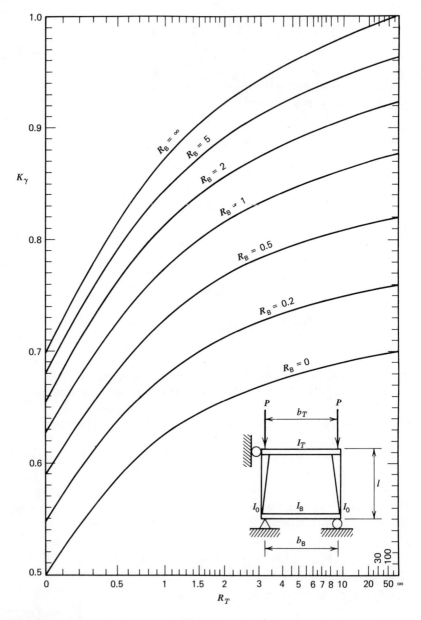

**Figure 9.17** Effective length factors $\gamma = 0.0$. $R_B = b_B I_0 / l I_B$; $R_I = b_r I_0 / l I_T$; $P_{er} = \pi^2 E I_0 / (K_\gamma l)^2$. *Courtesy AISC.*

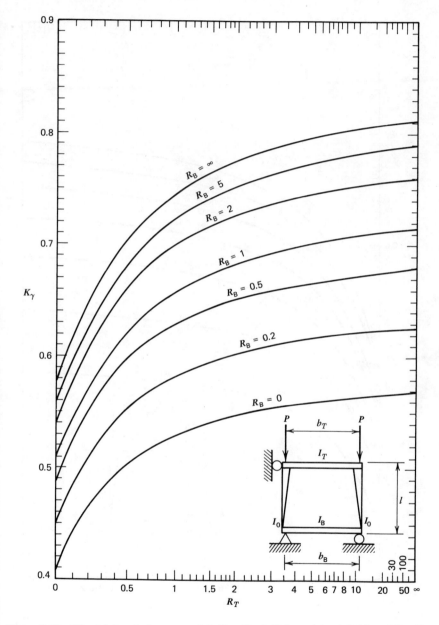

**Figure 9.18** Effective length factors $\gamma = 0.5$. (See Fig. 9.18 legend for definition of $R_B$, $R_I$ and $P_{cr}$.) *Courtesy AISC.*

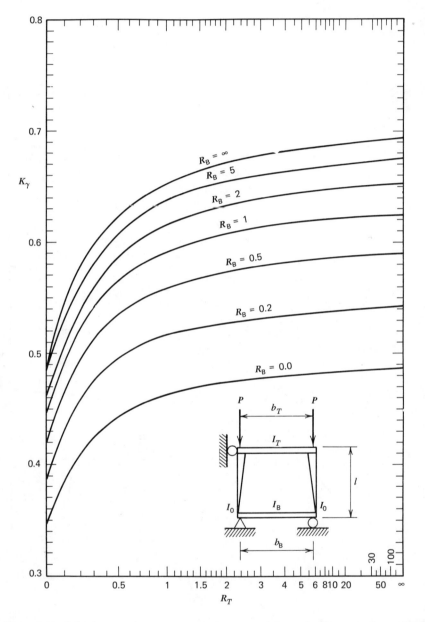

**Figure 9.19** Effective length factors $\gamma = 1.0$. See Fig. 9.18 legend for definition of $R_B$, $R_I$ and $P_{cr}$.

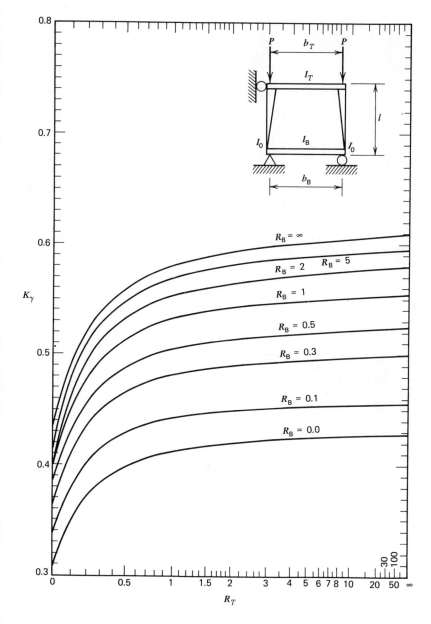

**Figure 9.20** Effective length factors $\gamma = 1.5$. See Fig. 9.18 legend for definition of $R_B$, $R_I$ and $R_{cr}$.

**317**

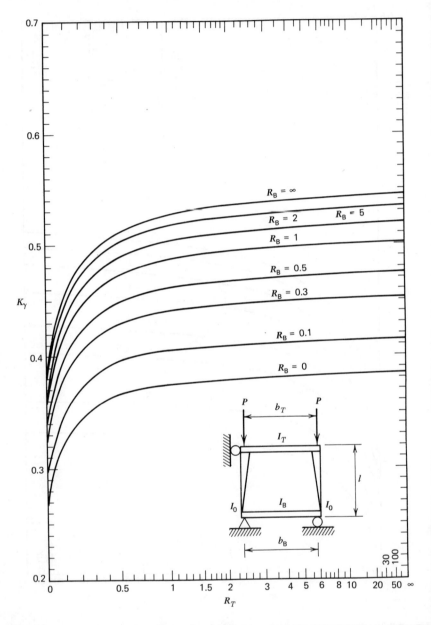

**Figure 9.21** Effective length factors $\gamma = 2.0$. See Fig. 9.18 legend for definition of $R_B$, $R_I$ and $P_{cr}$.

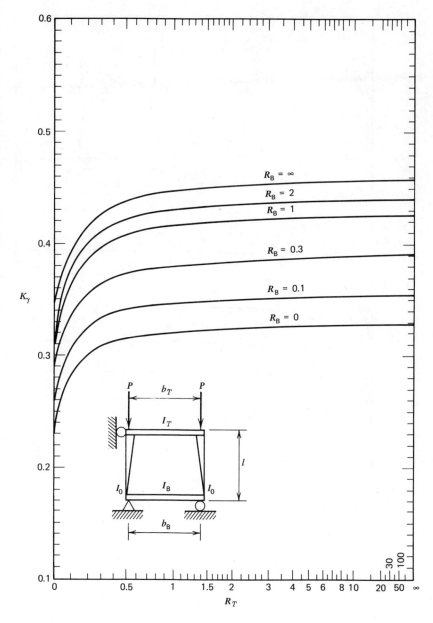

**Figure 9.22** Effective length factors $\gamma = 3.0$. See Fig. 9.18 legend for definition of $R_B$, $R_I$ and $R_{cr}$.

**319**

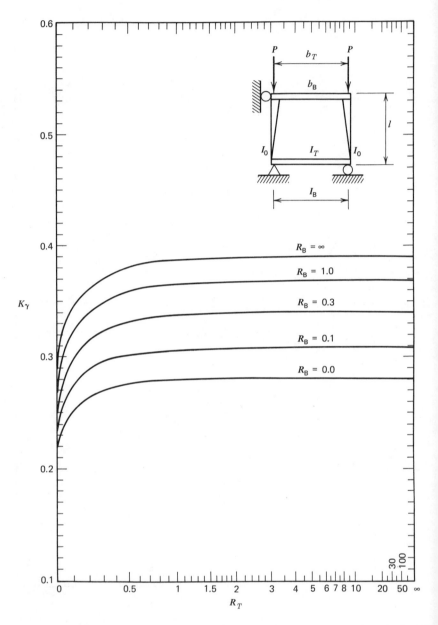

**Figure 9.23** Effective length factors $\gamma = 4.0$. See Fig. 9.18 legend for definition of $R_B$, $R_I$ and $P_{cr}$.

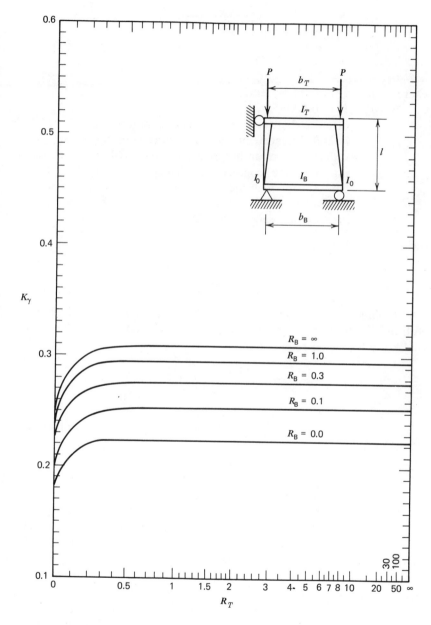

**Figure 9.24** Effective length factors $\gamma = 6.0$. See Fig. 9.18 legend for definition of $R_B$, $R_I$ and $P_{cr}$. *Courtesy AISC.*

In the above equations,

$$F_{s\gamma} = \frac{12 \times 10^3}{h_s L d_0 / A_f}$$

$$F_{w\gamma} = \frac{170 \times 10^3}{(h_w L / r_{T0})^2}$$

where   $h_s = 1.0 + 0.0230_\gamma \sqrt{L d_0 / A_f}$
        $h_w = 1.0 + 0.00385_\gamma \sqrt{L / r_{T0}}$

   $L$ = distance between cross sections braced against twist or lateral displacement of the compression flange (in.)

   $r_{T0}$ = radius of gyration of a section at the smaller end, considering only the compression flange plus one-third of the compression web area, taken about an axis in the plane of the web

   $A_f$ = area of the compression flange

and where $B$ is given by

$$B = \frac{1.75}{1.0 + 0.25\sqrt{\gamma}}$$

4. *Combined Stress Specifications.* Tapered members, subjected to both axial compression and bending stresses should be proportioned to satisfy the following requirements:

$$\frac{f_a}{F_a} + \frac{C_m}{1 - (f_a/F'_e)}\left(\frac{f_b}{F_b}\right) \leq 1.0$$

and

$$\frac{f_a}{0.472 F_y} + \frac{f_b}{F_b} \leq 1.0 \qquad \text{(at points of support)}$$

where   $F_a$ = axial stress that would be permitted if axial force alone existed

   $F_b$ = compressive bending stress that would be permitted if bending moment alone existed

   $F'_e = 135{,}000{,}740.0/(K_\gamma L/r_{b0})^2$ where $L$ is the actual unbraced length in the plane of bending, $r_{b0}$ is the corresponding radius of gyration at its smaller end, and $K_\gamma$ is as determined from Figs. 9.17 through 9.24.

   $f_a$ = computed axial stress at the smaller end of the member or unbraced segment, as applicable

   $f_b$ = computed bending stress at the larger end of the member or unbraced segment, as applicable

Using these specifications, the bridge leg is now designed. The following are assumed:

$$d_0 = 15 \text{ in.} \quad \text{and} \quad \gamma = 2.67$$

Therefore at $z = L$

$$d_z = d_0\left(1 + \gamma\frac{z}{L}\right) = 15(1 + 2.67) = 55 \text{ in.}$$

The corresponding depths between $d_0$ and $d_L$ are as given in Table 9.12. The section properties at pinned end ($E$) and moment end ($B$) are also listed.

*Compressive Stress.* Allowable stress is given by

$$F_a = 16,980.0 - 0.53\left(\frac{KL}{r}\right)^2$$

The slenderness ratio $KL/r$ is computed as

weak axis: $\left(\dfrac{KL}{r_{0y}}\right)$    $K = 0.80$    $L = 235.6$ in.    $r_{0y} = 2.58$ in.

$$\left(\frac{KL}{r_{0y}}\right) = \frac{0.8 \times 235.6}{2.58} = 73.1$$

strong axis: $\left(\dfrac{K_y L}{r_{0x}}\right)$    $L = 235.6$ in.    $r_{0x} = 7.64$ in.

**Table 9.12  Leg Properties**

| Location | E | 0.2 | 0.4 | 0.6 | 0.8 | B |
|---|---|---|---|---|---|---|
| $d$ (in.) | 15.0 | 23.0 | 31.0 | 39.0 | 47.0 | 55.0 |
| $b_f$ (in.) | 10.0 | 10.0 | 10.0 | 10.0 | 10.0 | 10.0 |
| $t_f$ (in.) | 1.5 | 1.5 | 1.5 | 1.5 | 1.5 | 1.5 |
| $t_w$ (in.) | 0.5 | 0.5 | 0.5 | 0.5 | 0.5 | 0.5 |
| $I_x$ (in.$^4$) | 2188.5 | — | — | — | — | 30,874.0 |
| $I_y$ (in.$^4$) | 250.0 | — | — | — | — | 250.0 |
| $S_x$ (in.$^3$) | 265.3 | — | — | — | — | 1064.0 |
| $A$ (in.$^2$) | 37.5 | — | — | — | — | 57.5 |
| $r_{0x}$ (in.) | 7.64 | — | — | — | — | 23.17 |
| $r_{0y}$ (in.) | 2.58 | — | — | — | — | 2.09 |

The effective length factor $K_\gamma$ is computed using Fig. 9.23, $\gamma = 2.67$ (use 3.0); the coefficients $R_B$ and $R_T$ are given by

$$R_B = \frac{b_B I_0}{L I_B} = \frac{1435.6 \times 2188.5}{235.6 \times 0} = \infty$$

$$R_T = \frac{b_T I_0}{L I_T} = \frac{1200 \times 2188.5}{235.6 \times 12{,}410.0} = 0.898$$

where $I_T = 12{,}410.0$ in.$^4$, which represents the midspan girder stiffeners at section C. Using Fig. 9.23, with $R_B = \infty$, $R_T = 0.898$, and $\gamma = 3.0$, gives $K\gamma = 0.45$. The slenderness ratio is computed as

$$\left(\frac{K\gamma L}{r_{0x}}\right) = \frac{0.45 \times 235.6}{7.64} = 13.88$$

Therefore the weak axis governs. Hence the stress $F_a$ is

$$F_a = 16{,}980.0 - 0.53(73.1)^2$$
$$F_a = 14.15 \text{ ksi}$$

The actual compressive stress is given by

$$f_a = \frac{P}{A},$$

where $A = 37.5$ in.$^2$, and the stress is located at the small end. Thus

$$f_a = \frac{212.2}{37.5} = 5.66 \text{ ksi}$$

*Bending Stress.*    Allowable bending stress is

$$F_b = \frac{2}{3}\left[1.0 - \frac{F_y}{6BF_{s\gamma}^2 + F_{w\gamma}^2}\right]F_y \leq 0.55F_y$$

where $F_{s\gamma} = \dfrac{12 \times 10^3}{h_s L d_0 / A_f}$

$\gamma = 2.67$
$L = 235.6$ in.
$A_f = 15$ in.$^2$.
$d_0 = 15$ in.

$$h_s = 1.0 + 0.023\gamma\sqrt{L d_0 / A_f}$$
$$L d_0 / A_f = \sqrt{(235.6 \times 15)/15} = 15.35$$

Then

$$h_s = 1.0 + 0.023(2.67)(15.35) = 1.94$$

$$F_{sy} = \frac{12 \times 10^3}{1.94 \times 235.6} = 26.25 \text{ ksi}$$

The term $F_{wy}$ is computed as

$$F_{wy} = \frac{170 \times 10^3}{(h_w L/r_{T0})^2}$$

where  $L = 235.6$ in.
   $r = 2.67$ in.

$$h_w = 1.0 + 0.00385 \, \gamma \, \sqrt{L/r_{T0}}$$

and $r_{T0}$ is computed from considering the flange plus one third of the web area,

$$I = \tfrac{1}{12}(1.5)(10)^3 = 125 \text{ in.}^2$$

$$A_T = 1.5 \times 10 + \frac{(15 \times 0.5)}{3} = 17.5 \text{ in.}^2$$

$$r_{T0} = \sqrt{\frac{125}{17.5}} = 2.67 \text{ in.}, \qquad \frac{L}{r_{T0}} = \frac{235.6}{2.67} = 88.2$$

The term $h_w$ is therefore

$$h_w = 1.0 + 0.00385(2.67) \sqrt{88.2}$$
$$h_w = 1.097$$

Then

$$F_{wy} = \frac{170 \times 10^3}{(1.097 \times 88.2)^2}$$

$$F_{wy} = 18.2 \text{ ksi}$$

The last term in the $F_b$ equation to be computed is $B$, which is given by

$$B = \frac{1.75}{1.0 + 0.25 \sqrt{\gamma}}$$

$$B = \frac{1.75}{1.0 + 0.25 \sqrt{2.67}} = 1.24$$

The allowable stress $F_b$ is now computed as

$$F_b = \frac{2}{3}\left[1.0 - \frac{F_y}{6B\sqrt{F_{s\gamma}^2 + F_{w\gamma}^2}}\right]F_y \leq 0.55F_y$$

$$F_b = \frac{2}{3}\left[1.0 - \frac{36.0}{6 \times 1.24[26.25^2 + 18.2^2]^{1/2}}\right]36.0$$

$$F_b = 24.0\left(1.0 - \frac{6}{1.24 \times 31.9}\right)$$

$$F_b = 20.4\,\text{ksi} > 20.0\,\text{ksi} \qquad \text{Say OK.}$$

The actual bending stress is given by

$$f_b = \frac{M}{S}$$

where $S = 1064.0$ in.$^3$ and the stress is located at the larger end. Then

$$f_b = \frac{350.07 \times 12}{1064.0} = 4.0\,\text{ksi}$$

*Combined Stresses.* At the support $B$:

$$\frac{f_a}{0.472F_y} + \frac{f_b}{F_b} \leq 1.0$$

$$\frac{5.66}{0.472(36.0)} + \frac{4.0}{20.0} \leq 1.0$$

$$0.33 + 0.25 \leq 1.0$$
$$0.58 \leq 1.0 \qquad \text{Section O.K.}$$

SHEAR CONNECTOR

The shear connector size and spacings are governed by two criteria: (1) fatigue and (2) ultimate strength. The shear connector requirements are chosen for fatigue and checked for ultimate strength. The following discussion details these criteria.

$\phi$ studs of 4 in. $\times \frac{3}{4}$ in. are utilized. The allowable horizontal shear is therefore

$$Z_r = \alpha d^2$$

where

$$\alpha = 10,600.0 \quad \text{for} \quad 500,000 \text{ cycles}$$

Then

$$Z_r = 10,600.0(\tfrac{3}{4})^2 = 5.96 \text{ k per stud}$$

The range of horizontal shear is determined from

$$S_r = \frac{V_r Q}{I}$$

where  $Q$ = statical moment
  $V_r$ = range of $LL$ shear
  $I$ = moment of inertia

From Fig. 9.10, the shear envelope, the value of $V_r$ along the span, is tabulated below

| Joint | Section A | $V_r$(k) | Reference |
|-------|-----------|----------|-----------|
| 2 | $0.1L_1$ | 45.0 | $L_1 = 45.0$ ft |
| 3 | $0.2L_1$ | 38.1 | |
| 4 | $0.3L_1$ | 31.6 | |
| 5 | $0.5L_1$ | 25.0 | |
| 6 | $0.5L_1$ | 20.3 | |

| Joint | Section D | $V_r$(k) |
|-------|-----------|----------|
| 6 | $0.5L_1$ | 20.3 |
| 7 | $0.6L_1$ | 15.6 |
| 8 | $0.7L_1$ | 11.6 |
| 9 | $0.8L_1$ | 8.4 |
| 10 | $0.9L_1$ | 5.6 |

| Joint | Section B | $V_r$(k) | Reference |
|-------|-----------|----------|-----------|
| 10 | $0.9L_1$ | 5.6 | |
| BA | Support B | 3.5 | |
| BC | Support B | 46.3 | |
| 17 | $0.1L_2$ | 46.3 | $L_2 = 100$ ft |
| 18 | $0.2L_2$ | 41.8 | |

| Joint | Section C | $V_r$(k) |
|-------|-----------|----------|
| 18 | $0.2L_2$ | 41.8 |
| 19 | $0.3L_3$ | 37.1 |
| 20 | $0.4L_2$ | 32.3 |
| 21 | $0.5L_2$ | 27.6 |

The section locations are shown in Fig. 9.11 and the properties $I_x$ and $Q$ are given below.

| Section | $I$ (in.$^4$) | $Q$(in.$^3$) |
|---------|--------------|--------------|
| A | 15,564.8 | $(\frac{96}{10} \times 8) \times$ 5.6 |
| D | 17,539.1 | $(\frac{96}{10} \times 8) \times$ 6.1 |
| B | 29,862.8 | $(\frac{96}{10} \times 8) \times$ 9.7 |
| C | 47,211.7 | $(\frac{96}{10} \times 8) \times 12.1$ |

It should be noted that sections D and B are negative-moment regions and thus composite action is generally not assumed. However, the values listed above $(I, Q)$ at these sections are for the composite section. These quantities have been determined to examine the shear-connector spacing requirement for composite action.

Using the tabulated $I$ and $Q$ values and the maximum shear in the section gives the following design requirements.

For section A:

$$S_r = \frac{V_r Q}{I} = \frac{45(9.6 \times 8)5.6}{15564.8} = 1.25 \text{ k/in.}$$

$$\text{pitch} = \frac{Z_r}{S_r} = \frac{5.9}{1.25} = 4.7 \text{ in. per stud}$$

It is assumed there are three studs per section: therefore, $P = 14.1$ in. Pitch values at the remaining locations are:

$$\text{At } 0.2L_1: P = 14.1 \text{ in.}$$
$$\text{At } 0.3L_1: P = 16.5 \text{ in.}$$
$$\text{At } 0.4L_1: P = 20.1 \text{ in.}$$
$$\text{At } 0.5L_1: P = 25.5 \text{ in.}$$

For section D:

$$S_r = \frac{V_r Q}{I} = \frac{20.3(9.6 \times 8)6.1}{17,539.1} = 0.54 \text{ k/in.}$$

$$P = \frac{5.9}{0.54} = 11.04 \text{ in. per stud or } 33.12 \text{ in. for three studs}$$

The pitch values at the remaining sections are:

$$\text{At } 0.6L_1: P = \ \ 43.2 \text{ in.}$$
$$\text{At } 0.7L_1: P = \ \ 58.0 \text{ in.}$$
$$\text{At } 0.8L_1: P = \ \ 80.1 \text{ in.}$$
$$\text{At } 0.9L_1: P = 119.0 \text{ in.}$$

For section C:

$$S_r = \frac{V_r Q}{I} = \frac{41.8(9.6 \times 8)12.1}{47,211.7} = 0.82$$

$$P = \frac{5.9}{0.82} = 7.3 \text{ in. per stud or } 21.9 \text{ in. for three studs}$$

The pitch values at the remaining sections are:

At $0.3L_2$: $P = 24.6$ in.
At $0.4L_2$: $P = 28.2$ in.
At $0.5L_2$: $P = 33.1$ in.

The resulting shear connector spacing, using this information, is as shown in Fig. 9.25.

The ultimate strength conditions are examined for sections A and C and require the following minimum number of connectors.

For section A the material properties are

$$f'_c = 3 \text{ ksi} \qquad E_c = 33w^{3/2}(f'_c)^{1/2}$$

where $w = 150 \text{ lb/ft}^3$. Therefore

$$E_c = 33(150)^{3/2}(3000)^{1/2} = 3.3 \times 10^6 \text{ psi}$$

The ultimate strength of the shear connector is

$$S_u = 0.4d^2(f'_c E_c)^{1/2}$$

$$S_u = 0.4(\tfrac{3}{4})^2(3000 \times 3.3 \times 10^6)^{1/2} = 22.46 \text{ k}$$

The maximum force in the slab is governed by

$$H_1 = A_s F_y$$

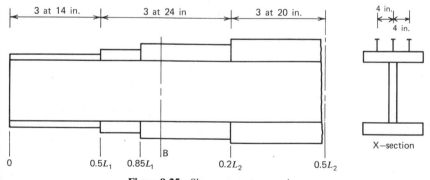

**Figure 9.25**  Shear connector spacing.

or

$$H_2 = 0.85 f'_c bc$$

where $b$ = slab width
$c$ = slab thickness

$$H_1 = A_s F_y = (25.5)36 = 915 \text{ k  Governs}$$

$$H_2 = 0.85 \times 3 \times 96 \times 8 = 1958.4 \text{ k}$$

The number of studs $(N)$ given by

$$N = \frac{H}{\phi S_u} = \frac{915}{0.85 \times 22.40} = 48 \text{ between support and } 0.5L_1$$

For section C:

$$S_u = 22.46 \text{ k}$$

$$H_1 = A_s F_y = 77.5 \times 36 = 2230 \text{ k}$$

$$H_2 = 0.85 f'_c bc = 1958.4 \text{ k}  \qquad \text{Governs}$$

$$N = \frac{H}{\phi S_u} = 1958.4/0.85 \times 22.46 = 103 \text{ between } 0.2L_2 \text{ and } 0.8L_2 \text{ of the midsp}$$

In section A a stud spacing of 14 in. is used which gives

$$N = \frac{0.5(45)12}{14} \times 3 \text{ studs per section}$$

$$N = 58 > 48 \qquad \text{O.K.}$$

In section C a stud spacing of 20 in. is used which gives

$$N = \frac{0.6(100) \times 12}{20} \times 3$$

$$N = 108 > 103 \qquad \text{O.K.}$$

WEB STIFFENERS

*Longitudinal Stiffeners.*  The following equation is used to check web plate thickness:

$$t \geq \frac{D\sqrt{f_b}}{23,000}$$

if longitudinal stiffeners are neglected.

$$t \geq \frac{36\sqrt{20,000}}{23,000} = 0.22 \text{ in.}$$

The actual $t = \frac{3}{8}$ in. $> 0.22$ in. Therefore longitudinal stiffeners are not required.

*Transverse Stiffeners.* Spacing of stiffeners should not be greater than

$$d = \frac{11,000t}{f_v}$$

nor greater than girder depth $D$ which is 36 in. the web shear $f_v$ is computed for all sections, as given in Table 9.13, where:

$$A = \frac{3}{8} \times 36 = 13.5 \text{ in.}^2$$

Knowing the shear $f_v$ and $t = \frac{3}{8}$ in., the required stiffener spacing $d$ is evaluated. As seen by these results the spacings all exceed 36 in., thus 36 in. governs using the above equation, and is also listed in Table 9.13.

**Table 9.13  Web Shear Stresses and Stiffener Spacing**

| Location | $V_{max}$ (k) | $f_v = V/A$ (k) | Stiffener Spacing $d$ (in.) |
|---|---|---|---|
| Support A | 59.8 | 4.4 | 62.0 |
| $0.1L_1$ | 59.8 | 4.4 | 62.0 |
| $0.2L_1$ | 46.8 | 3.5 | 70.1 |
| $0.3L_1$ | 34.3 | 2.5 | 81.9 |
| $0.4L_1$ | 23.8 | 1.8 | 98.4 |
| $0.5L_1$ | 33.6 | 2.5 | 82.7 |
| $0.6L_1$ | 44.5 | 3.3 | 71.8 |
| $0.7L_1$ | 56.3 | 4.2 | 63.9 |
| $0.8L_1$ | 69.0 | 5.1 | 57.7 |
| $0.9L_1$ | 80.5 | 6.0 | 53.4 |
| Section BA | 88.2 | 6.5 | 51.0 |
| Section BC | 106.9 | 7.9 | 46.4 |
| $0.1L_2$ | 100.2 | 7.4 | 47.9 |
| $0.2L_2$ | 82.1 | 6.1 | 52.9 |
| $0.3L_2$ | 63.9 | 4.7 | 59.9 |
| $0.4L_2$ | 45.8 | 3.4 | 70.8 |
| $0.5L_2$ | 27.6 | 2.0 | 91.3 |

Web stiffeners of 5 in. $\times \frac{3}{8}$ in are assumed. The stiffness is

$$I = \tfrac{1}{3}(\tfrac{3}{8})(5)^3 = 15.6 \text{ in.}^4$$

The minimum stiffness is

$$I = \frac{d_0 t^3 J}{10.92}$$

where

$$J = 25 \frac{D^2}{d^2} - 20.$$

For

$$d_0 = 36 \text{ in.} \quad \text{and} \quad d = 36 \text{ in.}, \quad J = 5.$$

Then

$$I = \frac{36(\tfrac{3}{8})^3 5}{10.92} = 0.87 \text{ in.}^4$$

The width of the plate is 5.0 in. and must be greater than

$$5 \text{ in.} \geqq 2 + \frac{D}{30}$$

$$= 2 + \frac{36}{30}$$

$$= 3.2 \qquad \text{O.K.}$$

Also,

$$5 \text{ in.} = \frac{\text{width of girder flange}}{4.0}$$

$$= \frac{12.0}{4.0} = 3 \text{ in.} \qquad \text{O.K.}$$

And the thickness of the plate stiffener is

$$\frac{3}{8} \text{ in.} \geqq \frac{\text{width}}{16} = \frac{5}{16} \text{ in.} \qquad \text{O.K.}$$

Therefore one 5 in. $\times \frac{3}{8}$ in. stiffener spaced at every 36 in.

BEARING STIFFENERS

The bearing stiffeners should be designed as columns. At support A two $4 \text{ in.} \times \frac{3}{8} \text{ in.}$ stiffeners are tried:

$$I = \frac{3}{8} \times \frac{8.275^3}{12} = 18.357 \text{ in.}^4$$

$$A = \frac{3}{8} \times 4 \times 2 + \frac{3}{8} \times \frac{3}{8} \times 18 = 5.53 \text{ in.}^2$$

$$r = \sqrt{\frac{I}{A}} = \sqrt{\frac{18.357}{5.53}} = 1.822 \text{ in.}$$

The allowable compression stress is

$$F_a = 16.98 - 0.00053\left(\frac{36}{1.822}\right)^2 = 16.77 \text{ ksi}$$

The actual stress is

$$f_a = \frac{59.78}{5.53} = 10.81 \text{ ksi}$$

The bearing stress is

$$f_b = \frac{59.78}{2 \times \frac{3}{8} \times (4.0 - 0.75)} = 24.53 < 29.0 \text{ ksi} = F_b = 0.8F_y$$

A web fillet weld of $\frac{3}{4}$ in. is assumed. The minimum thickness is checked:

$$t = \frac{b'}{12}\sqrt{\frac{F_y}{33}} = \frac{4}{12}\sqrt{\frac{36}{33}} = 0.348 \text{ in.} \qquad \text{Use} \quad t = \frac{3}{8}$$

Therefore use two $4 \text{ in.} \times \frac{3}{8} \text{ in.}$ plates.

## REFERENCES

1  C. P. Heins and R. Wang, "Influence Lines for Slant Legged Rigid Frame Highway Bridges," Civil Engineering Report, University of Maryland, College Park, Maryland, June 1976.

2  H. L. Kinnier and F. W. Barton, "A Study of a Rigid Frame Highway Bridge in Virginia," Virginia Highway and Transportation Research Council, Department of Highways, Charlottesville, Virginia, April 1975, VHTRC 75-R47.

3  J. C. Hacker, "A Simplified Design of Composite Bridge Structures," J. Struct. Div. Am. Soc. Civil Eng., Proc. paper 1432, November 1957.

4  AASHTO Interim Specifications Bridges 1974, American Association of State Highway and Transportation Officials, Washington, D. C., 1974.

5  AISC Supplement No. 3, American Institute of Steel Construction, New York, 1974.

**10**

# Substructure Design

## 10.1 BEARINGS

The bearings of a bridge are mechanical devices capable of transmitting vertical loads from the superstructure to the substructure, as well as in some cases transmitting lateral loads between these two segments of a bridge. In some bridges the lateral loads are transmitted by devices other than bearings. With regard to longitudinal action, there are two types of bearings—those classified as fixed and those classified as expansion. The fixed type resist any longitudinal forces and thus allow no longitudinal movement of the superstructure with respect to the substructure. The expansion type are designed to allow movement of the superstructure with the least possible restraining force.

Bearings have been as simple as two steel plates and as complex as precisely machined devices containing numerous parts. For several decades bearings were made exclusively of steel, however, in recent years elastomers such as Neoprene have been used, and for sliding bearings steel plates coated with Teflon are common.

As stated above, bearings must resist forces and they also must permit longitudinal rotations and translation. Any restraint from rotation and translation induces forces and moments in both the superstructure and substructure. The bridge must be designed for any such forces or moments, and minimizing such results in a more economical structure.

**334**

However, the bearings should be designed so that there is sufficient restraint to maintain the stability of the superstructure. Failure of bearings to adequately tie the superstructure to the substructure when subjected to earthquake forces resulted in the collapse of several bridges in southern California in 1971.

In bridges with span lengths of less than 50 ft, provision for deflection is not required. For spans equal to or greater than 50 ft the bearings must have hinges, curved plates, elastomeric pads or pin arrangements for deflection purposes.

There are a number of commercially available bearings for short-span bridges, especially bearings of elastomeric pads and Teflon-coated plates. Catalogues are available to aid the bridge engineer in the proper selection of bearings. Teflon has been incorporated into bridge bearings in recent years. Its primary attraction as a bearing material is its low coefficient of friction, only 0.06 as a maximum value. It is also corrosion resistant and has ample compressive strength. More sophisticated bearings as required for long-span bridges are individually designed. Because long-span bridges have considerable total expansion, rockers or rollers are utilized at the bearings to provide this total amount. When large reactions are produced, sets of rollers are necessary to keep the contact pressure between the rollers and the plates to the allowable values.

The following portion of this chapter is a detailed discussion of the design of typical bridge bearings.

## 10.1.1 Elastomeric Bearing Pads

Such pads are made of synthetic rubber, usually Neoprene. This material has adequate strength to support bridge beams. AASHTO permits a bearing pressure of 500 psi under dead load and 800 psi for dead plus live load, not including impact.

Neoprene is resistant to the elements and, if properly designed, the pad should last for a long time. Manufacturers claim Neoprene bearings will last the life of the bridge.

The elastomeric bearing compresses under application of load. The compressive strain under dead plus live load should not exceed 7% of the thickness. The amount of compressive strain depends on the durometer hardness, the compressive stress, and the shape factor. The shape factor $SF$ is defined as

$$SF = \frac{LW}{2t(L + W)} \tag{10.1}$$

where $L$ is the length of the bearing pad in the direction of the axis of the beam and $W$ is the width. The value of $t$ is the thickness of the plain pad, or when the bearing pad is composed of layers of elastomer laminated with steel plates, $t$ is the thickness of an individual layer. The amount of compressive strain decreases with an increase in shape factor. It is therefore obvious that dividing a bearing pad into two laminates instead of one single thickness decreases the total vertical compressive strain. The greater the hardness, the less the compressive strain.

For plain bearings, the minimum width (perpendicular to beam axis) or length should not be less than five times the thickness, and for laminated bearings, the minimum width should not be less than twice the effective rubber thickness ($ERT$), and the minimum length not less than three times the effective rubber thickness.

The bearing pad acts as an expansion bearing up to a total movement of one-half the effective rubber thickness. However, expansion produces a longitudinal force in the beam as a result of the shearing resistance of the pad. The modulus of elasticity of Neoprene in shear depends on the hardness (durometer) and temperature. Table 10.1 gives the shear modulus values. The shear modulus is the force per square inch of pad (contact area) to deflect the pad a distance equal to its thickness.

The durometer numbers used for Neoprene are 50, 60, and 70. Durometer 70 should not be used in laminated bearings. The compressive strain decreases with increasing durometer number. Figure 10.1 shows curves of compressive strain with increases in pressure and changes in shape factor. Each durometer number has a different set of curves. With long-time loading, the elastomer creeps. The amount of creep depends on the durometer number. In 10 years 50 hardness creeps about 25%, 60 hardness will creep 35%, and 70 hardness will creep 45%. Most of this creep occurs in the first 10 days. The final roadway wearing surface is usually not laid until after the 10-day period when most of the creep has taken place.

**Table 10.1   Modulus of Elasticity in Shear (Typical Neoprene Bearing Compositions)**

| 50 Hardness (psi) | Temp. (°F) | 60 Hardness (psi) | Temp. (°F) | 70 Hardness (psi) | Temp. (°F) |
|---|---|---|---|---|---|
| 110 | 70 | 160 | 70 | 215 | 70 |
| $1.1 \times 110$ | 20 | $1.1 \times 160$ | 20 | $1.1 \times 215$ | 20 |
| $1.25 \times 110$ | 0 | $1.25 \times 160$ | 0 | $1.25 \times 215$ | 0 |
| $1.9 \times 110$ | $-20$ | $1.9 \times 160$ | $-20$ | $1.9 \times 215$ | $-20$ |

LOG-LOG PLOT OF
SHAPE FACTOR VS. % COMPRESSION
50 DUROMETER NEOPRENE

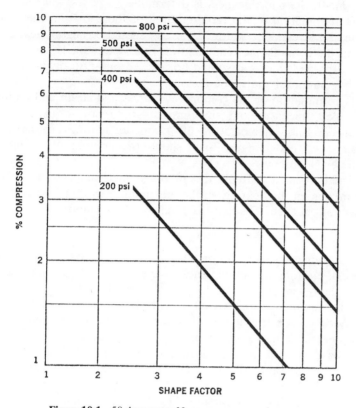

**Figure 10.1** 50 durometer Neoprene compression curve.

A typical elastomeric bearing pad design is shown in Design Example 10.1, which includes:

1. Determination of the dead- and live-load reactions on the beam.
2. Design for the range of temperature of +100 to −20°F. The total horizontal movement of the beam, resulting from expansion and contraction, is determined. The bearing pad must withstand this amount of shear strain or a portion thereof. If the beam is set on the pad at a temperature other than the upper or lower limit, then only that portion of total horizontal movement need be calculated. In Design Example 10.1, it is

considered that the beam is set at 100°F in the middle of a hot summer day. If the approximate date of erection were at a time of lower temperatures, then this could be scaled down. The thickness of the bearing pad should be at least twice the expected horizontal deflection. From this requirement, the pad thickness is determined.

3. Determination of the required area of the pad from the allowable compressive stress. The width of the pad should be slightly less than the width of the beam flange. The length is then determined. To prevent any wobble, the pad thickness should not exceed one-fifth of the length.

4. Calculation of the shape factor.

5. Determination of the compressive stress and strain and adjustment of the pad thickness and size until a strain not to exceed 0.07 is determined. Elastomer of 50 to 60 durometer hardness should be used for laminated pads. Figure 10.1 shows the amount of strain for 50 durometer Neoprene.

6. There is an acceptable design range that matches the dead load (*DL*) in kips on the bearing, to the maximum travel (inches) required by temperature change. This limit is given by the equation

$$DL \gtrless 4.8(\Delta)^2 \tag{10.2}$$

Only for longer spans with greater expansion does equation (10.2) limit the use of elastomeric bearing pads. If the dead-load reaction is less than $4.8(\Delta)^2$ the beam will slip on the pad. A second limit to prevent slippage is for the minimum compressive stress to be greater than 200 psi.

7. Calculation of the longitudinal force due to traffic from AASHTO specifications. To also prevent slippage, the shear force on a pad should not exceed one-fifth the dead-load reaction.

8. Determination of the force on the piers and/or abutments is determined due to the shear resistance of the pad when it is under strain due to thermal expansion or contraction. This force *F* is equal to

$$F = \frac{\text{shear modulus (area) }(\Delta)}{(ERT)}$$

where $\Delta$ is the expected horizontal travel of beam from the position when the beam is set in place. The value of *F* determined in the design example could be reduced if the beam were set in place at a temperature below the upper limit of temperature range. The longitudinal force is applied to the supporting structure at each end of the beam. At piers, these horizontal forces coming from each adjacent span oppose each other and do not stress the pier in bending. However, the abutment is subjected to this force. If this force is too great for a reasonable abutment design, it can be reduced considerably by incorporating Teflon sliding plates into the

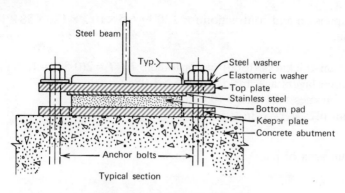

**Figure 10.2** Anchor bolt detail.

bearing. The load plate can be made wider so that anchor bolts can extend from the substructure as shown in Fig. 10.2. Such an arrangement would be necessary in regions of seismic activity. The holes in the load plate should be slotted to allow for any longitudinal adjustment. If the abutment is level and relatively smooth, the keeper plate is not necessary. In situations where it may be necessary to have holes in the elastomeric pad for anchor bolts, and so forth, an adjustment should be made in calculating the shape factor. The shape factor is calculated according to

$$SF = \frac{(L \times W) - N(\pi r^2)}{2[(L + W) + (2\pi r)N]t} \tag{10.3}$$

The terms in equation 10.3 are the same as those in 10.1, with the additions: $r$ = radius of holes and $N$ = number of holes.

Manufacturers of bridge bearings have design material readily available for use by engineers.

### Design Example 10.1

ELASTOMERIC BEARING

Total dead load per beam: 1.17 k/ft

Total beam length: 57 (center-to-center of bearings) + 1 = 58 ft
Dead-load reaction: 1.17 × 29 = 34 k
Live-load reaction: 43 k
No impact is required for elastomeric bearings
Total reaction: 57 k

Total expansion and contraction for 120°F: $6.5 \times 10^{-6} \times 120 \times 58 \times 12 = 0.54$ in.

Minimum thickness (total) of elastomer: $ERT = 2(0.54) = 1.08$ in.
For three layers at $\frac{3}{8}$ in.: $T = 1.125 > 1.08$ in.
Plate thickness: 14 gauge (0.075 in.) or thicker
Bottom ply: $\frac{1}{2}(\frac{3}{8}) = \frac{3}{16}$ in.

Maximum area of pad: $\dfrac{34}{0.2} = 170$ in.$^2$

Minimum area of pad: $\dfrac{34}{0.5} = 68$ in.$^2$. or $\dfrac{57}{0.8} = 72$ in.$^2$.

Flange width of beam: $11\frac{1}{2}$ in.

A thick load plate of approximately $1\frac{1}{4}$ in. is welded to the bottom flange of the beam to distribute the reaction. This plate can be wider than the beam flange. $W = 10$ in. and $L = 7\frac{1}{2}$ in., $t = \frac{3}{8}$ in., are tried.

Shape factor: $SF = \dfrac{7.5(10)}{2(0.375)(7.5 + 10)} = 5.7$

Compressive stress: $57/75 = 0.76$ ksi $< 0.8$.

Vertical strain: $\sim 5\%$ (Fig. 10.1)
Total pad thickness: see Fig. 10.3

Using equation 10.2 gives

$$4.8\,(\Delta)^2 = 4.8(0.54)^2 = 1.4\,\text{k} < DL = 34\,\text{k}$$

For a bridge which has five beams and is carrying two lanes of traffic, the longitudinal force is

$$\frac{0.64(58)(2) + 18(2)}{5}(0.05) = 0.72\,\text{k}$$

$$0.72 < \tfrac{34}{5} = 6.8\,\text{k (no slippage)}$$

$$F = \frac{\text{shear modulus } (L \times W)(\Delta)}{ERT}$$

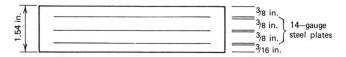

**Figure 10.3**   Elastomeric bearing pad.

Shear modulus: $1.9(110) = 209$ psi

$$F = \frac{209(75)(0.54)}{1.31} = 6460\,\text{lb}$$

Substructure must be designed for a total longitudinal force of
$6.46 + 0.72 = 7.18$ k
Side jacket: $\frac{1}{8}$ on all sides
Bearing dimensions: $10\frac{1}{4}$ in. $\times 7\frac{3}{4}$ in.
Load plate thickness: $1\frac{1}{4}$ in.
Size: $\frac{1}{2}$ in. greater than bearing all around
Minimum load plate size: $11\frac{1}{4}$ in. $\times 8\frac{3}{4}$ in. $\times 1\frac{1}{4}$ in.

## 10.1.2   Sliding Bearings

In the not too distant past short-span beam-bridge designs allowed for
expansion and contraction by the sliding of one steel plate on another.
The rotation of the end of the beam was neglected for very short spans or
the rotation was taken care of by providing for a curved surface on one of
the plates. Such designs did result in a longitudinal force in the substructure due to the friction of steel on steel, and a considerable force
developed when the steel plates rusted.

Such steel on steel sliding bearings are no longer considered a good
design since the advent of the chemical composition Teflon (TFE). This
material has the lowest coefficient of friction of any solid material. Bridge
bearings are commercially available with a sliding surface of Teflon
combined with other materials to give strength reinforcement. The objective of such material formulation is to produce moderate compressive
strength, chemical inertness, and low friction. This material is bonded to
special backing plates of carbon steel, stainless steel, and Neoprene.
When Teflon sliding surfaces are combined with a Neoprene pad, both
sliding and rocking are possible.

The sliding surfaces can consist of Teflon on Teflon or one surface of
Teflon in contact with a surface of steel (preferably stainless). This latter
combination was the matching sliding surface that made possible the

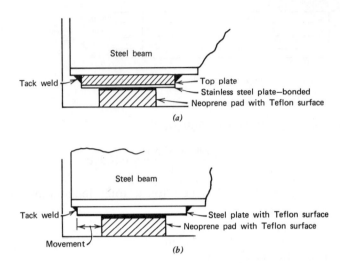

**Figure 10.4** Expansion bearing (a) one thickness of Teflon, (b) two thicknesses of Teflon.

156-ft. lateral movement of the 2000-ft. Oberkassel Bridge in Dussel-dorf, W. Germany, in 1976.

If there is danger of rusting on the sliding stainless steel plate, then both contact surfaces can be Teflon coated. Both systems are shown in Fig. 10.4.

The usual thickness of the Teflon coating bonded to the steel backing plates is $\frac{3}{32}$ in. The static coefficient of friction can be taken as 0.06 for calculation of longitudinal forces on the substructure. The design value of maximum compression on the Teflon varies from 1 to 2 ksi. If a Neoprene pad is used under the Teflon, the dimensions of the Teflon are controlled by the length and width dimensions of the neoprene.

To restrict side motion, anchor bolts can be made to extend through the top plate, in an arrangement similar to that shown in Fig. 10.2. The hole in the top plate should be slotted and the nut only slightly snug.

### 10.1.3  Rolling Expansion Bearings

For long-span steel bridges the loads are too large and longitudinal movements too great for the simple bearings previously described in this chapter. Bearings consisting of steel rollers rolling on heavy steel plates are required. For fixed bearings (no longitudinal movement), a large steel pin bearing is required.

When the total reaction is quite large, a set of segmental rollers may be necessary (Fig. 10.5). Using a single roller results in too large a diameter for the one roller.

A plague of all steel bearings is moisture that causes rusting. To prevent rusting several schemes have been used: (1) coating all steel surfaces with oil; (2) using steels that are more resistant to rusting; (3) enclosing the rollers and contact surfaces in a flexible rubber housing (Fig. 10.6). States that use salt for icy roads have had major problems with rusting of steel bearing parts. It is necessary to design the drainage systems of bridges so that roadway moisture is prevented from coming into contact with the bearings.

When the bearing consists of only one roller, the round surface of the roller takes care of the rocking action, as well as the longitudinal movement. With two or more rollers a separate pin or rounded surface must be provided to permit the rotation of the end of the bridge due to deflection of the structure. A tie bar connects the rollers and one roller is

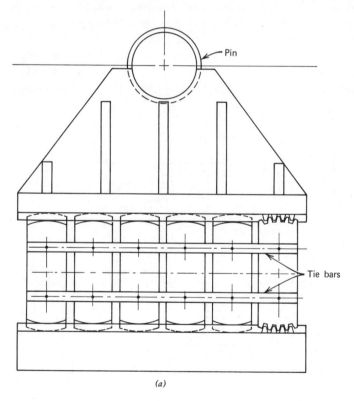

Pin

Tie bars

*(a)*

**Figure 10.5**  Rolling expansion bearings. (*a*) Expansion bearing; (*b*) fixed bearing.

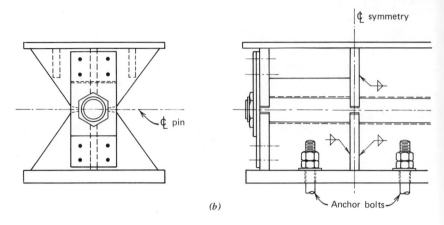

(b)

**Figure 10.5** *Continued*

geared to the top and bottom plate. These two provisions keep the rollers in proper position and orientation (Fig. 10.5). The Specifications give the minimum size of rollers as 4 in. in diameter. They also designate the allowable stresses for the various steels used in steel bridge bearings.

Design Example 10.2 shows the steps in the design of a roller bearing. The bearing consists of four individual parts. The masonary plate rests on

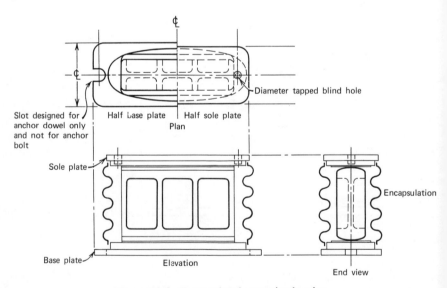

**Figure 10.6** Encapsulated expansion bearing.

the concrete abutment seat or on the pier cap. There is usually an area of concrete ground smooth and level to seat the base on. Segmental rollers permit the longitudinal translation of the bridge superstructure with respect to the substructure. On top of the rollers is the bottom bearing shoe, which transmits the reaction through a pin to the top bearing shoe, which is bolted to the bottom of the bridge superstructure. The forces on the assembly are just the vertical reaction for the expansion bearing.

The fixed bearing consists of the top and bottom bearing shoes and the interconnecting pin.

The first step in the design is to select the material and size of rollers. Other dimensions are most likely controlled by the roller length, thickness, and diameter. The Specifications give an allowable bearing stress on the rollers that is a function of the yield stress of the materials and the diameter of the roller. The number of rollers, length of rollers, and diameter of rollers are selected to meet the value of the reaction divided by the allowable contact bearing stress.

The masonary plate is sized on the basis of allowable bearing stress on the concrete and sufficient stiffness to distribute the reaction force uniformly over that area. This is assured by limiting the bending stress, assuming uniform bearing pressure, to within the allowable. Design Example 10.2 covers the expansion requirement first. The Specifications give an upper limit of $1\frac{1}{4}$ in. per 100 ft of span. Next the rollers are sized, followed by determination of the length–width dimensions. The bending stress is calculated at the contact line of the outside rollers. The cantilever portion is 4 in. Room must be sufficient to allow for the anchor bolts.

The design of the pin is the next step. Since the top and bottom shoe ribs are opposite each other, there is no bending stress, but only bearing stress in the pin. The pin diameter times the total width of ribs times the allowable bearing stress on pins subject to rotation should be greater than the total reaction on the bearing assembly.

The bottom bearing shoe must be of sufficient width and length to provide for full contact with the rollers. The thickness of the plate and depth of bearing shoe are controlled by the bending stress at a section through the centerline of the pin. The height of the shoe should be such that a 45° line from the contact of the center line of the outer rollers with the bottom plate of the shoe touches the horizontal diameter of the pin at the edge of the pin. This assures that the load to the rollers is divided uniformly.

The top bearing shoe usually matches the bottom shoe. Dimensions can vary to match the connection to the superstructure.

Stiffeners can usually be of $\frac{3}{8}$ in. thickness. The shoes can be cast or welded assemblies.

When reactions are sufficiently small to require only one moderate size roller, the bottom shoe can serve as a roller by curving the bottom plate of the bottom shoe. This type of bearing is then classified as a rocker.

If uplift forces are possible through earthquake, and so forth the top and bottom bearing shoes can be connected by bolted side plates.

## Design Example 10.2

ROLLER BEARING

Given:

1. Total reaction: 450 k.
2. Total length of bridge from fixed bearing to expansion bearing: 330 ft.
3. Temperature range: −20 to 100°F.
4. Rollers are set for vertical position at 50°F.
5. Movement: $6.5 \times 10^{-6}(70)(330)(12) = 1.8$ in.
6. For safety, bearings are designed for a movement of ±2 in. $1\frac{1}{4}/100$ ft.
7. Slope of the bridge at the bearing: as much as $\frac{1}{2}°$.

DESIGN OF ROLLER

Using A235 class G material, $F_y = 50,000$ psi. The allowable bearing/per in. is

$$p = \frac{F_y - 13,000}{20,000}(600\ d) = \frac{50 - 13}{20}(600\ d) = 1110(d)\text{lb/in.}$$

For three rollers each 15 in. long:

$$d = \frac{450,000}{1110(15)(3)} = 9 \text{ in.}$$

In order to decrease the length of the bearing, three segments will be used.

The required circumference: 4 in. (total movement ±2 in.)

The angle of rotation: $\dfrac{2}{4.5} = 0.444 \text{ rad} = 25.5°$

The minimum width: $2(4.5)(\sin 25.5°) = 3.87$; use 4 in.

Required area of masonary plate: $450/0.3(3) = 500$ in.$^2$

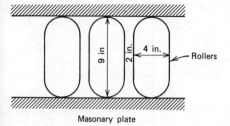

Masonary plate

**Figure 10.7**  Roller bearing.

Minimum size to accommodate rollers: 22 in. × 20 in. = 440 in$^2$ with con-
crete extending a minimum of 6 in. on each side of masonary plate;
  $A_2 = 34(32) = 1088$ in.$^2$
  For $A_1 = 440$ in.$^2$, $(A_2/A_1)^{\frac{1}{2}} = (1088/440)^{\frac{1}{2}} = 1.57 < 2.0$
  Allowable $F_b = 0.3(3)(1.57) = 1.41$ ksi (AASHTO 1.5.26(3)(a))
  Minimum area: $450/1.41 = 319 < 22 \times 20$
  Masonary plate: make 22 in. × 20 in.
  Thickness: determined from bending stresses assuming uniform pres-
sure under plate.
  Bending moment at section 1–1 (Fig. 10.8): $1.02(22)(4)(2) = 180$ in. k
  For A235 class G steel: $F_b = 40$ ksi; $s = M/F_b = 180/40 = 4.50$ in.$^3$; $d =$

$\sqrt{6(4.50)/22} = 1.11$; make $1\frac{1}{4}$ in. thick
  Masonary plate: 22 in. × 20 in. × $1\frac{1}{4}$ in. (Fig. 10.9)
  Four $1\frac{1}{2}$ in. diameter bolts in corners to anchor plate are required as
shown in Fig. 10.9.

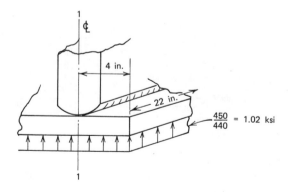

**Figure 10.8**  Bearing pressure.

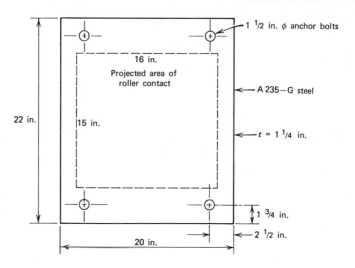

**Figure 10.9**   Masonary plate.

The general arrangement of the roller bearing is shown in Fig. 10.10.   The details of the bearing shoe will now be discussed.

Bearing plate (Fig. 10.10): $1\frac{1}{2}$ in. thick

Allowable bearing on pin: $0.4F_y = 0.4(50,000) = 20,000$ psi

Bearing area required: $\frac{450}{20} = 22.5$ in.$^2$

Diameter of pin: $22.5/3(1.5) = 5$ in.

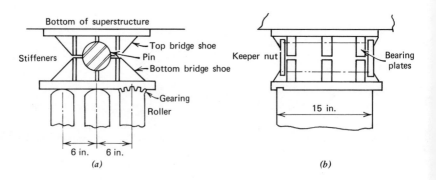

**Figure 10.10**   Roller bearing. (*a*) Side view. (*b*) End view.

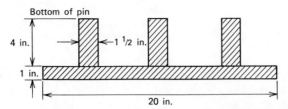

**Figure 10.11**   Cross section in bending—$\bar{\mathbb{C}}$ pin.

To check the strength of the bottom bearing shoe (see Fig. 10.11 for dimensions):

maximum bending at $\bar{\mathbb{C}}$ of pin $= \frac{450}{3}(6) = 900$ in. k

$$I = 84.9 \text{ in.}^4 \qquad C_{max} = 3.316 \text{ in.}$$

$$f_b = 900(3.316)/84.9 = 35.2 < 0.8F_y = 40 \text{ ksi}$$

Stiffener plates should be $\frac{1}{2}$ in. thick and top bearing shoe should match bottom bearing shoe.

## 10.2   PIERS AND ABUTMENTS

The substructure of a bridge consists of the abutments, and where multiple spans are used, piers. In the early age of highway bridges, dressed stone was used in the construction of piers and abutments. Today reinforced concrete is used in almost all cases. Stone is sometimes used for the facing when it is considered it would add to the appearance or when increased resistance to stream flow abrasion is required. The use of dressed stone in most localities is quite expensive. Brick has also been used.

For many years, the substructure was designed without any concern for appearance. Bridge designers are now aware of the effect of the design of the abutments and piers on the appearance of a structure. Concrete can be formed in any shape and thus lends itself to the aesthetic concepts of the bridge engineer. Piers allow more freedom in appearance selection than do abutments. Tall piers have been designed in unusual shapes such as those of the San Diego–Coronado bridge, or in extreme simplicity, such as those of the Widdern Bridge in West Germany (Fig. 10.12). Both of these examples are very attractive. Cost of the substructure is always a consideration in the selection of shape or form.

Steel piers have recently been used on occasion where they were not situated in a body of water.

**Figure 10.12** (a) Widdern Bridge (W. Germany) tall piers. (b) San Diego–Coronado Bridge.

350

Piers and abutments are subjected to many different kinds of loadings, including:

1. Live load from the superstructure.
2. Dead load from the superstructure.
3. Dead load of substructure.
4. Soil pressure.
5. Wind load on both superstructure and substructure.
6. Wind load on the live load.
7. Stream flow pressure.
8. Ice pressure.
9. Earthquake.

Any reasonable combination of the above loads can be applied in the design of the substructure. The AASHTO Specifications gives the various combinations of loads to use in the design. Because the probability of the peak value of the different items of loading occurring simultaneously varies with different load combinations, increases in allowable stresses of different amounts are permissible for the different load combinations. The substructure is normally designed for Groups I, II, and III loadings, as defined in the AASHTO Specifications. Where earthquake is likely, Group VII is also used, and where ice pressure is possible, the design must meet the Group VIII and IX load combination.

The superstructure must be attached to the substructure in such a way that lateral and longitudinal forces, as well as vertical forces, on the superstructure are transmitted to the substructure. This must be done while allowing expansion and contraction motion of the superstructure. This force transmission has to be accomplished by the proper design of the bridge bearings. Bearings are designed such that the lateral forces are transmitted at all the abutments and piers. However, the longitudinal force can only be transmitted at one point if expansion and contraction are not restricted. The most economical point of longitudinal restraint for most bridges is at one of the abutments. By their nature, abutments are more suitable for resisting the longitudinal forces on the superstructure. Longitudinal forces of an appreciable magnitude applied to the top of a pier may increase its size by a considerable amount (1).

The substructure is designed for stability, strength, and limit of soil or rock presure. One load combination may be most critical for stability while another loading may produce maximum stresses in the concrete or reinforcing steel, and a third combination may give the greatest soil pressure. If a portion of the pier is submerged, the buoyancy effect must be considered in determining the stability of the pier.

Because of the many possible load combinations, substructure design is not a simple task. But since the integrity of the superstructure depends on a properly designed substructure, a careful design is mandatory.

The stability, earth pressures, and permissible soil or rock pressures depend on a careful soil analysis. Therefore a proper soil investigation must be made and a laboratory study of sufficient soil samples must be performed. The bridge design engineer must work closely with the geotechnical engineers to achieve the desired results. The geotechnical specialist should be informed of the types of bridges under consideration. The soil condition in the area of the bridge location may prohibit the use of certain types of bridges and indicate the desirability of other types.

If the bridge is to be located in an earthquake region, it may be well to consult a geologist. Locations of fault zones, slide regions, or any possible large unstable ground areas should be known before the bridge location and type are selected.

## 10.3  ABUTMENTS

The bridge abutment supports the end of the bridge superstructure and also retains the earth so that there is a smooth transition from the roadway to the bridge deck. Thus the loads on the abutment come from the bridge superstructure and earth pressure.

The abutment must be designed for several different combinations of loads as given in the AASHTO Specifications. Group Loadings I, II, and III are used for the design of an abutment. It is possible for one part of an abutment to receive maximum stress under one loading while another part is most highly stressed under a different group loading. If the bridge is located in an earthquake zone, then the abutment should be checked for Group VII loading. Bridges should be so located that the abutments are not subjected to floating ice pressure.

An abutment consists of four parts (Fig. 10.13); the bridge seat, retaining backwall, wing wall, and footing or pile foundation. These four parts are all tied together to form an integrated unit. The wing walls may be brought back parallel with the roadway or extended out perpendicular to the roadway. Which method is used depends on the terrain at the location of the abutment, aesthetics, and the type of abutment that is required for the bridge.

Abutments may be classified as retaining type or spill-through type (Fig. 10.14). Which one is used depends on the bridge location. Both types may be supported on spread footings or if the soil conditions near

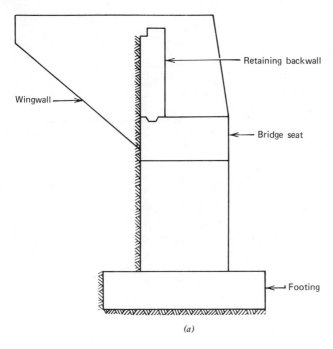

*(a)*

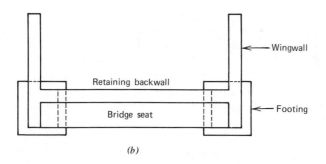

*(b)*

**Figure 10.13** Abutment details. (*a*) Cross section of abutment; (*b*) plan view of abutment.

the surface are not adequate, they can be supported on pile foundations. The spill-through abutment costs less since it involves less volume of concrete. However, if the slope of the soil in front of the abutment is relatively flat, the spill-through abutment may result in a longer span for the bridge. The total cost of the bridge has to be considered when the type of abutment is selected.

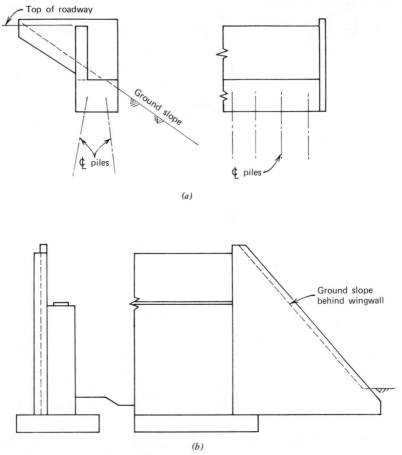

**Figure 10.14** Abutment types: ($a$) spill-through abutment; ($b$) retaining-type abutment.

### 10.3.1 Design of Spill-Through Abutment

After the superstructure of the bridge has been designed, the design of the abutment can be started. It is necessary to have the values of the dead- and live-load bearing reactions. The AASHTO Specification does not require impact loads for the design of abutments. In addition to the dead- and live-load reactions it is also necessary to determine the reactions due to the wind load on the superstructure and wind load on the live load (Group III loading).

The dimensional layout of the abutment is the first step in the design of this part of the bridge. The slope line is extended back from the toe of the slope in front of the abutment. The maximum slope is based on soil stability and erosion. This slope line intersects the top of roadway at some horizontal distance from the toe of the slope. The abutment is then placed so that it fits this slope line. The front edge of the beam seat must be above the slope line and if a side wall covering the area above the beam seat is not used, the back edge of the beam seat also must be above the slope line. The width of the beam seat must be sufficient to accommodate the bearings under each beam, girder, or truss and the edge of the bridge seat should extend at least 3 in. beyond the edge of bearing. The total length, perpendicular to the roadway, of the abutment depends on roadway and sidewalk widths. If the abutment is to be supported by piles, the width and length of bridge seat also have to be sufficient to accommodate the piles. The bridge seat must be sufficiently strong to carry, by beam action, the loads from the bridge superstructure to the supporting structure under the bridge seat.

The retaining back wall keeps the soil under the approach roadway in place. The concrete approach slab is carried over the top of the back wall. The design forces for this back wall are the soil pressures and possibly a surcharge. If no concrete approach slab is used, and there is a slight settlement at the edge of the back wall, a horizontal component of force on the top of the back wall may be produced by the traffic. The amount of this force is quite indeterminate. A concrete approach slab is recommended to eliminate this impact force, as well as to provide a smooth transition from the roadway to the bridge deck. Space between the face of the back wall and the end of the bridge deck is necessary. At the abutment supporting the fixed bearing, this space can be very small (not more than 1 to 2 in.). However, at the abutment supporting the expansion end, the space must be as much as the calculated expansion of the bridge span. This gap has to be bridged by a deck-expansion device of some type. For expansion up to 4 in. rubber devices are used. For expansions greater than 4 in., steel-toothed or other devices are required.

When steel-toothed expansion devices are used, a catch channel must be provided underneath to drain the salt-laden water away from steel members and bearings.

New patented deck-expansion devices are available that permit expansions larger than 4 in. yet give a closed roadway surface so that water cannot get to the bridge bearings. These devices consist of a series of expanding or contracting cells with walls of Neoprene and reinforced with steel sections to carry the wheel loads. Such devices are "off-the-shelf" items and only the forces and bridge movement are required to select the proper item.

The wing-walls are subjected to earth pressures and must have sufficient bending resistance to withstand these earth pressures. If the wing-walls are short, they can be cantilevered out from the abutment. For longer wing-walls, support to the ground is usually provided at the end. Sufficient cover for the reinforcing steel must be provided on the face exposed to the earth. On the other face, temperature steel must be adequate.

## Design Example 10.3

ABUTMENT

A spill-through abutment for the two-span bridge shown in Fig. 10.15 is designed. The detailed calculations for stability only are shown.

First the general dimensions of the abutment are determined. The superstructure consists of a steel beam with a total depth of 51 in. The $1\frac{1}{2}$ in. top flange is buried in the $7\frac{1}{2}$ in. concrete slab. A $1\frac{1}{2}$ in. asphalt wearing surface covers the slab. A traffic barrier weighing 0.39 k/ft (Fig. 10.16) is used.

The cross section of the superstructure is shown in Fig. 10.17a and the beam seat and columns are also shown. The elevation of the abutment is shown in Fig. 10.18. The ground slope is at 2 to 1. The beam seat is set at 6 in. above the ground line. The wing-walls are parallel to the roadway. The cross section of the abutment shows a width of beam seat of 4 ft, which is about the minimum for bearing space as well as for edge distances and clearances. The retaining back wall is selected as 1 ft 3 in. This value should be adequate for structural strength but should be checked.

The top of the wing-wall is selected as the same elevation as the top of the roadway at the edge of the approach slab. The slope of the bridge deck is satisfied by a haunch over the top of beams (not shown on sketch). If the ground slope is 2 to 1, and the rise from the front edge is $6+3+49.5+9=67.5$ in., the ground surface intersects the top of the wing-wall at $2(67.5)=135$ in. $=11.25$ ft from the front of the abutment.

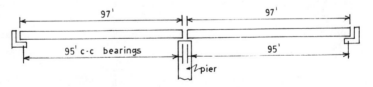

**Figure 10.15**   Two span bridge.

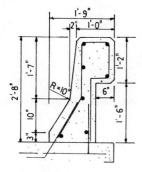

**Figure 10.16** Traffic barrier.

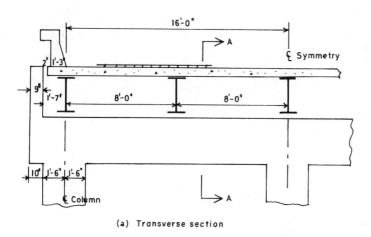

(a) Transverse section

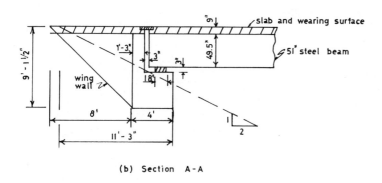

(b) Section A-A

**Figure 10.17** Spill-through abutment: (a) longitudinal elevation; (b) section A-A.

357

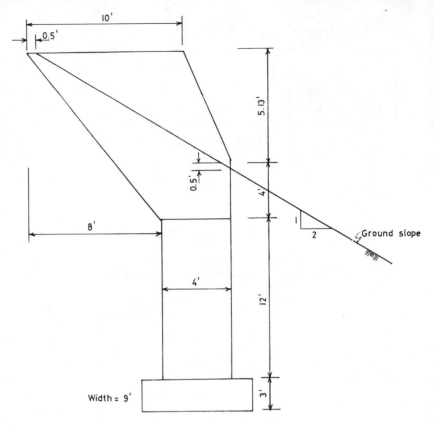

**Figure 10.18** Elevation of abutment.

The wing-wall should extend 6 to 12 in. beyond that point. It is made to extend 12 ft back from the front of the abutment to retain the fill behind the abutment. The front edge of the wing-wall is sloped back 2 ft for appearance only.

Three columns are used to support the beam seat (Fig. 10.17*a*). Three of the bridge beams are supported directly by the columns while the other two beams produce bending in the bridge seat.

The next step in the design after the general dimensions of the abutment have been selected is to determine all the forces that the abutment will be subjected to in service. The dead loads are calculated first.

1. *Dead Loads from Superstructure* (see Fig. 10.19).

Interior beam

asphalt: $\dfrac{1.5}{12} \times 8 \times 0.14$          0.14 k/ft

concrete: $\dfrac{7.5}{12} \times 8 \times 0.15$        0.75

steel beam                             0.21

                                      $\overline{1.10 \text{ k/ft}}$

$$\text{Reaction} = 1.10 \times \frac{97}{2} = 53.4 \text{ k}$$

Outside beam

asphalt: $\dfrac{1.5}{12} \times 4 \times 0.14$          0.07 k/ft

concrete slab: $\dfrac{7.5}{12} \times 5.42 \times 0.15$     0.51

barrier                               0.39

steel beam                            0.21

                                      $\overline{1.18 \text{ k/ft}}$

$$\text{Reaction} = 1.18 \times \frac{97}{2} = 57.2 \text{ k}$$

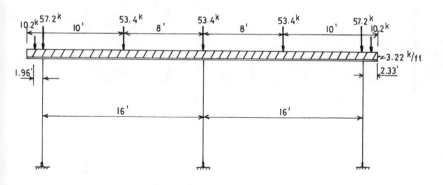

**Figure 10.19** Dead load.

2. *Live Loads from Superstructure* (see Fig. 10.20). The live loads can be shifted transversly across the width of the bridge. For maximum bending in the beam seat the vehicle is centered over beams 2 and 4. The 32 ft width of roadway in the bridge is suitable for high-speed limited-access highways. With this width the design vehicle can be placed in several different positions. The vehicle placement is not critical for stability, however, for stresses in different elements of the abutment it is necessary to place the live load in several possible positions. For maximum live-load reaction on the abutment the rear axle of the vehicle is placed directly over the beam bearing. The value of $P$ is then $16(1+81/95)+4\times67/95=32.5$ k. This value is somewhat conservative, since the wheel loads out on the span will be distributed transversely and will not be concentrated as implied in the calculation of the 32.5 k for $P$. This consideration is, however, not really necessary.

3. *Dead Load of Abutment.*

*Beam seat:* $4\times4\times0.15=2.4$ k/ft
*Back wall:* $1.25\times4.38\times0.15=0.82$ k/ft
*Wing-wall:* $[12\times9.13-(2\times5.13)/2-(8\times9.13)/2]0.15=10.2$ k each
*Column:* $4\times3\times12\times0.15=21.6$ k each
*Footing:* $3\times10\times9\times0.15=40.5$ k each

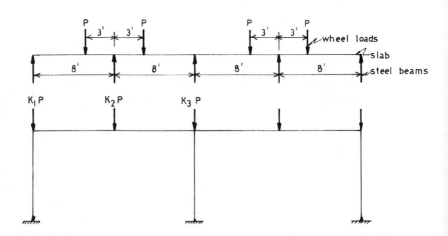

**Figure 10.20** Live load.

4. *Longitudinal Force from Traffic.* Following AASHTO Specifications this is taken as 5% of two lanes of live load. The longitudinal movement of the bridge is restricted at the abutment with expansion at the pier where there are Teflon-coated bearings. The force is given by

$$F_L = 2(0.64 \times 97 + 18)0.05 = 8.0 \text{ k}$$

5. *Longitudinal Force from Friction on Bearings due to Expansion and Contraction.* This force is due only to dead-load reaction. It is given by

$$F_L = \frac{97}{2}(1.10 \times 3 + 1.18 \times 2)0.06 = 6.9 \text{ k}$$

The coefficient of friction of the Teflon sliding bearings is equal to approximately 0.06. Since the longitudinal forces due to friction oppose each other at the piers, these forces are transmitted through the beams to the anchor bolts at the abutment bearings.

6. *Earth Pressures.* For stability calculations the soil pressures on the abutment must be determined. With a spill-through abutment, the soil pressure behind the beam seat and columns is partially balanced by the soil pressure in front of the abutment. This would not be so for a retaining type of abutment. Soil studies indicate that the lateral soil pressure is equal to an equivalent liquid pressure of 35 lb/ft³. Since there is an approach slab, the Specifications say that a surcharge soil pressure need not be considered. The lateral soil pressure diagram on the abutment is shown in Fig. 10.21.

Table 10.2 gives the total dead load plus soil pressure loads acting on the abutment. The total longitudinal and vertical forces as well as the overturning and righting moments are calculated and shown in the table.

The factor of safety against sliding and overturning can now be calculated. The longitudinal force of 76.3 k is resisted by friction of the footing on the soil underneath. The coefficient of friction of concrete on soil depends on the type of soil and ranges from about 0.25 to 0.6, for the design example a value of 0.4 is used. The factor of safety against sliding is then

$$FS = \frac{0.4(594.5)}{76.3} = 3.1$$

The value of 3.1 is conservative, since the longitudinal force due to traffic of 8.0 k was included in the total lateral force but no vertical live load was included. If the 8-k force is deleted, the factor of safety increases to 3.5.

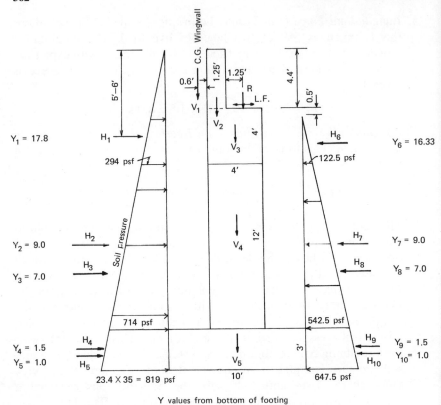

Y values from bottom of footing

**Figure 10.21** Forces on abutment.

There is additional safety in the condition that the footing is completely buried and potential additional restraining soil pressures are present.

The factor of safety against overturning is checked next. This value is

$$FS = \frac{2928}{1149} = 2.5$$

Here again, if the 8-k longitudinal force due to traffic is neglected, the factor of safety is

$$FS = \frac{2928}{997} = 2.9$$

**Table 10.2  Moments about Toe of Footing**

| Type of Force | Magnitude | Magnitude (k) | Arm (ft) | Moment (ft-k) |
|---|---|---|---|---|
| $H_1$: back soil pressure | $\dfrac{0.294}{2}(36.67)(8.4)$ | 45.3 | 17.8 | +806 |
| $H_2$: back soil pressure | $0.294(3\times3)(12)$ | 31.8 | 9.0 | +286 |
| $H_3$: back soil pressure | $(0.714-0.294)(\frac{9}{2})(12)$ | 22.7 | 7.0 | +159 |
| $H_4$: back soil pressure | $0.714(3)(3\times9)$ | 57.8 | 1.5 | + 87 |
| $H_5$: back soil pressure | $(0.819-0.714)(\frac{3}{2})(27)$ | − 4.3 | 1.0 | + 4 |
| $H_6$: front soil pressure | $0.1225\left(\dfrac{3.5}{2}\right)(36.67)$ | − 7.9 | 16.33 | −128 |
| $H_7$: front soil pressure | $0.1225(3\times3)(12)$ | −13.2 | 9.0 | −119 |
| $H_8$: front soil pressure | $(0.5425-0.1225)(\frac{9}{2})(12)$ | −22.7 | 7.0 | −159 |
| $H_9$: front soil pressure | $0.5425(3)(3\times9)$ | −43.9 | 1.5 | − 66 |
| $H_{10}$: front soil pressure | $(0.6475-0.5425)(\frac{3}{2})(27)$ | − 4.2 | 1.0 | − 4 |
| Longitudinal force | $(8.0)+6.9$ | +14.9 | 19.0 | +283 |
| Total longitudinal force | | 76.3 | | |
| Total overturning moment | | | | 1149 |
| $V_1$: weight of wing-walls | $2(10.2)$ | − 20.4 | 7.6 | −155 |
| $V_2$: weight of back wall | $0.82(35.17)$ | − 28.8 | 6.4 | −184 |
| $V_3$: weight of beam seat | $2.4(35.17)$ | − 84.4 | 5.0 | −422 |
| $V_4$: weight of columns | $21.6\times3$ | − 64.8 | 5.0 | −324 |
| $V_5$: weight of footings | $40.5\times3$ | −121.5 | 5.0 | −607 |
| $R$: dead-load reaction | $3\times53.4+2\times57.2$ | −274.6 | 4.5 | −1236 |
| Total vertical force | | −594.5 | | |
| Total righting moment | | | | −2928 |

For the condition of dead load only, a factor of safety of 1.5 is adequate. With live load added, the additional righting moment is

$$32.5\times4\times5=650\,\text{k-ft}$$

and the factor of safety for this condition is

$$FS=\frac{2928+650}{1149}=3.1$$

A value equal to or greater than 2.0 is recommended when live load is included.

There is also a question of lateral stability. However, lateral stability of an abutment is seldom critical, since the lateral loads are usually only wind load (possible earthquake forces) and in a short-span bridge the width of abutment is sufficient so that the factor of safety for lateral overturning is very high.

The longitudinal overturning moment about the center line of the footing is equal to the overturning moment due to the lateral soil pressure and longitudinal force minus the moment caused by the weight of the wing-wall and back wall plus the added moment of the beam reactions. This resultant moment is then

$$M_L = 1149 - 20.4(2.6) - 28.8(1.375) + (57.2 + 130)(0.5) = 1150 \text{ k-ft}$$

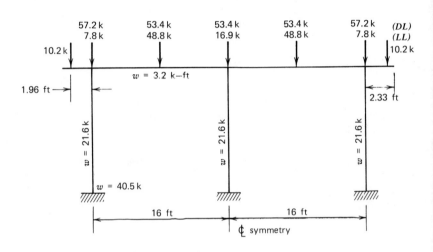

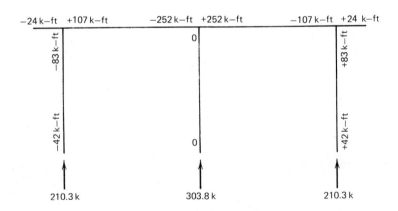

**Figure 10.22** Frame analysis.

Before actual overturning this moment can be equally divided among the three footings:

$$\frac{M_L}{\text{footing}} = \frac{1150}{3} = 383 \text{ ft-k}$$

Referring to Fig. 10.20 the beam reactions are calculated, for a $P$ value at 32.5 k, treating the slab as a continuous beam on rigid supports. The values of $k_1$, $k_2$, and $k_3$ can be determined from any method of structural analysis of statically indeterminate structures:

$$k_1 = 0.24 \qquad k_1 P = \phantom{0}7.8 \text{ k}$$
$$k_2 = 1.50 \qquad k_2 P = 48.8 \text{ k}$$
$$k_3 = 0.52 \qquad k_3 P = 16.9 \text{ k}$$

To determine the vertical loads on the footings, a frame analysis is necessary. The loading and results of the analysis are shown in Fig. 10.22. The loading is for full dead plus two lanes of traffic for live load. The live load is symmetrically placed to produce maximum reaction on the beams midway between columns.

This symmetrical loading produces no lateral moment on the middle column, but a vertical load of 303.8 k. The outer columns have a moment of 42 ft-k, but a lesser vertical force of 210.3 k. The soil pressure under the footings can now be calculated.

For the outside column:

$$p = \frac{P}{A} \pm \frac{M_L}{S_L} \pm \frac{M_T}{S_T} \qquad \text{(subscripts denote longitudinal and transverse directions)}$$

$$A = 9 \times 10 = 90 \text{ ft}^2$$

$$S_L = \frac{9(10)^2}{6} = 150 \text{ ft}^3$$

$$S_T = \frac{10(9)^2}{6} = 135 \text{ ft}^3$$

$$p = \frac{210}{90} \pm \frac{383}{150} \pm \frac{42}{135}$$

$$= 2.33 \pm 2.55 \pm 0.31 = +5.19 \text{ or } -0.53 \text{ psf}$$

For the inside column:

$$p = \frac{303.8}{90} \pm \frac{383}{150} = 3.38 \pm 2.55 = +5.93 \text{ psf}$$

The calculations show that there is an uplift pressure on the footing of the outside column. Therefore the calculation for maximum soil pressure may not be considered correct since there cannot be any tension developed between the bottom of the footing and the soil. However, since there is soil above the rigid footing the weight of this soil acts against this negative pressure of 0.53 psf. It would take only a small height of overburden to equal a weight of 0.53 psf.

The maximum pressure of 5.93 psf is well below the allowable value for most soils. Except for soft soils, the size of the footing can probably be reduced. If the footing dimensions are reduced, a recheck of the stability is necessary.

Some engineers might feel that the inner column should take more of the longitudinal moment than the outside columns. If a conservative assumption were made that the inside column resisted twice the overturning moment that an outside column did, then the maximum soil pressure would be

$$p = \frac{303.8}{90} \pm \frac{575}{150} = 3.38 \pm 3.83 = +7.21 \quad \text{or} \quad -0.45 \, \text{psf}$$

Live load in one lane only would produce a greater lateral bending moment at the bottom of the outside column (by about 50% more). There would be a moment at the bottom of the inside column of approximately 20 ft-k. The total vertical load on the outside column would be increased slightly, but this value for the inside column would be reduced. The resultant soil pressures would be changed slightly, but the maximum value would be that pressure under the inside column footing when both lanes were loaded. Other possible load conditions should be investigated. In fact, all possible loading conditions should be investigated for minimum stability, maximum concrete and steel stresses, and maximum soil pressures.

The final steps in the abutment design are checks of size of beam seat, back wall, columns, and footing, and the amount of reinforcing steel. The procedure would follow standard analysis of reinforced concrete sections. Either an allowable stress analysis or an ultimate strength approach is permitted by AASHTO. If the ultimate strength method is used, then the load factors must be used to determine axial forces and bending moments.

Such stress calculations are shown in the following section on pier design.

## 10.4 PIERS

Piers are subjected to various combinations of forces (3.11). All possible loading combinations and conditions should be considered. Some loading combinations are most critical for stability and others for maximum stresses.

The stability of the pier should be checked for the condition without superstructure, as well as with superstructure. The stability of the pier with wind on the front face is critical without the superstructure.

The first step in the design is to select the types of piers that are most likely the best from a standpoint of economics and appearance. Soil conditions are a major factor in the selection of type of pier. It is necessary to know whether spread footings or piles are required to carry the weight of the pier. If piles are necessary then the type of pile, and thus the bearing capacity of the pile or pile group, must be determined. This may involve test piles being driven and load tested. Therefore very early in the design process it is necessary to perform a soil investigation. Soil borings are required near the beginning of the project to determine the optimum location of the piers and maybe even the feasibility of piers. Span lengths, and thus the type of bridge selected, is dependent on pier locations. A more detailed discussion of piles and pile foundations is found later in this chapter.

There are a wide variety of types of piers that have been used in bridge construction. Figure 10.23 shows four basic types of piers. Type (a) is a gravity pier. This pier resists all forces to which it is subjected by the mass of the pier. This pier is simple in design and construction and is generally used for relatively low heights. The percent of reinforcing steel is low. The reinforcing consists of main vertical bars around the periphery of the concrete column. Horizontal steel of a quantity to meet shrinkage and temperature requirements and to tie the vertical steel together is also required. If the pier is subjected to stream flow the ends are usually rounded or pointed. If the pier is not subjected to flow of water, the ends can be flat, or rounded if the latter is more pleasing in appearance.

The gravity pier can sit on a spread footing or on a pile cap if piles are used. Formwork is simple and economical and erection of the gravity pier is relatively rapid. Although it is a good design for low heights—up to about 25 feet—it has been used at much higher elevations. When the height required becomes greater, it may be better to use other types of piers or to make the rectangular gravity pier with a hollow center. The Winnigen bridge over the Moselle River in West Germany is an example of a bridge with very high rectangular hollow piers. The tallest pier is 124.3 m above the pile cap. These piers have three cells with overall

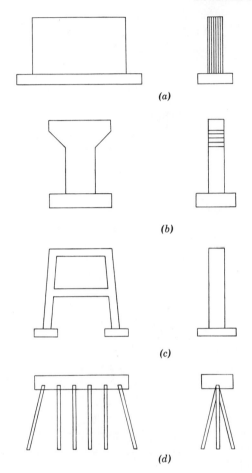

**Figure 10.23** Common types of piers: (*a*) gravity pier, (*b*) hammer head pier, (*c*) frame pier, (*d*) pile bent pier.

dimensions at the base of 7.30 m × 19.00 m. The maximum wall thickness is 0.3 m, or approximately 12 in.

When the required pier heights are above that for economy of solid gravity piers, a favorite type of many engineers is the hammerhead pier (Fig. 10.23*b*). This type is similar in action to the gravity pier but results in less total concrete, and thus less weight, on the soil or piles. It is well suited for multiple-beam superstructures. It can be used with superstructures composed of two girders to two trusses, but the cantilever moments and shears on the head are much greater.

The dimensions of the footing or pile cap can be selected so that there is sufficient stability in both directions. The main reinforcing in the pier consists of horizontal steel in the top of the head to resist the cantilever moment, and stirrups in the head to resist the shear. Vertical steel around the periphery of the column and ties linking this vertical steel together are required. Since the column is designed as a tied column subjected to axial forces and bending moments (possibly about both axes), the column steel area should be limited to a minimum of 1% of the area of the column in cross section. This minimum percent of steel is sufficient for most moderate-height piers. The vertical steel in the columns is lapped to vertical dowels extending out of the footing. Temperature and shrinkage reinforcement are needed on the faces of the hammerhead.

When pier heights above the upper limit of economy for hammerhead piers are required, the rigid-frame pier (Fig. 10.23c) is usable. The rigid-frame pier can obtain stability without using very large volume of concrete. This type also lends itself to superstructures of just two plate girders or trusses. The tops of the columns can be centered under the girders or trusses and the columns can be vertical or sloped slightly if necessary to obtain sufficient distance between the footings to meet stability requirements. The thickness of the legs, as seen in the side elevation, can be constant or can increase from the top to the bottom of the pier. For aesthetic reasons the variation in thickness should be small. The horizontal struts between legs should be spaced to meet maximum length-to-thickness ratios of the legs. Uniformity of appearance should be carefully considered in selecting strut spacing, as well as column sizes, when several piers of varying height are present in a bridge.

Here, as in other piers, the footings can be designed as spread footings on soil or rock or as pile caps. Whether the analysis of the frame is based on pin-ended, fixed, or partially fixed column bases depends on the foundation conditions and design.

The analysis of the pier is that of a rigid frame. As such there are axial forces and bending moments in the columns and horizontal struts. Major values of bending moments come from lateral loads. Longitudinal (in direction of bridge axis) forces can be minimized on tall piers by using expansion bearings. Thus the only longitudinal force on the pier is a result of any friction in the expansion bearings. The pier can be subjected to a wind force on the front face. The reinforcing-steel details for this frame type of pier are more complex than those for the gravity or the hammerhead type, but straight bars can be used in all places. Forming is more involved for the frame types than in the other two types.

When piles are required and beam spans are short, the piles can be left extended above ground and interconnected with a reinforced concrete pile cap, forming a pier (Fig. 10.23$d$). Such piers are economical for moderate height. Some of the piles are battered to resist longitudinal and lateral forces. Battering of piles minimizes bending. If steel piles are used they should be protected, where moisture is present, by being encased in concrete. Specifically, this would be near ground line or where stream flow would cause wetting and drying action.

Design Example 10.4 gives the details of the design of a hammerhead pier. The first step is to establish the pier dimensions from the known dimensions of the bridge. The total pier height is controlled by the deck elevation and elevation at the bottom of the footing. The pier should be kept as short as possible, but the footing must rest on firm material. The minimum width and thickness of the pier head is a function of the requirements of space for the bridge bearings.

The magnitude and position of the loads from the superstructure are established next. The pier head is the first item designed. The minimum depth at the outer edge is usually 3 ft. The depth is increased at a slope that is pleasing to the eye. The greater the depth, the less main steel needed in the top of the head. The projecting portion is treated as a cantilever beam. The first detail is to calculate the shear-steel requirements. Shear steel can be reduced by making the head deeper. However, some vertical steel is necessary. Number 3 bars at 8-in. spacing is quite minimal. With the depth at column face established, the bending stress in the concrete is calculated. The area of steel is determined and seven no. 11 bars are selected.

With the dimensions of the head fixed, a column size can be selected, as well as a trial size for the footing. In fact, a column dimension has to be selected before the cantilever moment in the head is calculated. With the dimensions selected the stability of the pier can be determined. It is first checked for Group I loading (Chapter 3). The only lateral loads in this group are stream flow, and centrifugal force when the bridge is curved. A factor of safety of 2 is used for this condition. AASHTO does not set factors of safety for stability. The engineer should select his values after careful consideration of the particular conditions of his structure. A factor of safety of 2 for Group I loading would generally be minimum.

The pier is then checked for the Group II and III loadings, which include wind forces and longitudinal forces. A factor of safety for wind on the front face of the pier without any superstructure is determined first. A factor of safety of 1.5 is adequate for this temporary condition. The pier is then checked for Group II with superstructure in place. The soil pressures are also calculated. Under Group II loading, the allowable stresses can be increased 25%.

The Specifications state that for girder and slab bridges having a maximum span length of 125 ft, the wind loading on the superstructure can be 50 psf transverse and 12 psf longitudinal with both forces applied simultaneously. The overturning moments are then added vectorally to obtain the total overturning moment. The righting moment is taken as the total vertical force multiplied by the distance from the corner to the center of the pier.

The soil pressure is calculated by the addition of $P/A \pm M_x y/I_x \pm M_y x/I_y$. The calculations show that the sum of the moment terms is greater than the direct stress term so there would be uplift and the value of 9.4 psi is not correct. Since there can be no tensile stresses between the footing surface and the ground, only the area of footing subject to pressure can be included in the $I$ values. A new method of calculation must be followed if uplift is to be permitted. Many engineers size a footing so that no uplift may occur. The value of 9.4 psf is close to the allowable of 10 psf, and the correct procedure, using a reduced section, would quite likely result in soil pressures greater than 10 psf. Therefore the size of the footing is increased sufficiently to eliminate uplift.

The calculations of stability and maximum soil pressure for Group III follow the same procedure as used previously with different loads. In Group III the wind on the structure is reduced and a wind force is applied to the live load.

From the three loading groups it may be possible to ascertain by inspection which loading or loadings would cause the greatest stress on the pier column. The minimum amount of reinforcing steel is 1%. In most cases this is sufficient. In Design Example 10.4 the stresses are checked by both the allowable stress method and the ultimate strength (load factor design) procedure. It is noted that the working stresses are well below the allowable values. The column cannot be reduced in size to any appreciable amount because of stability requirements.

After the column is designed, the footing is designed. The footing is designed for shear and bending at the face of the column. The Specifications are explicit on the design procedure for footings.

There is not space in this book to cover the detailed design of other types of piers. The loading systems would be similar to Design Example 10.4. The design procedures would also be simple and can be summarized as follows:

1. Determination of general dimensions and estimate size of pier.
2. Determination of pier forces and location of application of loads.
3. Check of pier for stability under all applicable group loadings.
4. Check of soil pressures under footings, or pile loads if piles are used.
5. Determination of required reinforcing steel quantities and concrete stresses.
6. Adjustment in dimensions and preparation of final calculations.

## Design Example 10.4

HAMMERHEAD PIER

*Given:*

1. Pier is to support three simple spans of steel beam superstructure.
2. Beams will be fixed at abutments.
3. $f'_c = 3000$ psi; reinforcing steel, grade 40.
4. Rock is located 40 ft below elevation of roadway surface; soil covering is 3 ft over rock.
5. Rock is good for 8 ksf.

The reaction (Fig. 10.26) is given by

$$R = W = 16 + \frac{44}{58}(20) = 31.2 \text{ k}$$

DESIGN OF PIER HEAD

The pier is assumed to be 4 ft thick. Shear is now determined. At Section $1-1 (D = 4 \text{ ft})$ (Fig. 10.27):

$$V = 19.6 + 62.6 + 4(3)(3.5)(0.15)$$
$$= 88.5 \text{ k}$$
$$v = \frac{V}{bd} = \frac{88.5}{48(45)} = 0.041 \text{ ksi}$$
$$v_c = 0.95\sqrt{3000} = 52 \text{ psi} > 41$$

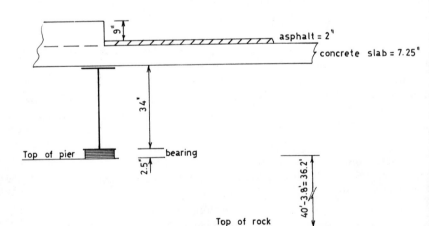

**Figure 10.24**   Top of rock.

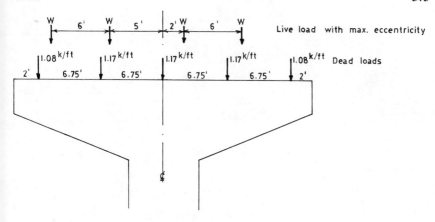

**Figure 10.25**   Live loading on pier.

At section 2–2 ($D = 6$ f):

$$V = 82.2 + 102.6 + 4(9)(4.5)(0.15)$$
$$= 209.1 \text{ k}$$
$$v = \frac{209.1}{48(69)} = 0.063 \text{ ksi} \qquad \text{Shear steel is required.}$$

At section 3–3 ($D = 7$ ft):

$$V = 184.8 + 4(11.5)(5)(0.15) = 219.3 \text{ k}$$
$$v = \frac{219.3}{48(81)} = 0.056 \text{ ksi}$$

Vertical stirrups ($\frac{3}{8}$ in. $\phi$) are used from first interior beam inward;

$$S = \frac{A_v f s}{(v - v_c)b} = \frac{0.22(20)(10^3)}{(63 - 52)(48)} = 8.3 \text{ in.}$$

Stirrups are placed as shown in Fig. 10.28.

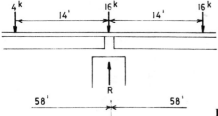

**Figure 10.26**   Live-loading reaction at pier.

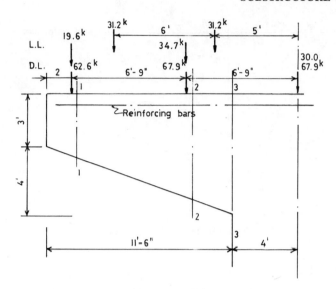

**Figure 10.27**    Pier-head loading.

DESIGN OF TENSION REINFORCING IN HEAD

The bending moment at section 3–3 is:

$$M = 82.2(9.5) + 102.6(2.75) + 20.7(5.75) + 13.8(3.83)$$
$$= 1235 \text{ ft-k}$$

For 2-in. cover and no. 10 bars

$$d = 84 - 2.63 = 81.4 \text{ in.}$$

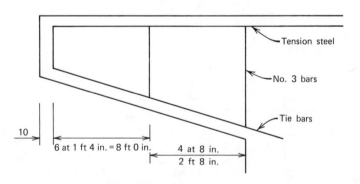

**Figure 10.28**    Stirrup placement.

Then

$$f_c = \frac{2(1235)(12)}{48(81.4)^2(3/8)(7/8)} = 0.28 \text{ ksi} < 1.2$$

$$A_s = \frac{1235(12)}{7/8(81.4)(20)} = 10.4 \text{ in.}^2$$

No. 10 bars = 8.2 (say 9) = 38.7 #/ft
No. 9 bars = 10.4 (say 11) = 37.4 #/ft
No. 11 bars = 6.7 (say 7) = 37.2 #/ft

Use No. 11 bars at 7-in. spacing (seven bars)

DESIGN OF PIER FOR STABILITY

*Group* I *Loading* (Fig. 10.29).

$$W_1 = [31 \times 3 + 11.5 \times 4 + 29.7 \times 8] \, 4(0.15)$$
$$= 226 \text{ k}$$
$$W_2 = 14 \times 10 \times 4 \times 0.15$$
$$= 84 \text{ k}$$

Total vertical load: 764 k
Overturning moment: 125(1.5) = 188 ft-k
Righting moment: 764(7) = 5348 ft-k
$FS = 5348/188 > 10 > 2.0$

Maximum soil pressure:

$$\frac{P}{A} + \frac{M}{S} = \frac{764}{140} + \frac{188(6)}{10(14)^2} = 5.45 + 0.58 = 6.0 \text{ ksf} < 8.0$$

Checking for only one lane loaded gives:

$$P = 701 \text{ k}$$
$$M = 62.4(8) = 499 \text{ ft-k}$$
$$FS > 2.0$$
$$p_{max} = \frac{701}{140} + \frac{499}{327} = 5.0 + 1.5 = 6.5 \text{ ksf} < 8.0 \text{ ksf}$$

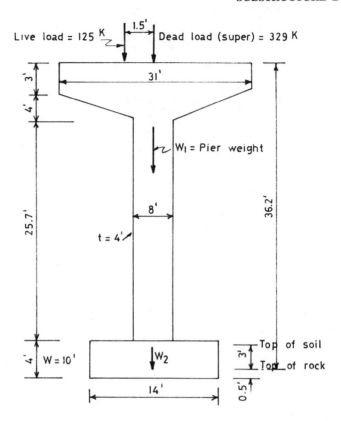

**Figure 10.29**  Group I loading.

*Group* II *Loading.*  Footing size for wind (40 psf) on pier face without superstructure is now checked.

| Wind Force (k) | Moment arm (ft) | Moment on footing (ft-k) |
| --- | --- | --- |
| $3 \times 31 \times 0.04 = 3.72$ | 35.2 | 131 |
| $4 \times 11.5 \times 0.04 = 1.82$ | 34.9 | 64 |
| $8 \times 29.7 \times 0.04 = 9.50$ | 18.9 | 180 |
| | | $\sum = 375$ |

The total applied vertical load is

$$P = 226 + 84 = 310 \text{ k}$$

Righting moment $= 310\left(\dfrac{10}{2}\right) = 1550 \text{ ft-k}$

The factor of safety and maximum pressure are completed as

$$FS = \frac{1550}{375} = 4 + > 1.5$$

$$p_{max} = \frac{310}{40} + \frac{375(6)}{14(10)^2} = 2.21 + 1.16 = 3.82 \text{ ksf} < 8 \times 1.25$$

Conditions with the superstructure in place are given below.

Wind on superstructure

Structure area: Curb, slab, and beam: $\dfrac{51.25}{12}(58)$ $\quad$ 248 ft$^2$

Parapet and railing: $\dfrac{9+6}{12}(58)$

$$\Sigma = \frac{73}{321 \text{ ft}^2}$$

Wind force: $321(0.05) = 16 \text{ k}$

Overturning moment: $16\left(\dfrac{5.4}{2} + 37\right) = 635 \text{ ft-k}$

Uplift force: $32.2(58)(0.02) = 37.3 \text{ k}$

Overturning moment: $37.3\left(\dfrac{32.2}{4}\right) = 300 \text{ ft-k}$

Wind on substructure: $33.2(4)(0.04) = 5.3 \text{ k}$

Overturning moment: $5.3\left(\dfrac{33.2}{2} + 3.5\right) = 107 \text{ ft-k}$

Total vertical force: $329 + 226 + 84 - 37 = 602 \text{ k}$

Total overturning moment: $635 + 300 + 107 = 1042 \text{ ft-k}$

Righting moment: $602(\tfrac{14}{2}) = 4214 \text{ ft-k}$

$$FS = \frac{4214}{1042} = 4.0 > 2.0$$

$$p_{max} = \frac{602}{140} + \frac{1042}{327} = 4.3 + 3.2 = +7.5 < 8.0(1.25)$$

In addition to the above lateral forces, a longitudinal force due to wind on the superstructure should be included:

$$\text{overturning moment} = 635\left(\frac{12}{50}\right) = 152 \text{ ft-k}$$

The wind on the substructure can be separated into lateral and longitudinal components:

lateral overturning moment $= 107(0.5) = 53$ ft-k
longitudinal overturning moment $= 375(0.866) = 325$ ft-k
total lateral overturning moment $= 988$ ft-k
total longitudinal overturning moment $= 477$ ft-k

These overturning moments are divided into components along a line from the center of the footing to a corner (Fig. 10.30):

$$M = \sqrt{(477)^2 + (988)^2} = 1097 \text{ ft-k}$$

$$\text{righting moment} = 602(8.6) = 5177 \text{ ft-k}$$

$$FS = \frac{5177}{1097} = 4.7 > 2.0$$

The pressure is computed as

$$p_{\max} = \frac{602}{140} + \frac{988}{327} + \frac{477}{233} = +4.3 + 3.0 + 2.1 = 9.4 < 8.0(1.25) \text{ ksf}$$

Since $3.0 + 2.1 > 4.3$ there is an uplift on the footing. It is necessary to increase footing size to eliminate uplift.

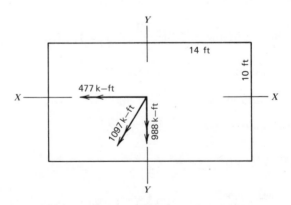

**Figure 10.30**  Overturning moments—Group II.

For length $= 16.5$ ft and width $= 12$ ft, the pressure is

$$p_{max} = \frac{602}{198} + \frac{988}{544.5} + \frac{477}{396} = 3.04 + 1.81 + 1.20 = 6.05 < 10 \text{ ksf}$$

*Group* III *Loading* (Fig. 10.31).

*Longitudinal forces*

Wind on live load: $5.8(0.4) = 2.3$ k
Wind on superstructure: $4.8(\frac{12}{50}) = 1.2$ k
Traffic: $(0.64 \times 58 + 18)2 \times 0.05 = 6$ k

No force from bearing friction since forces from adjacent spans balance

Lateral overturning moment (Fig. 10.32):

$$5.8(46.5) + 4.8(39.6) + (63 + 11.2)8 + 1.6(18.3) = 1083 \text{ ft-k} = M_{yy}$$

Longitudinal overturning moment (Fig. 10.32):

$$36.7(2.3 + 1.2 + 6.0) = 349 \text{ ft-k} = M_{xx}$$

Vertical load: $63 + 329 + 310 - 11 = 691$ k

Wind on pier could be divided into components but effect would be insignificant. The maximum soil pressure is given by

$$p_{max} = \frac{691}{140} + \frac{1083}{327} + \frac{349}{233} = 4.94 + 3.31 + 1.5 = 9.75 < 8(1.25) \text{ psf}$$

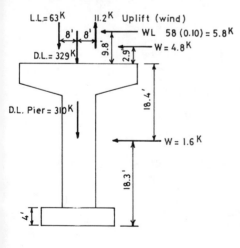

**Figure 10.31** Group III loading.

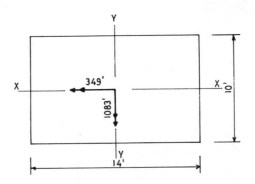

**Figure 10.32**   Overturning moments—Group III.

Since $3.31 + 1.5 < 4.94$ there will be pressure over the entire footing so the above $p_{max}$ is correct. Then

$$\text{overturning at corner} = \sqrt{(1083)^2 + (349)^2} = 1138\,\text{ft-k}$$

$$FS = \frac{691 \times 8.6}{1138} = 5.2 > 2.0$$

STRESS ANALYSIS OF COLUMN (Fig. 10.33)

From observation Group III loading will produce the maximum stresses in the pier shaft. The moments at the top of the footing are

$$M_{yy} = 5.8(42.5) + 4.8(35.6) + (74.2)8 + 1.6(14.3) = 1034\,\text{ft-k}$$

$$M_{xx} = 32.7(9.5) = 311\,\text{ft-k}$$

$$P = 691 - 84 = 607\,\text{k}$$

For load factor design the above values should be multiplied by the load factor 1.3:

$$M_y = 1034 \times 1.3 = 1344\,\text{ft-k}$$

$$M_x = 311 \times 1.3 = 404\,\text{ft-k}$$

$$p_u = 607 \times 1.3 = 789\,\text{k}$$

Using approximately 1% vertical steel, $A_s = 4608(0.01) = 46\,\text{in.}^2$ Using No. 10 bars,

$$\text{no. of bars} = \frac{46}{1.27} = 36$$

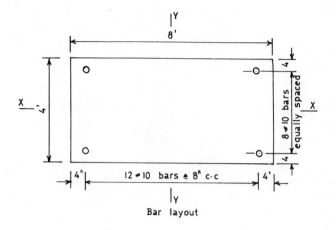

Bar layout

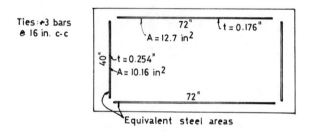

Figure 10.33   Column bar arrangement.

Therefore use 36 No. 10 bars, $A_s = 45.72 \text{ in.}^2$
The Elastic Analysis of the section can now be performed.

$$e_x = \frac{1034}{607} = 1.70 \text{ ft}$$

$$e_y = \frac{311}{607} = 0.51 \text{ ft}$$

$$\frac{e_x}{t_x} + \frac{e_y}{t_y} = \frac{1.7}{8.0} + \frac{0.51}{4.0} = 0.34 < 0.5$$

The section can be considered uncracked. For the transformed section

$$A_t = 4608 + (10 - 1)(45.72) = 5020 \text{ in.}^2$$

Then

$$I_x = 2(9)(12.7)(20)^2 + 96\left(\frac{48^3}{12}\right) + 2(9)\left(0.254 \times \frac{40^3}{12}\right) = 1 \times 10^6$$

$$I_y = 2(9)(0.176)\left(\frac{72^3}{12}\right) + 48\left(\frac{96^3}{12}\right) + 2(9)(10.16)(44)^2 = 3.99 \times 10^6$$

$$f_c = \frac{607}{5020} + \frac{1034(48)(12)}{3.99 \times 10^6} + \frac{311(24)(12)}{1 \times 10^6} = 0.37 \text{ ksi} < 1.2 \times 1.25$$

$$f_s = 10\left[-0.121 + \frac{1034(44)(12)}{3.99 \times 10^6} + \frac{311(20)(12)}{1 \times 10^6}\right] = 1.0 \text{ ksi} < 20 \times 1.25$$

LOAD FACTOR DESIGN

For biaxial bending

$$P_u = 789 \text{ k} < 0.1 f'_c \text{ Ag} = 0.1(3)(4608) = 1382 \text{ k}$$

$$\frac{M_x}{M_{ux}} + \frac{M_y}{M_{uy}} \le 1.0$$

For bending about $x$–$x$ axis

$$\frac{As}{bd} = \zeta = \frac{12.7}{4608} = 0.003$$

$$M_{ux} = 0.7\left[A_s F_y d\left(1 - 0.6\zeta \frac{F_y}{f'_c}\right)\right] = 0.7\left[12.7(40)(44)\left(1 - 0.6(0.003)\left(\frac{40}{3}\right)\right)\right]$$

$$= 0.7[21,816] = 15,270 \text{ k-in.} = 1273 \text{ ft-k}$$

For bending about $y$–$y$ axis

$$\zeta = \frac{10.16}{4608} = 0.0022$$

$$M_{uy} = 0.7\left[10.16(40)(96)\left(1 - 0.6(0.0022)\left(\frac{40}{3}\right)\right)\right]$$

$$= 0.7[38,328] = 26,830 \text{ in-k} = 2236 \text{ ft-k}$$

$$\frac{M_x}{M_{ux}} + \frac{M_y}{M_{uy}} = \frac{404}{1273} + \frac{1344}{2236} = 0.32 + 0.60 = 0.92 < 1.0$$

In the above calculations for $M_u$ the compression reinforcing steel and the side steel are neglected. Therefore the above value of 0.92 is more than the real value.

The pier is now checked for a condition with stream flow. Water at depth of 10 ft and flowing at a rate of 5 ft/sec is considered.

$$p = KV^2 = 1.375(25) = 34.4 \text{ psf}$$

$$\text{total force (lateral)} \qquad = 34.4(4 \times 10) = 1375 \text{ lb}$$

$$\text{moment at bottom of footing} = 1.375(5+4)$$
$$= 12.4 \text{ ft-k}$$

The weight of the pier is reduced by the buoyant effect of the water; reduction is given by

$$(8 \times 4 \times 10 + 14 \times 10 \times 4)0.063 = 55 \text{ k}$$

Overturning moment:

| | |
|---|---|
| Wind on superstructure | $= 635$ k-ft |
| Wind uplift | $= 300$ |
| Stream flow | $= \phantom{0}12$ |
| | $947$ ft-k |

Total vertical force $= 602 - 55 = 547$ k
Righting moment $= 547(7) = 3829$
$FS > 4.0 > 2.0$
$p_{max}$: $< 7.5$ psf (pressure under full wind—no stream flow)

One may want to consider full wind on that portion of the pier above the water surface. Even if this is a possibility, the pier is amply safe.

DESIGN OF FOOTING—GROUP I LOADING (Fig. 10.34)

Since the projection of 3 ft is greater than $\frac{2}{3}(4)$, bending must be considered:

$$\text{moment at face of pier} = 5.86(30)(1.5) + \frac{0.64}{2}(30)(2)$$

$$= 283 \text{ ft-k}$$

Then

$$f_c = \frac{283(12)(2)}{120(\frac{3}{8})(\frac{7}{8})(44.5)^2} = 0.09 \text{ ksi} < 1.2 \text{ ksi}$$

$$A_s = \frac{283(12)}{20(\frac{7}{8})(44.5)} = 4.36 \text{ in.}^2$$

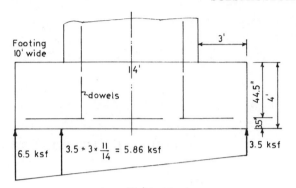

**Figure 10.34**   Footing pressure.

Number 8 bars are placed at 18-in. spacing for minimum reinforcement (seven bars = 7 in.²). Number 8 bars are used at 18-in. in both directions. Since $d = 44\frac{1}{2}$ in. $> 3.0$ ft, shear stresses are negligible.

The minimum length of No. 10 dowels in footing is

$$1.27 \times \frac{20,000}{3.99(207)} = 31 \text{ in.} < 44.5 \text{ in.}$$

Therefore there is sufficient room for dowels without bends.

DESIGN OF PILE FOUNDATION

The use of point bearing piles of HP $10 \times 42$ is now considered. $A = 12.4$ in.²; allowable load at pile tip, is

$$12.4(9) = 112 \text{ k}$$

If only vertical load is considered, the number of piles is

$$\frac{691}{112} = 6.2 \qquad \text{Try 10 piles}$$

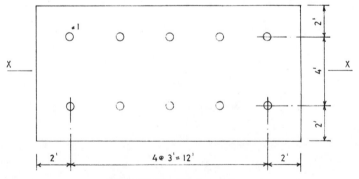

**Figure 10.35**   Pile foundation.

The spacing shown in Fig. 10.35 is tried where the

minimum spacing = 2 ft 6 in. as per AASHTO
minimum edge distance = 9 in. $<24-6=18$ in.

$$n = 10$$

$$I_x = 10(2)^2 = 40$$

$$I_y = 4(6)^2 + 4(3)^2 = 180$$

$$\frac{P}{n} = \frac{691}{10} = 69 \ k$$

$$\frac{M_x c}{I_x} = \frac{392(2)}{40} = 20 \ k$$

$$\frac{M_y c}{I} = \frac{1083(6)}{180} = 36 \ k$$

Total load on pile No. 1 = 69 + 20 + 36 = 125 k < 112(1.25). It would not be possible to reduce two piles and keep maximum pile load less than 140 k, so the above design is satisfactory. The batter is now checked:

lateral force $= 5.8 + 4.8 + 1.6(0.5) = 11.4$ k
longitudinal force $= 2.3 + 1.2 + 6.0 + 15(0.3)(0.866) = 13.4$ k

Two piles take longitudinal and lateral forces, batter in $x$ direction:

$$\frac{11.4}{2} \times \frac{12}{125} = 0.5 \qquad \text{say } \tfrac{1}{2} \text{ in. in 12 in.}$$

The batter in the $y$ direction is

$$13.4/2 \times 12/125 = 0.64 \qquad \text{say } \tfrac{5}{8} \text{ in. in 12 in.}$$

The total axial load is given by

$$\sqrt{(125)^2 + (5.7)^2 + (6.7)^2} = 125.4 \ k < 140$$

## REFERENCES

1  A. A. Witecki and V. K. Raina, "Distribution of Longitudinal Horizontal Forces Among Bridge Supports," Americal Concrete Institute Publication SP-23, pp. 803–815, 1969.

2  C. W. Dunham, *Foundation of Structures*, McGraw-Hill, 1962.

3  G. A. Leonards, *Foundation Engineering*, McGraw-Hill, 1962.

4  W. C. Teng, *Foundation Design*, Prentice-Hall, 1962.

CHAPTER **11**

# Bridge Rating

## 11.1 GENERAL

In the previous chapters the design of modern types of bridge structures is presented in detail. Such designs have required new analytical techniques and specifications to ensure safety of the structure. They offer the engineer new challenges and responsibility. However, the engineer has the greater responsibility of *maintaining* the safety of these new structures and, more important, of those bridges that were built in the early 1900s and still serve the public today.

Many of these early bridges still exist throughout the states and are quite serviceable when properly maintained. They are generally truss-type structures, as shown in Figures 11.1 through 11.4. The cost of replacing these bridges is prohibitive. Thus proper maintenance and rating of these bridges are necessary.

The rating of such bridges is performed by a combination of field inspection of the bridges and an analytical study, as guided by the AASHTO *Manual for Maintenance Inspection of Bridges*. This manual provides guidelines for (1) inspection of bridges, (2) records of such bridges, (3) rating of bridges, and (4) specifications for checking capacities of existing bridges.

**Figure 11.1**  Through-truss bridge.

**Figure 11.2**  Name tag on truss bridge.

**Figure 11.3**  Through-truss bridge.

The application of these guidelines is illustrated here through presentation of the rating of an actual truss bridge located in Frederick County, Maryland, as given in Fig.11.5

## 11.2  DESIGN CODE

In general, the AASHTO Specifications (1) apply in the case of loadings, impact, and distribution of loads. The primary change, relative to rating of existing bridges, is concerned with the allowable maximum unit stresses (2). The recommended values for (1) structural steel, (2) wrought iron, (3) reinforcing steel, (4) concrete, (5) prestressed concrete, and (6) timber are given below.

### 11.2.1  Structural Steel

Tables 11.1 and 11.2 give the allowable steel stresses based on inventory conditions. Following these tables the specifications (2) give the following section on batten plate columns.

**Figure 11.4**  Through-truss bridge.

*Batten Plate Columns.* To allow for the reduced strength of batten plate columns, the actual length of the column shall be multiplied by the following factor to obtain the value of $L/r$ to be substituted in the column formulae, given in Table 11.1.

For columns having a solid plate on one side and the batten plates on the other, the foregoing increase factors shall be reduced 50 percent.

Adjusted $L/r$ (batten plate both sides) = actual $L/r \times$ factor

adjusted $L/r$ (batten plate one side) = actual $L/r$

$$\times [1 + \tfrac{1}{2}(\text{factor} - 1)]$$

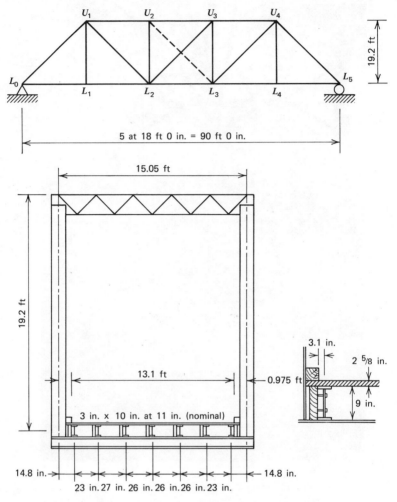

**Figure 11.5** Rating example—truss bridge details.

| Actual | Spacing Center-to-Center of Batten Plates[a] | | | |
|---|---|---|---|---|
| $L/r$ | Up to $2d$ | $4d$ | $6d$ | $10d$ |
| 40 | 1.3 | 2.0 | 2.8 | 4.5 |
| 80 | 1.1 | 1.3 | 1.7 | 2.3 |
| 120 | 1.0 | 1.2 | 1.3 | 1.8 |
| 160 | 1.0 | 1.1 | 1.2 | 1.5 |
| 200 | 1.0 | 1.0 | 1.1 | 1.3 |

[a] $d$ = depth of member perpendicular to battens.

# Table 11.1 Allowable Stresses (Inventory Rating)

Allowable unit stresses are shown in pounds per square inch. The modulus of elasticity of all grades of steel is assumed to be 29,000,000 psi and the coefficient of linear expansion 0.0000065 per degree Fahrenheit.

| | Date Built–Steel Unknown | | | | Carbon Steel | Silicon Steel | | | Nickel Steel |
|---|---|---|---|---|---|---|---|---|---|
| | Prior to 1905 | 1905 to 1936 | 1936 to 1963 | After 1963 | | | | | |
| AASHO Designation[a] ASTM Designation[b] | | | | | M-94 (1961) A7 (1967) | M-95 (1961) A94 (1966) t ≤ 1⅛ in. | M-95(1961) A94 (1966) 1⅛ in. < t ≤ 2 in. | M-95 (1961) A94 (1966) 2 in. < t ≤ 4 in. | M-96 (1961) A8 (1961) |
| Minimum tensile strength ($F_u$) | 52,000 | 60,000 | | | 60,000 | 75,000 | 72,000 | 70,000 | 90,000 |
| Minimum yield point ($F_y$) | 26,000 | 30,000 | 33,000 | 36,000 | 33,000 | 50,000 | 47,000 | 45,000 | 55,000 |
| Axial tension, net section | | | | | | | | | |
| Tension in extreme fiber of rolled shapes, girders and built-up sections, subject to bending ($0.55F_y$) | 14,000 | 16,000 | 18,000 | 19,000 | 18,000 | 27,000 | 25,000 | 24,000 | 30,000 |
| Axial compression, gross section: Stiffeners of plate girders | | | | | | | | | |
| Compression in splice material, gross section | | | | | | | | | |
| Compression in extreme fibers of rolled shapes, girders and built-up sections, subject to bending, gross section, when compression flange is: ($0.55F_y$) | | | | | | | | | |
| a. Supported laterally its full length by embedment in concrete. | 14,000 | 16,000 | 18,000 | 19,000 | 18,000 | 27,000 | 25,000 | 24,000 | 30,000 |
| b. Partially supported or unsupported[b] | $14{,}000 - 3.9\left(\frac{l}{b}\right)^2$ | $16{,}000 - 5.2\left(\frac{l}{b}\right)^2$ | $18{,}000 - 6.3\left(\frac{l}{b}\right)^2$ | $19{,}000 - 7.5\left(\frac{l}{b}\right)^2$ | $18{,}000 - 6.3\left(\frac{l}{b}\right)^2$ | $27{,}000 - 14.4\left(\frac{l}{b}\right)^2$ | $25{,}000 - 12.7\left(\frac{l}{b}\right)^2$ | $24{,}000 - 11.7\left(\frac{l}{b}\right)^2$ | $30{,}000 - 17.4\left(\frac{l}{b}\right)^2$ |
| with $\frac{L}{b}$ not greater than | 42 | 39 | 38 | 36 | 38 | 30 | 31 | 32 | 29 |
| Compression in concentrically loaded columns[c] with $C_c = \sqrt{\dfrac{2\pi^2E}{F_y}}$ | 148.4 | 138.1 | 131.7 | 126.1 | 131.7 | 107.0 | 110.4 | 112.8 | 102.0 |
| when $\dfrac{KL}{r} \le C_c$ $F_a = \dfrac{F_y}{FS}\left[1 - \dfrac{(KL/r)^2 F_y}{4\pi^2 E}\right]$ when $\dfrac{KL}{r} \ge C_c$ | $12{,}260 - 0.28\left(\frac{KL}{r}\right)^2$ | $14{,}150 - 0.37\left(\frac{KL}{r}\right)^2$ | $15{,}570 - 0.45\left(\frac{KL}{r}\right)^2$ | $16{,}980 - 0.53\left(\frac{KL}{r}\right)^2$ | $15{,}570 - 0.45\left(\frac{KL}{r}\right)^2$ | $23{,}580 - 1.03\left(\frac{KL}{r}\right)^2$ | $22{,}170 - 0.91\left(\frac{KL}{r}\right)^2$ | $21{,}230 - 0.83\left(\frac{KL}{r}\right)^2$ | $25{,}940 - 1.25\left(\frac{KL}{r}\right)^2$ |

## Table 11.1 (continued)

| | | Date Built—Steel Unknown | | | | Carbon Steel | Silicon Steel | Nickel Steel |
|---|---|---|---|---|---|---|---|---|
| | | Prior to 1905 | 1905 to 1936 | 1936 to 1963 | After 1963 | | | |
| Shear in girder webs, gross section | | 8,500 | 9,500 | 11,000 | 11,500 | 11,000 | 15,500 | 17,500 |
| Bearing on milled stiffeners and other steel parts in contact | | | | | | | | |
| Stress in extreme fiber of pins | $0.80\,F_y$ | 20,000 | 24,000 | 26,000 | 28,000 | 26,000 | 37,000 | 44,000 |
| Bearing on pins not subject to rotation | | 20,000 | 24,000 | 26,000 | 28,000 | 26,000 | 37,000 | 40,000 |
| Bearing on pins subject to rotation (such as rockers and hinges) | | 10,000 | 12,000 | 13,000 | 14,000 | 13,000 | 18,000 | 18,000 |
| Shear in pins | $0.40\,F_y$ | 10,000 | 12,000 | 13,000 | 14,000 | 13,000 | 18,000 | 22,000 |
| Bearing on power-driven rivets and high-strength bolts (or as limited by allowable bearing on the fasteners) | $1.22\,F_y$ | 31,000 | 36,000 | 40,000 | 44,000 | 40,000 | 57,000 | 67,000 |

$$F_a = \frac{\pi^2 E}{FS\left(\dfrac{KL}{r}\right)^2} = \frac{135,008.740}{\left(\dfrac{KL}{r}\right)^2}$$

with $FS = 2.12$

[a] Number in parentheses represents the last year these specifications were printed.

[b] $L$ = length, in inches, of unsupported flange between lateral connections, knee braces or other points of support.

[c] $b$ = flange width, in inches; $E$ = modulus of elasticity of steel; $r$ = governing radius of gyration; $L$ = actual unbraced length; $K$ = effective length factor. The formulae do not apply to members with variable moment of inertia. For values of $K$ and for allowable loads on columns eccentrically loaded and/or with transverse bending, see AASHTO *Interim Bridge Specifications, 1974.*

# Table 11.2  Allowable Stresses (Operating Rating)

Allowable unit stresses are shown in pounds per square inch. The modulus of elasticity of all grades of steel is assumed to be 29,000,000 psi and the coefficient of linear expansion 0.0000065 per degree Fahrenheit.

| | Date Built–Steel Unknown | | | | Carbon Steel | Silicon Steel over 2 in. or 4 in. incl. | Nickel Steel | 8 in. and Under | 1-⅛ in. and Under | Over 1-⅛ in. to 2 in. incl. |
|---|---|---|---|---|---|---|---|---|---|---|
| | Prior to 1905 | 1905 to 1936 | 1936 to 1963 | After 1963 | | | | | | |
| AASHO Designation[a] | | | | | M-94 (1961) | M-95 (1961) | M-96 (1961) | | | |
| ASTM Designation[a] | | | | A-7 (1967) | A94 (1966) | A-8 (1961) | A36 | A94 | A94 | |
| Minimum tensile strength $F_u$ | 52,000 | 60,000 | | | 60,000 | 70,000 | 90,000 | 58,000 | 75,000 | 72,000 |
| Minimum yield point $F_y$ | 26,000 | 30,000 | 33,000 | 36,000 | 33,000 | 45,000 | 55,000 | 36,000 | 50,000 | 47,000 |
| Axial tension, net section | | | | | | | | | | |
| Tension in extreme fiber of rolled shapes, girders and built-up sections, subject to bending | | | | | | | | | | |
| Axial compression, gross section:p' stiffeners of plate girders   $0.75\,F_y$ | 19,500 | 22,500 | 24,500 | 27,000 | 24,500 | 33,500 | 41,000 | 27,000 | 37,500 | 35,000 |
| Compression in splice material, gross section | | | | | | | | | | |
| Compression in extreme fibers of rolled shapes, girders and built-up sections, subject to bending, gross section, when compression flange is: | | | | | | | | | | |
| a. Supported laterally its full length by embedment in concrete:   $0.75\,F_y$ | 19,500 | 22,500 | 24,500 | 27,000 | 24,500 | 33,500 | 41,000 | 27,000 | 37,500 | 35,000 |
| b. Partially supported or unsupported[b]   $0.75\,F_y$ | $1.37\left[14,000-3.9\left(\dfrac{l}{b}\right)^2\right]$ | $1.37\left[16,000-5.2\left(\dfrac{l}{b}\right)^2\right]$ | $1.37\left[18,000-6.3\left(\dfrac{l}{b}\right)^2\right]$ | $1.37\left[19,000-7.5\left(\dfrac{l}{b}\right)^2\right]$ | $1.37\left[18,000-6.3\left(\dfrac{l}{b}\right)^2\right]$ | $1.37\left[24,000-11.7\left(\dfrac{l}{b}\right)^2\right]$ | $1.37\left[30,000-17.4\left(\dfrac{l}{b}\right)^2\right]$ | $1.37\left[20,000-7.5\left(\dfrac{l}{b}\right)^2\right]$ | $1.37\left[27,000-14.4\left(\dfrac{l}{b}\right)^2\right]$ | $1.37\left[25,000-12.7\left(\dfrac{l}{b}\right)^2\right]$ |
| with $\dfrac{l}{b}$ not greater than | 42 | 39 | 38 | 36 | 38 | 32 | 29 | 36 | 30 | 31 |
| Compression in concentrically loaded columns[c] | | | | | | | | | | |

## Table 11.2  (continued)

| | | Date Built—Steel Unknown | | | | Carbon Steel | Silicon Steel over 2 in. or 4 in. incl. | Nickel Steel | 8 in. and Under | 1⅛ in. and Under | Over 1⅛ in. to 2 in. incl. |
|---|---|---|---|---|---|---|---|---|---|---|---|
| | | Prior to 1905 | 1905 to 1936 | 1936 to 1963 | After 1963 | | | | | | |
| with $C_c = \sqrt{\dfrac{2\pi^2 E}{F_y}}$ when $\dfrac{KL}{r} \le C_c$ | | 148.4 | 138.1 | 131.7 | 126.1 | 131.7 | 112.8 | 102.0 | 126.1 | 107.0 | 110.4 |
| $F_a = \dfrac{F_b}{FS}\left[1 - \dfrac{(KL/r)^2 F_y}{4\pi^2 E}\right]$ when $\dfrac{KL}{r} \ge C_c$ $F_a = \dfrac{\pi^2 E}{FS\left(\dfrac{KL}{r}\right)^2} = \dfrac{168,363,840}{FS\left(\dfrac{KL}{r}\right)^2}$ with $FS = 1.70$ | | $15{,}290 - 0.35\left(\dfrac{KL}{r}\right)^2$ | $17{,}650 - 0.46\left(\dfrac{KL}{r}\right)^2$ | $19{,}410 - 0.56\left(\dfrac{KL}{r}\right)^2$ | $21{,}180 - 0.67\left(\dfrac{KL}{r}\right)^2$ | $19{,}410 - 0.56\left(\dfrac{KL}{r}\right)^2$ | $26{,}470 - 1.04\left(\dfrac{KL}{r}\right)^2$ | $32{,}350 - 1.55\left(\dfrac{KL}{r}\right)^2$ | $21{,}180 - 0.67\left(\dfrac{KL}{r}\right)^2$ | $29{,}410 - 1.28\left(\dfrac{KL}{r}\right)^2$ | $27{,}650 - 1.13\left(\dfrac{KL}{r}\right)^2$ |
| Shear in girder webs, gross section | $0.45F_y$ | 11,500 | 13,500 | 15,000 | 16,000 | 15,000 | 20,000 | 24,500 | 16,000 | 22,500 | 21,000 |
| Bearing on milled stiffeners and other steel parts in contact | $0.9\,F_y$ | 23,000 | 27,000 | 29,500 | 32,000 | 29,500 | 40,500 | 49,500 | 32,000 | 45,000 | 42,000 |
| Bearing on pins not subject to rotation | $0.90\,F_y$ | 23,000 | 27,000 | 29,500 | 32,000 | 29,500 | 40,500 | 49,500 | 32,000 | 45,000 | 42,000 |
| Bearing on pins subject to rotation (such as rockers and hinges) | $0.55F_y$ | 14,000 | 16,500 | 18,000 | 19,500 | 18,000 | 24,500 | 30,000 | 19,500 | 27,500 | 25,500 |
| Shear in pins | $0.55F_y$ | 14,000 | 16,500 | 18,000 | 19,500 | 18,000 | 24,500 | 30,000 | 19,500 | 27,500 | 25,500 |
| Bearing on power-driven rivets and high-strength bolts or as limited by allowable bearing on the fasteners) | $1.66F_y$ | 43,000 | 49,500 | 54,500 | 59,500 | 54,500 | 70,000 (74,500) | 91,000 | 58,000 (59,500) | 75,000 (83,000) | 72,000 (78,000) |

| | 1½ in. max. | ½ in. max. | Over 2½ in. 4 in. incl. | ¾ in. and under 4 in. and under (A588) | To 2½ in. incl. All thick (A517) | incl. (A588) over ¾ in. to 1½ in. incl. | 1½ in. max. | 1 in. max. | over 5 in. to 8 in. incl. (A588) over 1½ in. to 4 in. incl. | Over 4 in. to 8 in. incl. |
|---|---|---|---|---|---|---|---|---|---|---|
| AASHO Designation[a] ASTM Designation[a] | A572 | A572 | A514 | A242, A440, A441, A588 | A514/A517 | A242, A440, 441, A588 | A572 | A572 | A242, A440, A441, A588 | A441 |
| Minimum tensile strength | 60,000 | 80,000 | 105,000 | 70,000 | 115,000 | 67,000 | 70,000 | 75,000 | 63,000 | 60,000 |
| Minimum yield point | 45,000 | 65,000 | 90,000 | 50,000 | 100,000 | 46,000 | 55,000 | 60,000 | 42,000 | 40,000 |
| Axial tension, not section Tension in extreme fiber of rolled shapes, girders and built-up sections, to bending | 33,500 | 48,500 | 67,500 | 37,500 | 75,000 | 34,500 | 41,000 | 45,000 | 31,500 | 30,000 |
| Axial compression, gross section: stiffeners of plate girders Compression in splice material, gross section Compression in extreme fibers of rolled shapes, girders and built-up sections, subject to bending, gross section, when compression flange is: a. Supported laterally its full length by embedment in concrete: | 33,500 | 48,500 | 67,500 | 37,500 | 75,000 | 34,500 | 41,000 | 45,000 | 31,500 | 30,000 |
| b. Partially supported or unsupported[b] | $1.37\left[24,000 - 11.7\left(\frac{l}{b}\right)^2\right]$ | $1.37\left[35,000 - 24.4\left(\frac{l}{b}\right)^2\right]$ | $1.37\left[49,000 - 47\left(\frac{l}{b}\right)^2\right]$ | $1.37\left[27,000 + 14.4\left(\frac{l}{b}\right)^2\right]$ | $1.37\left[55,000 - 58\left(\frac{l}{b}\right)^2\right]$ | $1.37\left[25,000 - 12.2\left(\frac{l}{b}\right)^2\right]$ | $1.37\left[30,000 - 17.4\left(\frac{l}{b}\right)^2\right]$ | $1.37\left[33,000 - 20.7\left(\frac{l}{b}\right)^2\right]$ | $1.37\left[23,000 - 10.2\left(\frac{l}{b}\right)^2\right]$ | $1.37\left[22,000 - 9.2\left(\frac{l}{b}\right)^2\right]$ |
| with $\frac{l}{b}$ not greater than: | 32 | 27 | 23 | 30 | 21 | 32 | 29 | 28 | 33 | 34 |
| Compression in concentrically loaded columns[c] with $C_c \sqrt{\dfrac{2\pi^2 E}{F_y}}$ when $\dfrac{KL}{r} \leq C_c$ | 112.8 | 93.8 | 79.8 | 107.0 | 75.7 | 111.6 | 102.0 | 97.7 | 116.7 | 119.6 |

# Table 11.2  (continued)

$$F_a = \frac{F_y}{FS}\left[1 - \frac{(KL/r)^2 F_y}{4\pi^2 E}\right]$$

when $\dfrac{KL}{r} \geq C_c$

$$F_a = \frac{\pi^2 E}{FS\left(\frac{KL}{r}\right)^2} = \frac{168,363,840}{\left(\frac{KL}{r}\right)^2}$$

with $FS = 1.70$

| | $1\frac{1}{2}$ in max. | $\frac{1}{2}$ in. max. | Over $2\frac{1}{2}$ in. 4 in. incl. | $\frac{3}{4}$ in. and under 4 in. and under under (A588) | To $2\frac{1}{2}$ in. incl. (A511) All thick (A517) | incl. (A588) over $\frac{3}{4}$ in. to $1\frac{1}{2}$ in. incl. | $1\frac{1}{2}$ in. max. | 1 in. max. | over 5 in. to 8 in. incl. (A588) over $1\frac{1}{2}$ in. to 4 in. incl. | Over 4 in. to 8 in. incl. |
|---|---|---|---|---|---|---|---|---|---|---|
| | $26,470-$ $1.04\left(\frac{KL}{r}\right)^2$ | $38,240-$ $2.17\left(\frac{KL}{r}\right)^2$ | $52,940-$ $4.16\left(\frac{KL}{r}\right)^2$ | $29,410-$ $1.28\left(\frac{KL}{r}\right)^2$ | $58,820-$ $5.14\left(\frac{KL}{r}\right)^2$ | $27,060-$ $1.09\left(\frac{KL}{r}\right)^2$ | $32,350-$ $1.55\left(\frac{KL}{r}\right)^2$ | $35,290-$ $1.85\left(\frac{KL}{r}\right)^2$ | $24,710-$ $0.91\left(\frac{KL}{r}\right)^2$ | $23,530-$ $0.82\left(\frac{KL}{r}\right)^2$ |
| Shear in girder webs, gross section | 20,000 | 29,000 | 40,500 | 22,500 | 45,000 | 20,500 | 24,500 | 27,000 | 18,500 | 18,000 |
| Bearing on milled stiffeners and othr steel parts in contact | 40,500 | 58,500 | 81,000 | 45,000 | 90,000 | 41,000 | 49,500 | 54,000 | 37,500 | 36,000 |
| Stress in extreme fiber of pins; Bearing on pins not subject to rotation | 40,500 | 58,000 | 81,000 | 45,000 | 90,000 | 41,000 | 49,500 | 54,000 | 37,500 | 36,000 |
| Bearing on pins not subject to rotation (such as rockers and hinges) | 24,500 | 35,500 | 49,500 | 27,500 | 55,000 | 25,000 | 30,000 | 33,000 | 23,000 | 22,000 |
| Shear in pins | 24,500 | 35,500 | 49,500 | 27,500 | 55,000 | 25,000 | 30,000 | 33,000 | 23,000 | 23,000 |
| Bearing on power-driven rivet and high-strength bolts or as limited by allowable bearing on the fasteners) | 60,000 (74,500) | 80,000 (107,500) | 105,000 (149,000) | 70,000 (83,000) | 115,000 (166,000) | 67,000 (76,000) | 70,000 (91,000) | 74,500 | 63,000 (69,500) | 80,000 (66,000) |

[a] Number in parentheses represents the last year these specifications were printed.

[b] $l$ = length, in inches, of unsupported flange between lateral connections, knee braces or other points of support;

[c] $b$ = flange width, in inches, $E$ = modulus of elasticity of steel; $r$ = governing radius of gyration; $L$ = actual unbraced length; $K$ = effective length factor; ... bending, see AASHTO Interim Bridge Specifications, 1974.

## 11.2.2 Wrought Iron

The allowable maximum unit stress in wrought iron for tension and bending is 14,600 psi.

## 11.2.3 Reinforcing Steel

The following are the allowable unit stresses in tension for reinforcing steel. These are ordinarily used without reduction when the condition of the steel is unknown.

|  | Inventory Rating (psi) | Operating Rating (psi) |
|---|---|---|
| Structural or unknown grade | 18,000 | 25,000 |
| Grade 40 (intermediate) | 20,000 | 28,000 |
| Grade 50 (hard) | 20,000 | 32,500 |
| Grade 60 | 24,000 | 36,000 |

## 11.2.4 Concrete

The maximum allowable unit stresses in concrete in psi are given below. The value of $n$ should be varied approximately according to the following table:

| $f'_c$(psi) | $n$ |
|---|---|
| 2000–2400 | 15 |
| 2500–2900 | 12 |
| 3000–3900 | 10 |
| 4000–4900 | 8 |
| 5000 or more | 6 |

Compression due to bending is 1300 psi.*
Compression, short column, in which $L/D$ is 12 or less is $0.3f'_c$.
Compression, long columns, in which $L/D$ is greater than 12, where $L$ = unsupported length of column and $DL$ = least dimension of column, is equal to $0.3f'_c(1.3-0.3L/D)$. When compressive strength of concrete is not known, maximum $f'_c$ is to be taken as 3300 psi.

---

* Compression caused by bending may be increased up to 0.55 of the ultimate compressive strength of the concrete if investigation indicates that the concrete is sound and that contract plans and controls specified and furnished compressive strengths of 2400 psi or more.

Maximum safe axial load in columns is given by

$$P = f_c A_g + f_s A_s$$

where     $P$ = allowable axial load on column

             $f_c$ = allowable unit stress of concrete taken from $d$ or $e$

             $A_g$ = gross area of column

             $f_s$ = allowable stress of steel = $f_y(0.4 \times 1.38)$

                = 18,000 psi unless information on specifications of reinforcing steel is available and $f_y$ is greater than 33,000

             $f_y$ = yield strength of reinforcing steel

             $A_s$ = area of longitudinal reinforcing steel

Shear in beams showing no diagonal tension cracking is given by total unit shear = shear taken by steel + shear taken by concrete or

$$v = v_s + v_c = v_s + 0.05 f'_c$$

The maximum value of $0.05 f'_c$ to be used is 160 psi. Where severe diagonal tension cracking has occurred, $v_c$ may be considered as zero and all shear stress may be taken by the reinforcing steel.

### 11.2.5 Prestressed Concrete

For prestressed concrete members in which the reinforcement index, determined in accordance with Article 1.6.10 of the AASHTO Standard Specifications, does not exceed 0.30, the operating rating should result in moments not to exceed 75 percent of the ultimate moment capacity of the member (Article 1.6.9, AASHTO Standard Specifications). In situations of unusual design with wide dispersion of the tendons, the operating rating might further be controlled by stresses not to exceed 0.90 of the yield point stress in the prestressing steel in the layer of tendons nearest the extreme tension fiber of the member.

### 11.2.6 Timber

Determining allowable unit stresses for timber in existing bridges requires sound judgment on the part of the engineer making the field investigation. The maximum allowable unit stresses should not exceed 1.33 times the allowable unit design stresses for stress-grade lumber given in the current AASHTO *Standard Specifications for Highway Bridges*. Reduction from the maximum allowable stress depends on the grade and condition of the timber and is determined at the time of the inspection. Allowable unit loads in pounds per square inch of cross-sectional area of simple solid columns are determined by the following formulae, but the maximum unit load should not exceed 1.33 times the values for compression parallel to grain given in the design stress table of the standard specifications referred to above.

$$\frac{P}{A} = \frac{4.813E}{(L/r)^2}$$

where    $P$ = total load in pounds
$A$ = cross-sectional area in square inches
$E$ = modulus of elasticity
$L$ = unsupported overall length, in inches, between points of lateral support of simple columns
$r$ = least radius of gyration of the section

For columns of square or rectangular cross section, this formula becomes

$$\frac{P}{A} = \frac{0.40E}{(L/d)^2}$$

in which $d$ = dimension in inches of the face under consideration.

The above formula applied to long columns for $(L/d)$ over 11, but not exceeding 50.

For short columns, $L/d$ not over 11, the compression is determined by multiplying the allowable design unit stress in compression parallel to grain by 1.33 for the grade of timber used.

## 11.3  RATING EXAMPLE—THROUGH TRUSS

ALLOWABLE STRESSES

As shown in Fig. 11.5, a five-panel through-truss bridge is rated for live load. This bridge was built prior to 1905 and thus the following allowable steel and timber stresses are recommended (2) as given in Section 11.2.

For *steel and wrought iron* the bending stress for tension is given by

$$F_b = 14,000 \text{ psi (inventory)}$$

$$F_b = 14,000 \times (1.38) \text{ psi (operating)}$$

The bending stress for compression is

$$F_b = 0.55F_y\left[1 - \frac{(L/r)^2 F_y}{4\pi^2 E}\right]$$

where $r^2 = b^2/12$

$$E = 29 \times 10^6 \text{ psi}$$
$$F_y = 26,000 \text{ psi}$$

Substituting these values into $F_b$ gives

$F_b = 14,000 - 3.9(L/b)^2 \le 12,500$ psi (inventory) as shown in Table 11.1
$F_b \times (1.38) = $ (operating)

Axial stresses for steel and wrought iron are given by

Tension:   $F_a = 14,000$ psi (inventory)
           $F_a = 14,000$ psi (1.38 (operating)

Compression:   $F_a = \dfrac{F_y}{2.12}\left[1 - \dfrac{(L/r)^2 F_y}{4\pi^2 E}\right]$

$F_a = 12,260 - 0.28(L/r)^2$, as shown in Table 11.1.

For damaged and/or severely corroded compression members assume

$$F_a = 10,000 - 0.152\left(\frac{L}{r}\right)^2 \text{ (inventory)}$$

$$F_a \times (1.38) = \text{operating}$$

For *timber* the bending stress

$$F_b = 1000 \text{ psi (inventory)}$$
$$F_b = 1333 \text{ psi (operating)}$$

Horizontal shear stress for timber is given by

$$\tau = 90 \text{ psi (inventory)}$$
$$\tau = 120 \text{ psi (operating)}$$

The allowable stresses given above fall into two categories, operating and inventory. The operating rating permits an upper stress level not to exceed 0.75 of the yield point of the material. This stress level is used in conjunction with heavier than normal vehicle loads, as governed by special permits. The inventory rating is a lower stress level that cannot exceed 0.55 of the yield point of the material. This rating level is designated as that load that can safely utilize an existing structure for an indefinite period.

TIMBER DECK

As shown in Fig. 11.5, the bridge planking has the following dimensions:

> plank size $= 3$ in. $\times 9\frac{1}{2}$ in.
>
> average spacing $= 11$ in. on center
>
> average stringer spacing $= 25.2$ in.
>
> maximum distance between two stringers $= 27$ in.
>
> flange width of stringers $= 4\frac{1}{4}$ in.

Since field inspection of the planking indicates servere decay the planking is assumed to be only 70% effective. The allowable stresses are therefore reduced.

Allowable stresses

Bending: 850 psi
Horizontal shear: 75 psi

Dead load

Weight of planks: $\dfrac{3 \times 9.5}{144} \times 50 \, \text{lb/ft}^3 = 10 \, \text{lb/linear ft}$

$M_{DL} = \dfrac{10(2.25)^2}{8} = 6.33 \, \text{ft-lb}$

$S_x = \frac{1}{6}(9.5)(3)^2 = 14.25 \, \text{in.}^3$

$f_{b_{DL}} = \dfrac{6.33 \times 12}{14.25} = 5.33 \, \text{psi}$

Live load (AASHTO Article 1.3.4)

Span

Clear distance $+\frac{1}{2}$ flange width $= (27 - 4.25) + 2.125 = 24.875$ in. (controls)

Clear span $+$ floor thickness $= 22.75 + 3 = 25.75$

Using H15 loading, as shown in Fig. 11.6, the distribution of wheel load on timber flooring is

$$w = 12,000 \, \text{lb}/15 \, \text{in.} = 800 \, \text{lb/in.} = 9.6 \, \text{k/ft}$$

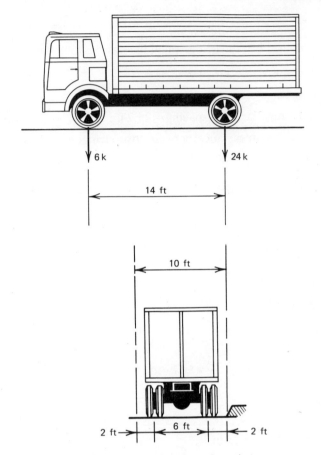

**Figure 11.6**   H15 truck configuration.

The distribution of load across the plank is as shown in Fig. 11.7. The maximum live-load moment is therefore

$$M_{LL_{(max)}} = \left[ 6 \times 4.94 + 7.5 \times \frac{6}{2} \right] = 4.34 \text{ k-ft}$$

$$f_{b_{LL}} = \frac{4.34 \times 12}{14.25} = 3.658 \text{ ksi}$$

Rating

Stress available for $LL$: $850 - 5 = 845$ psi

$LL$ rating of deck: $\dfrac{0.845}{3.66} \times 15 = \text{H3.5}$

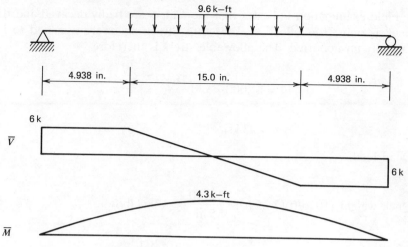

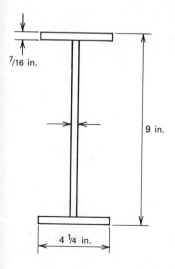

**Figure 11.7** Timber deck loading and forces.

STRINGERS

The properties of the stringer (Fig. 11.8), categorized as an American Standard Beam 9120, are given in reference (4).

The section properties are

$$\text{Flange width} = 4.25 \text{ in.}$$
$$S_{xx} = 17.55 \text{ in.}^3$$

**Figure 11.8** Stringer details.

Field examination indicates that the nailers are badly decayed and the deck is only 70% effective. Therefore the stringers are assumed to be laterally unsupported. The allowable stress is therefore

$$F_b = 14,000 - 3.9\left(\frac{18 \times 12}{4.25}\right)^2$$

$$F_b = 4.23 \text{ ksi}$$

Dead load

Deck weight $(10 \text{ lb/ft})(2.25 \text{ ft}): \dfrac{12 \text{ in./ft}}{9.5 \text{ in.}}$         28.4 lb/ft

Beam                                                                          20.0

Nailers: $4 \times 9.5 \times \dfrac{50}{144}$                                            $\underline{13.2}$

                                                                            61.6 lb/ft      Use 65 lb/ft

$$M_{DL} = \frac{65(18)^2}{8} = 2.6 \text{ k-ft}$$

$$f_{b_{DL}} = \frac{2.6 \times 12}{17.55} = 1.8 \text{ ksi}$$

Live load

Position: 12 k load at the center of an 18-ft span; then

$$M_{LL} = \frac{PL}{4} = \frac{12 \times 18}{4} = 54 \text{ k-ft}$$

$$M_{LL+I} = 54 \times 1.3 = 70.2 \text{ k-ft}$$

Distribution factor: $\dfrac{S}{4} = \dfrac{2.25}{4} = 0.563$

Design $M_{LL+I}$: $70.2 \times 0.563 = 39.49$ k-ft; then

$$f_{LL+I} = \frac{39.49 \times 12}{17.55} = 27 \text{ ksi}$$

Rating

Stress available for $LL$:   $4.2 - 1.8 = 2.43$ ksi

$$LL \text{ rating:} \qquad \frac{2.43 \times 15}{27} = \text{H}1.35$$

Inventory rating: H1.4
Operating rating: H1.8

FLOOR BEAM

The floor beam is shown in Fig. 11.9 and falls into the category of a 15I39 beam. The section properties are

$$S_x = 53.8 \text{ in.}^3$$
$$w = 39 \text{ lb/ft}$$

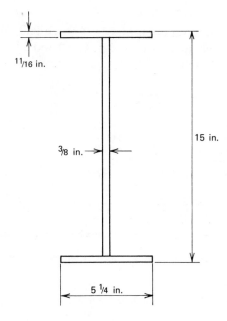

11/16 in.

15 in.

3/8 in.

5 1/4 in.

**Figure 11.9**  Floor-beam details.

From the deck and stringer information, the dead load on each stringer is 65 lb/ft, and it occupies a spacing of 2.1 ft on a span of 18 ft. Therefore $w_T = 65 \times 18 = 1170$ lb concentrated on the floor beam, or

$$w \text{ lb/ft} = \frac{1170}{2.1} = 557 \text{ lb/ft along the floor beam}$$

$$\text{beam weight} = \underline{\phantom{00}39}$$
$$596 \text{ lb/ft} \qquad \text{say 600 lb/ft on the floor beam}$$

Then the dead-load moment is

$$M_{DL} = \frac{0.6(15.05)^2}{8} = 16.98 \text{ k-ft}$$

and the induced stress is computed as

$$f_{b_{DL}} = \frac{16.98 \times 12}{53.8} = 3.79 \text{ ksi}$$

The *DL* reaction is

$$DL \text{ reaction} = \frac{0.6 \times 15.05}{2} = 4.52 \text{ k}$$

The live-load effect is formed by positioning the H15 truck laterally and longitudinally on the bridge deck, as shown in Figs. 11.10 and 11.11. The increase in the 12 k wheel load due to the 3 k load distribution is

$$R = \frac{12 \times 4}{18} \times 3 = 12.67 \text{ k}$$

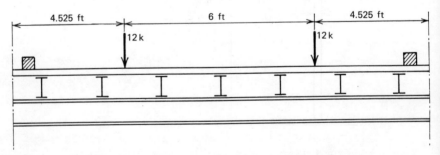

**Figure 11.10**  Transverse truck position for floor-beam analysis.

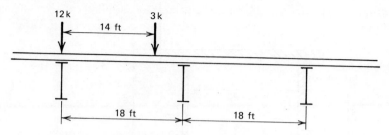

**Figure 11.11**  Longitudinal truck position for floor-beam analysis.

The maximum moment is now computed as

$$M_{max} = 4.52 \times 12.67$$
$$M_{max} = 57.3 \text{ k-ft}$$

The impact is given by

$$I = 0.30$$

and

$$M_{LL+I} = 1.3 \times 57.3 = 74.49 \text{ k/ft}$$
$$f_{b_{LL+I}} = \frac{74.49 \times 12}{53.8} = 16.6 \text{ ksi}$$

Assuming full bracing due to the stringers the allowable bending stress is 14.0 ksi. The stress available for live load is therefore $14.0 - 3.79 = 10.2$ ksi.

Rating

*LL* rating:      $\dfrac{10.2}{16.6} \times 15 = 9.2$

Inventory:      H9.2
Operating:      H12.7

TRUSS

In order to evaluate the proper rating of the truss, all truss elements must be examined. This necessitates evaluation of the entire dead load of the truss and influence lines for each truss element. The configurations of the various elements are shown in Figs. 11.12 through 11.17. The properties of these sections are given below.

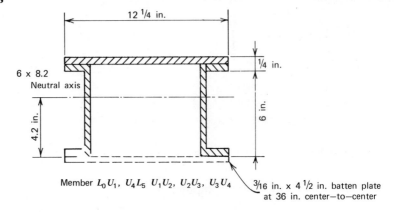

**Figure 11.12** End post and top chords.

For members $L_0U_1$, $U_4L_5$, $U_1$–$U_4$—end and top chords (Fig. 11.12)

1. *Centroid from Bottom of Channel.*

| Item | $A$ | $y$ | $A\bar{y}$ |
|---|---|---|---|
| Top plate | $\frac{1}{4} \times 12.25 = 3.06$ | 6.125 | 18.758 |
| Two channels | $2 \times 2.39 = 4.78$ | 3.0 | 14.34 |
| | 7.84 in². | | 33.09 in.³ |

From these data:

$$\bar{y} = \frac{33.09}{7.84} = 4.2 \text{ in.}$$

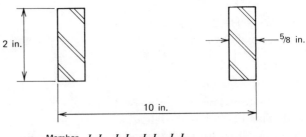

Member $L_0L_1$, $L_1L_2$, $L_3L_4$, $L_4L_5$

**Figure 11.13** Bottom chords.

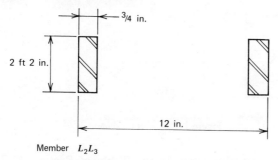

3/4 in.
2 ft 2 in.
12 in.
Member $L_2L_3$

**Figure 11.14** Bottom chord—center panel.

2. *Inertia*

| Item | $I_0$ | $A$ | $y$ | $Ay^2$ |
|---|---|---|---|---|
| Plate | 0.0160 | 3.06 | 1.91 | 11.11 |
| Channels | 26.0 | 4.78 | 1.22 | 7.12 |
| | 26.02 in.$^4$ | | | 18.23 in.$^4$ |

These data give:

$$I_T = 44.25 \text{ in.}^4$$

$$r = \left(\frac{I}{A}\right)^{1/2} = \left(\frac{44.25}{7.8}\right)^{1/2}$$

$$r = 2.38 \text{ in.}$$

$$\left(\frac{L}{r}\right)_{\text{end post}} = \frac{26.3 \times 12}{2.38} = 133$$

$$\left(\frac{L}{r}\right)_{\text{top chord}} = \frac{18.0 \times 12}{2.38} = 91$$

1/2 in.
2 in.
5 in.
Member $L_2U_1$, $L_3U_4$

**Figure 11.15** End diagonal.

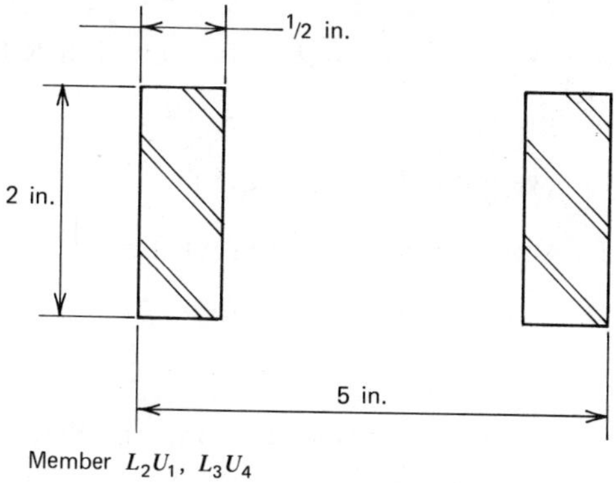

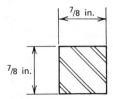

Member $L_2U_3$, $L_3U_2$       **Figure 11.16**   Center diagonals.

If the modified factor, given in the Table of Section 11.2, is used the value of $L/r$ is increased. For center-to-center batten plates equal to $6d$ and $(L/r) = 91$ and 133, the factors are 1.7 and 1.3 respectively. These factors are then multiplied by the $(L/r)$ values.

For this example the factors are not utilized, and the stresses are computed as follows.

3. *Allowable Stresses*

$$F_a = 10,000 - 0.152\left(\frac{L}{r}\right)^2$$

$$F_a = 7.3 \text{ ksi (end post)}$$
$$F_a = 8.7 \text{ ksi (top chord)}$$

4. *Weight* (steel wt = 3.4 lb/ft-in.$^2$ area)

Top plate: $(\frac{1}{4} \times 12.25)\left(\dfrac{3.4 \text{ lb/ft}}{\text{in}^2}\right)$     10.4

Two channels                 16.4
Batten plates                  <u>2.2</u>
                             29.0 (say 30.0 lb/ft)

For members $L_0L_1$, $L_1L_2$, $L_3L_4$, $L_4L_5$—bottom chords (Fig. 11.13):

$$\text{area} = 2 \times 2 \times \tfrac{5}{8} = 2.5 \text{ in.}^2$$
$$\text{weight} = 2.5 \times 3.4 = 8.5 \text{ lb/ft}$$

For member $L_2L_3$—bottom chord (Fig. 11.14):

$$\text{area} = 2 \times 2.5 \times \tfrac{3}{4} = 3.75 \text{ in.}^2$$
$$\text{weight} = 3.75 \times 3.4 = 12.76 \text{ lb/ft}$$

For members $L_2L_1$, $L_3U_4$—end diagonal (Fig. 11.15):

$$\text{area} = 2 \times 2 \times \tfrac{1}{2} = 2 \text{ in.}^2$$
$$\text{weight} = 2 \times 3.4 = 6.8 \text{ lb/ft}$$

For members $L_2U_3$, $L_3U_2$—center diagonal (Fig. 11.16):

$\frac{7}{8}$-in. bar with turn buckle

area $= 0.766$ in.$^2$

weight $= 0.766 \times 34 = 2.6$ lb/ft

For members $L_1U_1$, $L_2U_2$, $L_3U_3$, $L_4U_4$—verticals (Fig. 11.17):
1. *Inertia.*

$A = 3.9$ in.$^2$

$I_x = 2 \times 7.4 = 14.8$ in.$^4$

$I_y = 2 \times 0.48 + 3.9(3.24)^2 = 41.9$ in.$^4$

$$r = \left(\frac{I}{A}\right)^{1/2} = \left(\frac{14.8}{3.9}\right)^{1/2}$$

$$= 1.95 \text{ in.}$$

$$\left(\frac{L}{r}\right) = \frac{19.2 \times 12}{1.95} = 118.1$$

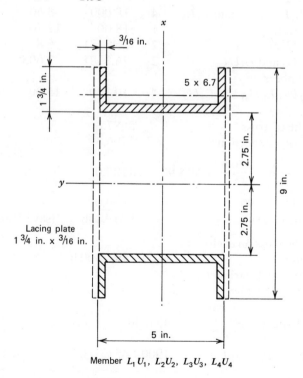

Member $L_1U_1$, $L_2U_2$, $L_3U_3$, $L_4U_4$

**Figure 11.17** Vertical members.

2. *Allowable Stresses.* Using $F_a = 10{,}000 - 0.152(L/r)^2$ gives

$$F_a = 7.88 \text{ ksi} \quad \text{for compression}$$
$$F_a = 14.0 \text{ ksi} \quad \text{for tension}$$

3. *Weight.*

| | |
|---|---|
| 2 Channels: $2 \times 6.7$ | 13.4 |
| Lacing bars | 1.6 |
| | 15 lb/ft |

*Weight of the Truss.* The factors involved in determining truss weight are tabulated below.

| Members | No. | Length | Weight per foot | Total weight |
|---|---|---|---|---|
| End post $L_0U_1$ and $U_4L_5$ | 2 | 26.3181 | 30.00 | 1579.1 |
| Top chord $U_1-U_4$ | 3 | 18.000 | 30.00 | 1620.0 |
| Bottom chord | | | | |
| $L_0L_1$, $L_1L_2$, $L_3L_4$, and $L_4L_5$ | 4 | 18.000 | 8.50 | 612.0 |
| Bottom chord $L_2L_3$ | 1 | 18.00 | 12.76 | 229.7 |
| Member $L_2U_1$ and $L_3U_4$ | 2 | 26.3181 | 6.8 | 358.0 |
| Member $L_2U_3$ and $L_3U_2$ | 2 | 26.3181 | 2.603 | 137.0 |
| Vertical members | | | | |
| $L_1U_1$, $L_2U_2$, $L_3U_3$, and $L_4U_4$ | 4 | 19.200 | 15.00 | 1152.0 |

| | |
|---|---|
| Dead weight of truss | 5687.8 |
| Add 30% for connection | 1706.3 |
| Wooden curb: $\dfrac{6 \text{ in.} \times 6 \text{ in.}}{144} \times 50 \times 90$ | 1125.0 |
| Railing 10 lb/ft at 90 | 900.0 |
| Total dead load | 9419.1 lb    (say 9500 lb.) |

Weight of the truss on each
panel point:      $9500 \div 5 = 1900 \text{ lb}$

Weight of the deck, stringers,
and floor beam which is
transferred by floor beam
at every panel point:      4515 lb

Total dead load at panel
point $L_1$, $L_2$, $L_3$, and $L_4$:      $1900 + 4515 = 6415$    Use $P$
                                                                        $= 6.42 \text{ k}$

Dead load at panel point
$L_0$ and $L_5$: $\qquad\qquad 1900 \div 2 = 950\,\text{lb}$

Applying these concentrated loads to the panel points gives the dead-load bar forces, as shown in Fig. 11.18.

LIVE LOADS

*Influence Line Diagrams.* The resulting influence line diagrams for each truss bar are shown in Figs. 11:19 through 11.28. These diagrams are now used to evaluate maximum live-load forces. It should be noted that the center-panel diagonals are only assumed to act in tension.

*Distribution of Live Loads.* Both truck load (H15) and lane load are considered.

1. *Truck Load.* The H15 truck is positioned laterally on the bridge such as to induce the maximum reaction on the truss, as shown in Fig. 11.29. The maximum live-load reaction is computed as

$$\text{reaction} = \frac{12(6.025 + 12.075)}{15.05} = 14.47\,\text{k}$$

$$\text{impact} = \frac{50}{90 + 125} = 0.2326$$

$$(LL + I)\ \text{reaction} = 14.47 \times 1.2326$$

$$= 17.84\,\text{k (rear axle)}$$

$$= 4.46\,\text{k (front axle)}$$

The resulting truck, which is now positioned along the truss, is shown in Fig. 11.30.

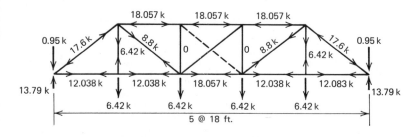

**Figure 11.18** Dead-load bar forces.

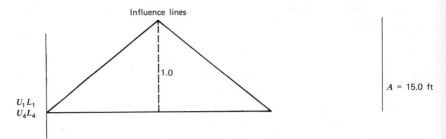

**Figure 11.19**   Influence line for members $U_1L_1$, $U_4L_4$.

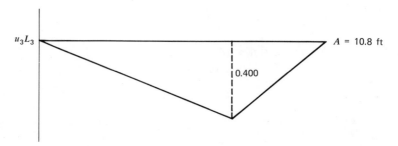

**Figure 11.20**   Influence line for member $U_1L_1$, $U_4L_4$.

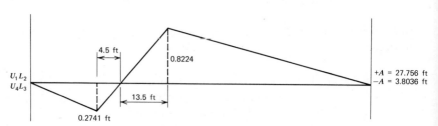

**Figure 11.21**   Influence line for member $U_3L_3$.

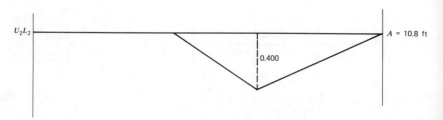

**Figure 11.22**   Influence line for member $U_2L_2$.

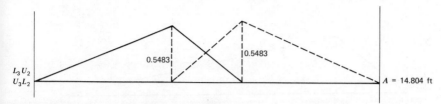

**Figure 11.23** Influence line for member $L_3U_2$, $U_3L_2$. $A$ is the area under the influence line diagram.

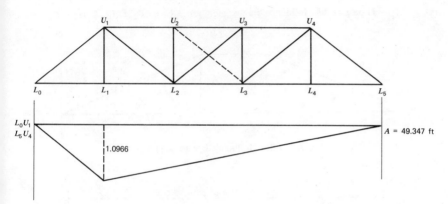

**Figure 11.24** Influence line for member $L_0U_1$, $L_5U_4$.

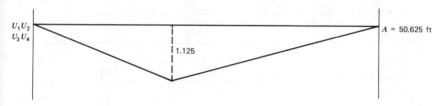

**Figure 11.25** Influence line for member $U_1U_2$, $U_3U_4$.

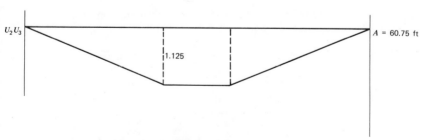

**Figure 11.26** Influence line for member $U_2U_3$.

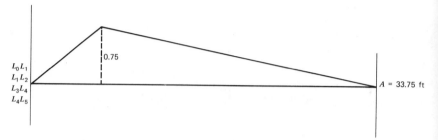

**Figure 11.27** Influence line for member $L_0L_1$, $L_1L_2$, $L_3L_4$, $L_4L_5$.

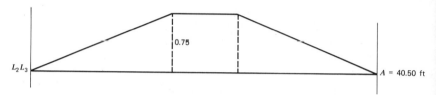

**Figure 11.28** Influence line for member $L_2L_3$.

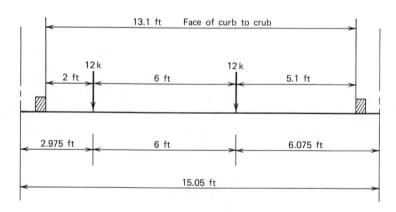

**Figure 11.29** H15 lateral truck position—truss loading

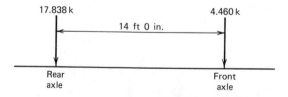

**Figure 11.30** H15 Truck—longitudinal loading on truss

**416**

2. *Lane Load.* The uniformly distributed load over a 10.0-ft lane gives $w = 480/10 = 48$ lb/ft. The concentrated load for shear distributed over 10-ft width of lane is $19.5/10 = 1.95$ k/ft. The uniform load $w$ is positioned as shown in Fig. 11.31. The maximum live-load reaction is computed as

$$\text{reaction} = \frac{10w \times 9.075}{15.05} = 6.03w$$

$$(LL + I) \text{ reaction} = 6.03w \times 1.2326 = 7.4325w$$

For the concentrated load

$$F = 7.4325 \times 1.95 = 14.493 \text{ k}$$

For the uniform load

$$R = 7.4325 \times \frac{48}{1000} = 0.357 \text{ k/ft}$$

*Forces in Various Members.* Applying the equivalent live loads (truck and lane) and the various influence lines gives the resultant forces, as shown in Tables 11.3 and 11.4.

*Stress Analysis.* Summing up the dead-load and live-load forces for each member and the corresponding properties, the proper rating for each member is obtained as shown in Table 11.5.

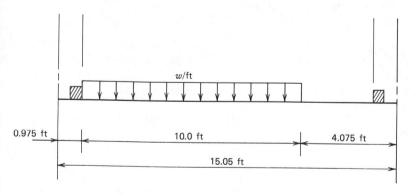

**Figure 11.31** Lane loading—truss loading

**Table 11.3  Forces in Truss Members due to Truck Loading**

| Member | Type | Force in the Member | Force (k) |
|---|---|---|---|
| $L_0U_1$ | Compression | $17.838 \times 1.0966 + 4.460 \times 0.8838$ | $-23.501$ |
| $L_5U_4$ | | | |
| $U_1U_2$ | Compression | $17.838 \times 1.1250 + 4.460 \times 0.8333$ | $-23.784$ |
| $U_3U_4$ | | | |
| $U_2U_3$ | Compression | $17.838 \times 1.1250 + 4.460 \times 1.1250$ | $-25.085$ |
| $L_0L_1, L_1L_2$ | Tension | $17.838 \times 0.7500 + 4.460 \times 0.6042$ | $16.073$ |
| $L_3L_4, L_4L_5$ | | | |
| $L_2L_3$ | Tension | $17.838 \times 0.7500 + 4.460 \times 0.7500$ | $16.724$ |
| $U_1L_1$ | Tension | $17.838 \times 1.000 \ \ + 4.460 \times 0.2222$ | $18.829$ |
| $U_4L_4$ | | | |
| $U_1L_2$ | Tension | $17.838 \times 0.8224 + 4.460 \times 0.6092$ | $17.387$ |
| $U_4L_3$ | | | |
| $U_1L_2$ | Compression | $17.838 \times 0.2741 + 4.460 \times 0.0609$ | $-5.161$ |
| $U_4L_3$ | | | |
| $U_2L_2$ | Compression | $17.838 \times 0.400 \ \ + 4.460 \times 0.2444$ | $-8.225$ |
| $U_3L_3$ | | | |
| $L_2U_3$ | Tension | $17.838 \times 0.5483 + 4.460 \times 0.3351$ | $11.275$ |
| $L_3U_2$ | | | |

**Table 11.4  Forces in Truss Members due to Lane Loading**

| Member | Force due to Uniform Load | Force due to Concentrated Load | Total Force (k) |
|---|---|---|---|
| $L_0I_1$ | $0.357 \times 49.347 = 17.617$ | $14.493 \times 1.0966 = 15.893$ | $-33.510$ |
| $L_5U_4$ | | | |
| $U_1U_2$ | $0.357 \times 50.625 = 18.073$ | $14.493 \times 1.125 = 16.305$ | $-34.378$ |
| $U_3U_4$ | | | |
| $U_2U_3$ | $0.357 \times 60.750 = 21.688$ | $14.493 \times 1.125 = 16.305$ | $-37.993$ |
| $L_0L_1, L_1L_2$ | $0.357 \times 33.750 = 12.049$ | $14.493 \times 0.750 = 10.870$ | $22.919$ |
| $L_3L_4, L_4L_5$ | | | |
| $L_2L_3$ | $0.357 \times 40.50 = 14.459$ | $14.493 \times 0.750 = 10.870$ | $25.329$ |
| $U_1L_1$ | $0.357 \times 15.00 = 5.355$ | $14.493 \times 1.00 = 14.493$ | $19.848$ |
| $U_4L_4$ | | | |
| $U_1L_2$ | $0.357 \times 27.756 = 9.909$ | $14.493 \times 0.8224 = 11.919$ | $21.828$ |
| $U_4L_3$ | | | |
| $U_1L_2$ | $0.357 \times 3.0836 = 1.101$ | $14.493 \times 0.2741 = 3.973$ | $-5.074$ |
| $U_4L_3$ | | | |
| $U_2L_2$ | $0.357 \times 10.80 = 3.856$ | $14.493 \times 0.400 = 5.797$ | $-9.653$ |
| $U_3L_3$ | | | |
| $L_2U_3$ | $0.357 \times 14.8041 = 5.285$ | $14.493 \times 0.5483 = 7.947$ | $13.232$ |
| $L_3U_2$ | | | |

**Table 11.5 Stress and Rating of Truss Members[a]**

| Member | DL force (k) | LL force (k) | Length (in.) | Area | r (in.) | L/r | DL Stress (ksi) | LL Stress (ksi) | Allowable Stress (ksi) | Available Stress for LL (ksi) | Inventory Rating |
|---|---|---|---|---|---|---|---|---|---|---|---|
| $L_0U_1$, $L_5U_4$ | −17.600 | −33.510 | 315.82 | 7.8425 | 2.3752 | 133 | −2.244 | −4.273 | 7.311[b] | 2.874 | H10 |
| $U_1U_2$, $U_3U_4$ | −18.057 | −34.378 | 216.00 | 7.8425 | 2.3752 | 91 | −2.302 | −4.384 | 8.741[c] | 4.254 | H14.6 |
| $U_2U_3$ | −18.057 | −37.993 | 216.00 | 7.8425 | 2.3752 | 91 | −2.302 | −4.845 | 8.741[c] | 4.254 | H14.6 |
| $L_0L_1$, $L_1L_2$, $L_3L_4$, $L_4L_5$ | 12.038 | 22.919 | 216.00 | 2.500 | — | — | 4.815 | 9.168 | 14.000 | 9.185 | H15 |
| $L_2L_3$ | 18.057 | 25.329 | 216.00 | 3.75 | — | — | 4.815 | 6.754 | 14.000 | 9.185 | H20.4 |
| $U_1L_1$, $U_4L_4$ | 6.42 | 19.848 | 230.40 | 3.90 | — | — | 1.646 | 5.089 | 14.000 | 12.354 | H36.4 |
| $U_1L_2$, $U_4L_3$ | 8.800 | 21.828 | 315.82 | 2.00 | — | — | 4.400 | 10.914 | 14.000 | 9.600 | H13.2 |
|  | 8.800 | −5.074 | 315.82 | 2.00 | — | — | 4.400 | −2.537 | — | — | Dead-load tension is greater than LL compression |
| $U_2L_2$, $U_3L_3$ | 0.000 | 9.653 | 230.40 | 3.90 | 1.95 | 118.2 | 0.000 | 2.475 | 7.878 | 7.878 | H47.7 |
| $L_2U_3$, $L_3U_2$ | 0.00 | 13.232 | 315.82 | 0.7656 | — | — | 0.000 | 17.283 | 14.000 | 14.000 | H12.2 |

[a] Positive values indicate compression and negative values indicate tension.
[b] Only 70% allowable stress is used for rating top chord, $L_0U_1$, because it is rusted and buckled.
[c] 25% allowable stress reduction for rust.

Timber Deck

Inventory rating: H4
Operating rating: H6

Steel Stringers

Inventory rating: H1.4
Operating rating: H1.8

Floor Beams

Inventory rating: H8.0
Operating rating: H10.0

Truss

Inventory rating: H10.0
Operating rating: H14.0

The bridge is therefore limited by the stringers. If the deck system is improved, the rating could be substantially increased (3).

## REFERENCES

1  American Association of State Highway and Transportation Officials, *Standard Specifications for Highway Bridges*, 11th Ed., 1973, Washington, D.C.
2  American Association of State Highway and Transportation Officials, *Manual for Maintenance Inspection of Bridges*, 1974, Washington, D.C.
3  C. P. Heins, W. S. Fout and R. Y. Wilkinson, "Old Truss Bridges—Replace or Repair," Transportation Research Record No. 607, January 1976, Washington, D.C.
4  *Cambria Steel–Handbook*, Philadelphia, Pennsylvania, 1919.

# CHAPTER 12

# Design of Steel
# Cable-Stayed Bridges

## 12.1 INTRODUCTION

There has not been a new bridge development as dramatic as the design of the cable-stayed bridge since the world-record suspension bridges of the 1930s. The development of the prestressed concrete bridge shortly after World War II aroused the interest of many bridge engineers, since this type of structure had relatively short spans and was within the design capabilities of the majority of design engineers. The cable-stay arrangement is logical for long spans only and as such falls in the domain of only a small number of engineers.

The first modern cable-stayed bridge was designed by West German engineers for a bridge at Stromsund, Sweden, completed in 1955. This bridge has modest spans of 74, 183, and 74 m. The success of this structure led to the design and construction of the North-bridge in Dusseldorf, West Germany in 1958 (later named Theodore Heuss Brücke). The central span of 260 m gave ample room for the heavy barge traffic on the Rhine River. The clear span requirements on the Rhine are such that this bridge type is the most logical. Many cable-stayed bridges have been built over this famous waterway and other rivers in Europe. Canada, South America, and Japan are also the locations of several interesting structures, In 1977 the United States had one modest span structure in Alaska and one long-span cable-stayed bridge under construction in Pasco, Washington. Several are presently on the drawing boards.

To date (1978) the longest clear span is the St. Nazaire Bridge in France, which measures 404 m. (1325 ft). The cable-stayed bridge is well suited for clear spans from 400 to 2000 ft. It can have girders of steel or prestressed concrete.

A very favorable aspect of this type of bridge is that it can be erected with no false work in the main span. Over deep canyons or waterways this is a considerable advantage.

It is interesting to note that the engineering text, '*Theory and Practice of Modern Framed Structures*, by Johnson, Bryan, and Turneaure, printed in 1907, showed a picture of a cable-stayed bridge but gave no details of design or analysis. It is not certain what the authors had in mind for the stays.

The idea of the cable-stayed bridge goes back several centuries. A paper by the American Society of Civil Engineers Committee on Long-Span Steel Bridges (1) gives an extensive bibliography, as well as dimensions of all cable-stayed bridges at the date of publication.

## 12.2  COMPOSITION OF CABLE-STAYED BRIDGES

The cable-stayed bridge is not a single structural arrangement but has the possibility of a multiplicity of designs. There are a wide variety of features in the many cable-stayed bridges built to date (1). Variations in the construction material and in several other design aspects are possible.

The main feature of this type of bridge is that it has the girder supported not only at the abutments and piers, but also by cables radiating from the towers to the girders (Fig. 12.1). The various design alternatives besides span length are discussed below.

### 12.2.1  Cables

The cables are composed of strands made from steel wires. The wires can be arranged in a number of different ways. The Europeans have used a locked-coil cable, particularly in their earlier designs. This cable cross section (Fig. 12.2a) has the advantage of a smooth outside surface that can be painted more easily. The outside elements have a greater contact

**Figure 12.1** Cable-stayed bridge.

surface than round wires and thus show a greater impediment to moisture penetrating to the inner elements.

This type of cable has less strength per unit weight and also higher cost than either the wire bridge strand (Fig. 12.2b) or the parallel-wire cable (Fig. 12.2c). The locked-coil cable also has a smaller modulus of elasticity than the other two. It is doubtful that the locked-coil will be used in bridges in the United States. The advantage of resistance to moisture is being minimized by development of new methods of cable protection.

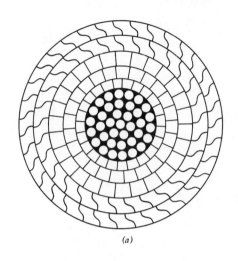

(a)

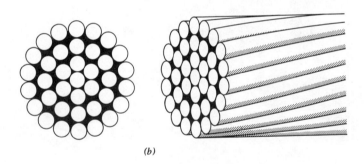

(b)

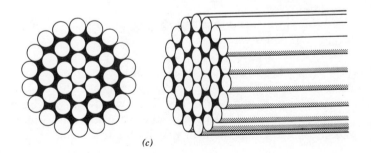

(c)

424

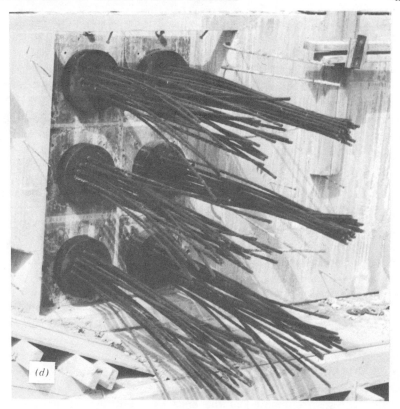

**Figure 12.2**  Types of cables; (*a*) locked oil; (*b*) 1×37 galvanized structural strand; (*c*) parallel wire; (*d*) prestressing strand.

The wire bridge strand consists of multiple wires of ultimate strength of approximately 250 ksi twisted into a compact unit. The size of the cable is determined by the number of wires. This type of cable has been used for the main cables of small suspension bridges, cable suspender ropes for large suspension bridges and arches, and cable-hung roof structures. This is the type of cable used on the Sitka Harbor Bridge, which was the first cable-stayed bridge in the United States. It has an effective modulus of elasticity of 23 to 24 million lb/in². Moisture penetration is of concern, and cables for bridges should have wires with at least a class A galvanized coating.

The bridge strand is easily shipped and handled. The wires do not separate if handled properly. A variety of end connections are available.

In large parabolic cable suspension bridges the main cables are composed of many small-diameter wires and are spun in place. It is now possible to obtain various size cables that are preassembled of parallel wires. In the United States the wires are usually $\frac{1}{4}$ in. in diameter and in Europe they are usually 7 mm. The strength of the parallel-wire cable is the highest per unit weight since there is no loss due to twisting. The modulus of elasticity of such a cable is about $28 \times 10^3$ ksi. To hold it compact during shipping and erection the cable is wrapped with an outside circumferential wire or a moisture-protection sheath or both.

The Mannheim Bridge over the Rhine River in West Germany was the first to be built with the vinyl moisture-protection sheath for the parallel-wire cable. The Pasco-Kennewick Bridge in Washington is the first with this sheath in the United States. When the moisture protection sheath is used a filling material has to be inserted in the air space. This is done after the cable is in place and finally tensioned. Cement grout and a mixture of polyurethane and zinc chromate have been used. The sheath must be well sealed at the anchorages.

Until recently the method for anchoring wires in an end fitting has been the splaying of the wires in a conical cavity in the fitting and the filling of the spaces with molten zinc. Overheating of the molten metal can result in a reduction in the strength of the wires. Tests have also proven that if there is any flexure at the fitting the fatigue strength of the wires is reduced at this location. A new process of end anchorage of wires was used at the Mannheim Bridge. This method is shown in Fig. 12.3. The parallel wires are anchored at the ends by a buttonhead bearing on a heavy steel plate. A space ahead of the anchor plate is filled by a mixture of small steel balls and an epoxy material. The space has a conical shape that provides a wedging action. This anchorage, which does not reduce the

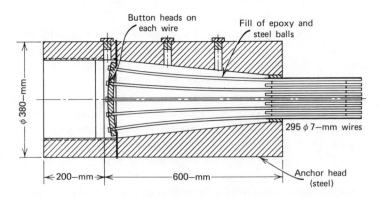

**Figure 12.3**   Anchorage head—first used in Mannheim bridge.

strength of the wires, has a higher fatigue strength than the old zinc method. The cables on the Pasco-Kennewick Bridge are anchored in this manner.

A fourth type of cable that has been used in the past (Pont Brotonne, France), and will most likely be used in the future, is composed of multiple strands of the standard seven-wire prestressing cable (Fig. 12.2d). The number of strands can be varied to fit the force requirement. Anchorage of each strand is by the usual end anchorage used in conventional prestressing. The wires in these strands have an ultimate strength of 270 ksi and since they have no galvanizing, they must be weather protected. This has been done on the Brotonne Bridge by using a thin-guage metal tube filled with grout. The Ruck-a-Chucky Bridge plans show this kind of cable with a polyurethane tube filled with grout. The various strands are held together by a wire wrapping in helix formation.

## 12.2.2  Cable Pattern

The designer of a cable-stayed bridge has many choices in cable pattern. The first choice is in the number of cables to use. The number of cables have varied from the single-cable Wye Bridge in England (Fig. 12.4a) to the 20-cable Rheinbrücke Bonn-Nord (Fig. 12.4b). When only a single cable is used the girder of course must be much larger than if several cables are used. Though the single-cable is much larger in diameter than the size required when multiple cables are used, the total weight of cable steel is greater for the multiple-cable design. There is an optimum solution to the question of the number of cables to use, but it is very difficult to achieve, since cost factors available to the design engineer are not sufficiently precise.

A single-cable design results in the difficulty of erecting a very heavy cable and the problems inherent in the design of the anchor at the girder and tower to resist the large cable force. However, a plus is that there is only one cable to erect.

Multiple cables simplify the erection of the girder, since as sections of girder are cantilevered out they can be connected to the cables and the end elevations adjusted. A single-cable system may call for temporary erection cables. The present trend is toward the use of multiple cables.

Another choice in cable pattern is whether to have all cables radiate from the top (or near the top) of the tower or to arrange parallel to each other and thus connect at even spacing up the tower. The former is called the radiating pattern and the latter the harp pattern. The Bonn-Nord Bridge is a good example of a third arrangement that has been used. The cables are not parallel but do not all connect to the tower at the top. This

**Figure 12.4** Cabled-stayed bridges with variation in number of cables: (a) Wye Bridge, England; (b) Bonn Nord Bridge, West Germany.

has been called the fan type arrangement. A fourth arrangement, called the star, has been used a few times. Figure 12.5 shows these four cable arrangements.

The higher in elevation the cable meets the tower, the less force in the cable, since the cable has a steeper slope. Leonhardt and Zellner (2) have indicated that there is a slight economy in the radiating and fan patterns over the harp pattern. A detail difficulty in the radiating pattern is that if many cables are used there is not sufficient space at the top of the tower to accommodate all the cables. Special tower heads such as those used on the Ludwigshafen Bridge can be used to meet this requirement (Fig. 12.6).

Which of the cable arrangements is more aesthetically pleasing is still a matter of personal preference. Whichever pattern is used, the cable-stayed bridge is very striking in appearance. The functional truth of this bridge type is obvious to both engineer and the casual observer.

Bridges have been built with both single and double vertical planes of cables. The single-plane cables lie down the center line of the bridge and connect to a single pylon tower. Such single-plane designs must have a torsionally rigid box girder. Many major long-span cable-stayed bridges have been built with the single-plane system. Again there is the choice of fewer cables of larger size for the single-plane system versus more cables of smaller diameter for the two-plane system. Most of the concrete girder bridges have been designed with the two-plane system. However, the Brotonne Bridge in France is an exception.

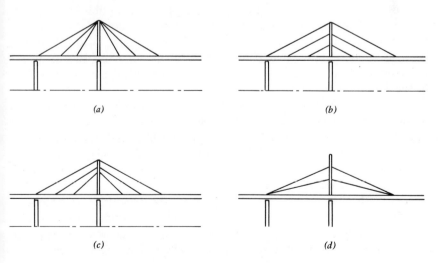

**Figure 12.5** Cable arrangements; (a) radiating; (b) harp; (c) fan; (d) star.

**Figure 12.6**   Ludwigshafen Bridge, West Germany.

## 12.2.3  Towers

The towers are a salient feature of the cable-stayed bridge. The design choices are many. The choice of single-plane or double-plane cable systems dictates the use of a single pylon tower, a double-pylon tower, the portal, or an A-frame tower (Fig. 12.7). The double-pylon tower can be composed of individual pylons, such as in the Kniebrücke, or the pylons may be tied together at the top to form a portal, such as was done in the Pasco-Kennewick (Fig. 12.8).

The Severin Bridge over the Rhine River at Cologne was the first (1960) cable-stayed bridge to use the A-frame tower. Several other bridges have been built with this type of tower, but it is less common than the pylon tower. With the A-frame tower the cables have to meet the tower near or at the top; otherwise considerable bending moment is introduced in the legs of the tower.

A few bridges have the single-pylon tower with a considerable length of tower extending above the location of the topmost cable. This design is for aesthetic reasons only. There is some question whether such treatment was successful, especially in consideration of the extra cost.

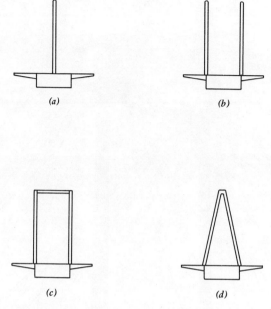

**Figure 12.7**  Various tower arrangements: (a) single pylon, (b) double pylon, (c) portal, (d) A-frame.

The cables can be terminated at the tower or they can be continuous over a tower saddle and terminate on the girder in the end span. If the cables run over a saddle, they must be clamped at the saddle so that no sliding of the cable takes place. Final adjustment of the cable tension has been achieved by vertical jacking of the saddle and shiming under the saddle. Space to perform this operation within the tower may be limited.

The tower can be fixed or pinned at the base. For construction purposes a fixed-base tower has advantages. For long-span bridges the selection would most likely be a tower with a fixed base. The only advantage in a pinned base is a reduction in bending moment in the tower.

The towers can either be steel or reinforced concrete. To date most of the cable-stayed bridges have been built with the steel tower. The concrete tower is very heavy, with heights approaching 100 ft or more. The Erskine Bridge in Scotland has a tower with a steel outer shell filled with concrete. Composite action is accomplished by internal shear studs. Several bridges with concrete girders have concrete towers.

Proper heights of towers are discussed in a later section on design parameters.

**Figure 12.8** Pasco-Kennewick Bridge.

432

## 12.2.4 Girders

The use of cable stays reduces the size of the girder a great amount over that required otherwise for a similar span. A recently designed plate girder bridge in California with continuous spans of 380, 480, and 380 ft has a depth of girder over the piers of 25 ft, in contrast with cable-stayed bridges of nearly 1000-ft central span and with box girders 8 to 10 ft deep.

Girders for cable-stayed bridges of both prestressed concrete and steel have been of various shapes and sizes. They have usually had a box shape with either sloping or vertical sides. Two long-span steel-girder bridges in West Germany, the Kniebrücke and the Rees, have plate girders. The smaller Sitka Harbor Bridge in Alaska also has plate girders. However, the majority of bridges have the torsionally stiff, more aerodynamically stable box girder. Wind tunnel tests in the United States, Canada, and Europe have shown that the box with sloping sides is preferable when moderate- to high-velocity winds are present. The steel box girder has exclusively used the steel orthotropic deck. This is necessary to obtain the torsional stiffness, as well as the reduction in dead load. Dead-load reduction saves in girder material, but more so in cable size, tower size, and foundation requirements. The cable-stayed bridge concentrates a large percentage of the dead weight of the superstructure at the piers.

## 12.2.5 Cable Anchorages

The design of the anchorage of the cable to the girder is a demanding feature of the cable-stayed bridge. The anchorage of the cable to the tower is also complex. If the bridge has two planes of cables, then it may be necessary to anchor the cable to the girder outside the fascia of the bridge, which results in a more complex arrangement than when the cable is brought some distance in from the outside edges of the girder. This latter arrangement is possible where there is only one plane of cables along the longitudinal center line of the structure or where wide cantilever sidewalks extend outside the tower legs. Considerable details of cable anchorages are shown in reference 3. It is very important that a thorough analysis of the connection of cable to girder be performed, since the cables transfer large forces at these locations. Proper design of fasteners and consideration of plate buckling is absolutely necessary.

The use of multiple cables reduces the magnitude of load being applied at any one location.

## 12.3 ECONOMIC PROPORTIONING OF STEEL CABLE-STAYED BRIDGES

There are few published data on the economics of cable-stayed bridges. References 2 through 5 discuss to some extent the relative costs in general terms of the variation of design parameters. Design organizations responsible for the design of such bridges have undoubtedly studied the economic effects of various design possibilities; however, little of this information is available in the literature.

The economics of the following design parameters are discussed to a limited extent in this section:

1. Arrangement and ratio of spans.
2. Height of towers.
3. Slope of cables.

### 12.3.1 Arrangement and Ratio of Spans

For cable-stayed bridges the number of spans and location of towers are primarily determined by the features of the terrain and waterway rather than the economics of the girder. This bridge type lends itself very readily to a multiplicity of features. The bridge can have single towers, double towers, or as in the case of the California Ruck-a-Chucky Bridge, no towers (Fig. 12.9). In this case the cables are anchored into the rock of the steep mountain at each end of the bridge.

A single tower, resulting in two spans, is quite common in bridges built to date. In some cases the tower is centrally located, but in most cases the tower can be located at the edge of deep water with the long span over the deep section of the river and the shorter span over a shallow section or flood plain, such as in the Kniebrücke and Oberkassel bridges at Dusseldorf, or the short span can be over approach roads on the bank, such as the Rhine River Bridge near Speyer. A study of span ratios (3) for existing two-span bridges shows that the long span varies from 0.6 to 0.8 of the total bridge length. For three-span bridges the ratio of center span length to total length is from 0.53 to 0.60, a limited variation. The designers of the Kniebrücke and Oberkassel Bridge used tiedown piers in the short span to anchor the cables in this section of the bridge (Fig. 12.10). These piers were in the flood plain and were designed to resist the vertical component of the cable tension. This resulted in a stiffer bridge in both spans.

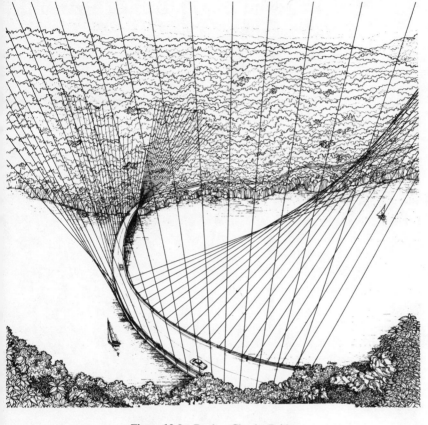

**Figure 12.9**    Ruck-a-Chucky Bridge.

## 12.3.2    Height of Towers

The cable forces and quantity of steel in the cables is influenced by the height of the tower. Making the tower higher reduces the force in the cables, since the cable is at a steeper angle to the girder. However, the cables are longer and the tower is more expensive because of greater height.

The ratio of tower height to length of longest span is given for several bridges in Table 12.1.

## 12.3.3    Cable Slope and Pattern

The radiating- and fan-cable patterns result in less cable steel. Leonhardt and Zellner (2) say this is true for ratios of tower height to span of up to

**Table 12.1   Dimensions of Existing Cable-Stayed Bridges**

| Bridge | Length of Maximum Span (ft) | Tower Height (ft) | Ratio |
|---|---|---|---|
| St. Nazaire (France) | 1325 | 220 | 0.167 |
| Erskine (Scotland) | 1000 | 125 | 0.125 |
| Kohlbrand (W. Germany) | 1060 | 322 | 0.304[a] |
| Ishikari (Japan) | 820 | 141 | 0.172 |
| Speyer (W. Germany) | 902 | 249 | 0.276[a] |
| Sitka Harbor (U.S.A.) | 450 | 100 | 0.222 |
| Hawkshaw (Canada) | 722 | 112 | 0.155 |
| Oberkassel (W. Germany) | 846 | 256 | 0.303 |
| Kniebrücke (W. Germany) | 1050 | 315 | 0.315 |
| Duisburg (W. Germany) | 1148 | 164 | 0.143 |
| LeVerkusen (W. Germany) | 918 | 148 | 0.161 |
| Karlsruhe (W. Germany) | 574 | 151 | 0.263 |
| Bonn-Nord (W. Germany) | 918 | 161 | 0.175 |
| Theodore-Heuss (W. Germany) | 853 | 131 | 0.154 |
| Severin (W. Germany) | 987 | 213 | 0.216[a] |

[a] A frame towers.

0.3 where the quantity of steel is essentially equal. Beyond 0.3 the radiating and fan patterns require more steel. This requirement appears reasonable since the length of cable becomes a greater factor than the steepness of the cable.

The single, vertical plane of cables of course results in fewer, but larger, cables. Of course, by having one plane of cables instead of two the total length of the cables is reduced by one-half, but since the area is doubled there is no change in total cable steel. However, this is not quite correct, because with two planes of cables, one set of cables is required, due to live-load eccentricity, to carry somewhat more than one-half the live load. Labor cost in handling and erecting favors a fewer number of cables. If a corrosion-resistant cover is used on the cable then the smaller number of cables also produces a cost benefit because less labor is required in installing the protective cover.

## 12.4   ANALYSIS

A cable-stayed bridge is a girder that is supported at several locations. The ends of the girder rest on abutments, which can usually be considered as fixed supports. The girder is usually supported at the tower pier, which is again nonyielding and it is generally free of any moment restraint at the tower.

The additional girder supports come from the cables emanating from the towers. These are yielding supports, since the cables change length under loads and the location of the connection of the cable and tower also moves in space because of the flexibility of the tower. The problem can then be modeled as a continuous beam on both rigid and flexible supports.

The analysis procedure is developed in the following sections.

### 12.4.1  General Behavior

The response of a cabled-supported structure can be readily analyzed by the general technique of consistent deformations. The positions or locations of the cables serve as restraints or supports along the main girder. These restraints act as springs at various locations and are dependent on the induced forces and geometry.

In determining the response of the main girder, various influence from the cables and tower must be considered. The following phases are assumed and are superimposed on these influences.

*Phase I:* rigid supports at cable location
*Phase II:* elastic cable effects at supports
*Phase III:* tower shortening
*Phase IV:* tower rotation
*Phase V:* combined effects

### 12.4.2  Rigid Supports at Cable Location

A double-cable system, shown in Fig. 12.11, is subjected to a load $P$. The column or tower is pinned at point 2 and has cables extending to locations 1 and 3, which are also pinned to the deck or girder. It is assumed that the cables at these locations are replaced by supports, as shown in Fig. 12.12. These supports create an indeterminate system to the third degree. Removing redundant reactions $R_1$, $R_2$, and $R_3$, as shown in Fig. 12.13, creates a determinate system. It is known, however, that the induced deformations $\Delta_1$, $\Delta_2$, and $\Delta_3$, created by removal of the reactions $R_1$, $R_2$, and $R_3$, must be zero in the real structure. It is therefore required to determine the magnitude of $R_1$, $R_2$, and $R_3$ such that compatibility is maintained; that is, $\Delta_n = 0$. These reactions are determined by examining

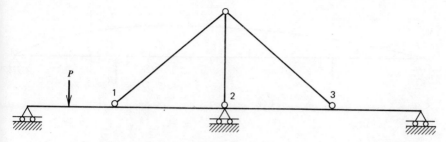

**Figure 12.11**  Cabled-Stayed bridge.

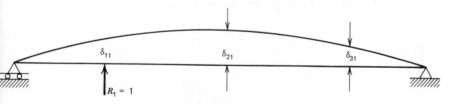

**Figure 12.12**  Equivalent continuous span.

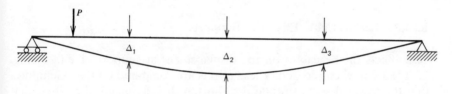

**Figure 12.13**  Simple span—$P$ loading.

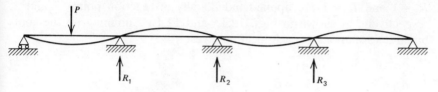

**Figure 12.14**  Simple span—unit loading $R_1$.

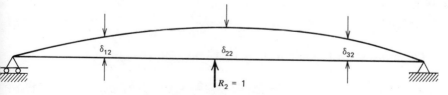

**Figure 12.15**  Simple span—unit loading $R_2$.

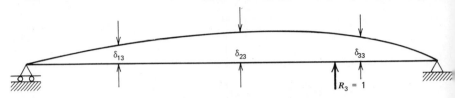

**Figure 12.16**  Simple span—unit loading $R_3$.

the effect of individual unit reactions on the system. As shown in Fig.
12.14, $R_1$ is set equal to one and induces $\delta_{11}$, $\delta_{21}$, and $\delta_{31}$. Similarly,
$R_2 = 1$ and $R_3 = 1$ are applied individually at reaction points 2 and 3,
respectively, as shown in Figs. 12.15 and 12.16. Summing up the combined deformations at each location gives:

$$R_1 \delta_{11} + R_2 \delta_{12} + R_3 \delta_{13} = \Delta_1 \tag{12.1}$$

$$R_1 \delta_{21} + R_2 \delta_{22} + R_3 \delta_{23} = \Delta_2 \tag{12.2}$$

$$R_1 \delta_{31} + R_2 \delta_{32} + R_3 \delta_{33} = \Delta_3 \tag{12.3}$$

### 12.4.3  Elastic Cable Effects at Supports

The effects of the cables on the displacement at supports 1 and 3 (Fig.
12.17) are now considered. These effects are computed on the assumption
that the tower does not rotate nor shorten. It is assumed, therefore, that
the cable will elongate and thus will permit a displacement of $\Delta v_1$, as
shown in Fig. 12.18. Assuming a small angle change such that $\theta_1 = \theta_2$ (Fig
12.19), the cable length change $\Delta L$ is

$$\Delta_L = \frac{T_c L_c}{A_c E_c} \tag{12.4}$$

and from geometry

$$\sin \theta = \frac{\Delta_L}{\Delta_v}$$

or

$$\Delta_v = \frac{\Delta_L}{\sin \theta} \tag{12.5}$$

Therefore substituting equation 12.4 into 12.5 gives

$$\Delta_v = \frac{T_c L_c}{A_c E_c \sin \theta} \tag{12.6}$$

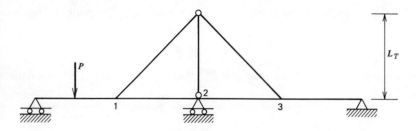

**Figure 12.17** Bridge system.

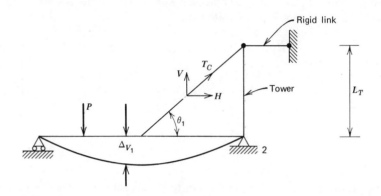

**Figure 12.18** Equivalent cable forces.

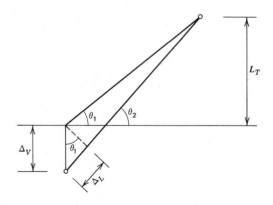

**Figure 12.19** Cable elongation.

The total force in the cable is $T_c$ and the vertical component $V$ given by

$$V = T_c \sin \theta$$

which is equivalent to the reaction $R_1$; therefore equation 12.6 is written as

$$\Delta_{v1} = \frac{R_1 L_c}{A_c E_c \sin^2 \theta} \tag{12.7}$$

This displacement coefficient times the redundant force at support point 1 is then subtracted from the resultant displacement $\Delta_1$ at point 1. A similar condition occurs at cable support 3. Therefore equations 12.1 through 12.3 are modified as follows:

$$R_1 \delta_{11} + R_2 \delta_{12} + R_3 \delta_{13} = \Delta_1 - \Delta v_1 \tag{12.8}$$

$$R_1 \delta_{21} + R_2 \delta_{22} + R_3 \delta_{23} = \Delta_2 \tag{12.9}$$

$$R_1 \delta_{31} + R_2 \delta_{32} + R_3 \delta_{33} = \Delta_3 - \Delta v_3 \tag{12.10}$$

### 12.4.4 Tower Shortening

As shown in Fig. 12.20, the tower shortens vertically as a result of strain in the tower caused by the vertical component of the cable force. This shortening is given by

$$\Delta_{vT_1} = R_1 \frac{L_T}{A_T E_T} \tag{12.11}$$

The total shortening also includes the effect of the cables connected at node 3; therefore

$$\Delta_{vT} = (r_1 + R_3) \frac{L_T}{A_T E_t} \tag{12.12}$$

This effect is not included in the general equations 12.8 through 12.10, but is shown by

$$R_1 \delta_{11} + R_2 \delta_{12} + R_3 \delta_{13} = \Delta_1 - \Delta v_1 - \Delta v_T \tag{12.13}$$

$$R_1 \delta_{21} + R_2 \delta_{22} + R_3 \delta_{23} = \Delta_2 \tag{12.14}$$

$$R_1 \delta_{31} + R_2 \delta_{32} + R_3 \delta_{33} = \Delta_3 - \Delta v_3 - \Delta v_T \tag{12.15}$$

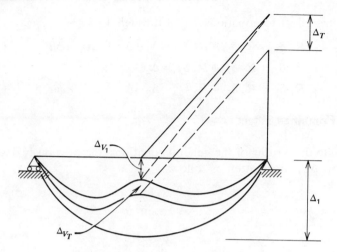

**Figure 12.20**  Bridge-system deformations.

## 12.4.5  Tower Rotation

If the loading is not symmetrical with respect to tower support 2, the tower rotates, as shown in Fig. 12.21. This rotation causes a vertical displacement $\Delta\phi$ in the girder, which is given by

$$\tan\phi = \frac{\Delta\phi}{h_1} \quad \text{or} \quad \Delta\phi = \phi \cdot h_1 \tag{12.16}$$

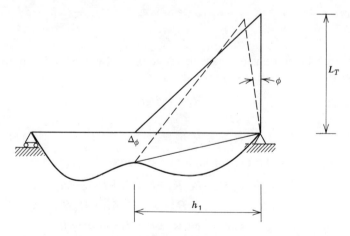

**Figure 12.21**  Tower rotation.

Adding this effect to equations 12.13 through 12.15

$$R_1\, \delta_{11} + R_2\, \delta_{12} + R_3\, \delta_{13} = \Delta_1 - \Delta v_1 - \Delta v_T - \phi h_1 \qquad (12.17)$$

$$R_1\, \delta_{21} + R_2\, \delta_{22} + R_3\, \delta_{33} = \Delta_2 \qquad (12.18)$$

$$R_1\, \delta_{31} + R_2\, \delta_{32} + R_3\, \delta_{33} = \Delta_3 - \Delta v_3 - \Delta v_T + \phi h_3 \qquad (12.19)$$

### 12.4.6   Combined Effect

To expedite the writing of the general equations, the previously described terms $\Delta v_1$, $\Delta v_T$ are written as follows:

$$\Delta v_1 = \frac{R_1 L_c}{A_c E_c \sin^2 \theta}$$

Given

$$\delta_c = \frac{L_c}{A_c E_c \sin^2 \theta} \qquad (12.20)$$

Therefore

$$\Delta v_1 = R_1\, \delta_c, \qquad \Delta v_3 = R_3 \delta_c$$

Also

$$\Delta v_T = \frac{R_1 L_T}{A_T E_T}$$

Given

$$\delta_T = \frac{L_T}{A_T E_T} \qquad (12.21)$$

Therefore;

$$\Delta v_T = (R_1 + R_3)\, \delta_T$$

Another useful relationship can be developed by summing moments about the tower support, as shown in Fig. 12.22:

$$R_1 h_1 = R_3 h_3$$

or

$$R_1 h_1 - R_3 h_3 = 0 \qquad (12.22)$$

Rewriting equations (12.17) through (12.19) gives

$$R_1\, \delta_{11} + R_2\, \delta_{12} + R_3\, \delta_{13} + \delta c_1 R_1 + \delta_T (R_1 + R_3) + \phi h_1 = \Delta_1$$

$$R_1\, \delta_{21} + R_2\, \delta_{22} + R_3\, \delta_{23} = \Delta_2$$

$$R_1\, \delta_{31} + R_2\, \delta_{32} + R_3\, \delta_{33} + \delta c_3 R_3 + \delta_T (R_1 + R_3) - \phi h_3 = \Delta_3$$

$$R_1 h_1 - R_3 h_3 = 0$$

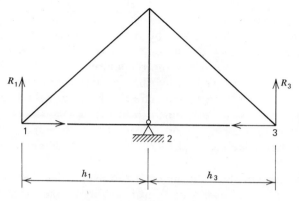

**Figure 12.22** Cable reactions.

Collecting terms and writing the equations in matrix form gives

$$
\begin{bmatrix}
(\delta_{11}+\delta c_1+\delta_T) & \delta_{12} & (\delta_{13}+\delta_T) & h_1 \\
\delta_{21} & \delta_{22} & \delta_{23} & 0 \\
(\delta_{13}+\delta_T) & \delta_{32} & (\delta_{33}+\delta c_3+\delta_T) & -h_3 \\
h_1 & 0 & -h_3 & 0
\end{bmatrix}
\begin{bmatrix}
R_1 \\
R_2 \\
R_3 \\
\phi
\end{bmatrix}
=
\begin{bmatrix}
\Delta_1 \\
\Delta_2 \\
\Delta_3 \\
0
\end{bmatrix}
\quad (12.23)
$$

Equation (12.23) can readily be solved for the redundant reactions $R_1$, $R_2$, $R_3$, and $\phi$ as for all unit displacements terms are known. The column matrix $\Delta$ values are also known quantities, as they represent the node point displacements for any loading, when no restraints are present.

If the bridge has multiple cables a matrix similar to (12.23) can be developed (6–8).

The previous analysis develops the basic equations for a single-cable-stayed bridge with the tower pinned at the base. Additional cables add additional terms to equation 12.23. If the tower is fixed at the base one additional redundant reaction is present.

More detailed analyses have been published (6–9), but space is not available in this book to present the variety of approaches. For final design a refinement considering the structure in three dimensions may be desirable.

Formalized computer programs, such as STRUDL, STRESS, and FRAN, have or can be used for anlysis of cable-stayed bridges. The Ruck-a-Chucky Bridge was analyzed from a finite element approach using the SAP-4 program.

An interesting and very important aspect of this type of bridge is that the dead-load bending moments and axial forces can be controlled by post-tensioning of the cables after the girder is erected. With one or two cables this is possible; however, with many cables the tensioning of one cable changes the tension in all the others. The cables can be pretensioned to some degree as the structure is built. This is the simplest and most economical time to induce the predetermined cable prestress.

As the cables are stressed during the erection, compression is induced into the girder. This compressive stress can easily be calculated from statics if the prestress is determined from the jacking forces.

Influence lines for cable tensions, girder thrusts, bending moments, and torsional moments are required for the design. In addition, the influence lines for tower bending moments are also necessary.

In the development of the analysis the support of the cable to the girder was considered as elastic. The flexibility of the cable was determined from equation 12.6:

$$\frac{T_c L_c}{A_c E_c \sin \theta}$$

The value of $E_c$ for a cable has to be considered as being less than the modulus of elasticity of the steel itself. Since a cable is composed of a combination of wires and/or strands, additional strain in the total cable, over and above the strain in the steel, results from the compacting of the wires and the strands. A steel wire has a modulus of close to $29.6 \times 10^6$ psi. A bridge strand that has been tensioned before erection has an effective modulus of $24 \times 10^6$ psi for cables less than $2\frac{5}{8}$ in. diameter and $23 \times 10^6$ psi for cables over $2\frac{5}{8}$ in. diameter. The bridge rope cable has an $E$ as low as $20 \times 10^6$ psi. These values are for cables with wire having a class A coating. For class B and C coatings the value of $E$ is reduced by $1 \times 10^6$ psi. Wire-type cable gives a less-flexible structure than the bridge strand. The seven-wire strand has an $E$ equal to approximately $27.5 \times 10^6$ psi. These strands are uncoated. When a cable is composed of multiple strands in a parallel arrangement, the modulus of elasticity remains equal to that of an individual strand.

Another factor affecting the effective modulus of elasticity of a cable in place is the sag in the cable caused by its own weight. If the cable has an outer steel or plastic shell with the inner space around the strands filled with cement grout for purpose of corrosion protection, then the cable can weigh considerably more than the weight of the steel strands only. The equivalent cable modulus (Fig. 12.24) as a result of the sag in the cable is

$$E_{eq} = \frac{E}{[1 + (wL)^2 AE/12T^3]} \qquad (12.24)$$

where  $E_{eq}$ = equivalent cable modulus
  $E$ = modulus of cable element
  $L$ = horizontal projected length of cable
  $w$ = weight per unit length of cable
  $T$ = cable tension
  $A$ = cross-sectional area of steel cable

A cable composed of 97 parallel wires of $\frac{1}{4}$-in. diameter and the following physical characteristics is considered as an example.

$$A = 0.0491(97) = 4.76 \text{ in}^2$$
$$T = 100(4.76) = 476 \text{ k.}$$
$$L = 200 \text{ ft}$$
$$E = 28 \times 10^6 \text{ psi}$$
$$w = 20 \text{ lb/ft with protection}$$

The value of the equivalent modulus is

$$E_{eq} = \frac{28 \times 10^6}{1 + (20 \times 200)^2 (4.76)(28 \times 10^6)/12(476,000)^3} = \frac{28 \times 10^6}{1.00165}$$
$$= 27.95 \times 10^6 \text{ psi}$$

The above example shows that the equivalent modulus is changed very little in this situation. However, the tension $T$ is for a stress of 100 ksi, which is the condition under its full dead load plus live load. At dead load only, the stress is in the region of 60 ksi or a $T$ of 285.6 k. The value of $E_{eq}$ before the live load is applied is then

$$E_{eq} = \frac{28 \times 10^6}{1.0076} = 27.79 \times 10^6 \text{ psi}$$

It is noted here that for this length and size of cable the equivalent $E$ is reduced by less than 1%. For all practical purposes the value of $E_{eq}$ could be taken as the average of the values at stress of 60 and 100 ksi for all live-load calculations. A linear analysis would be satisfactory.

If the value of $L$ is doubled to 400 ft with stress at 60 ksi, the equivalent modulus is $27.17 \times 10^6$ psi. At a stress of 100 ksi, $E_{eq} = 27.82 \times 10^6$ psi. The cable tension changes during the application of the live load. The equivalent modulus during a load increment can be calculated by the following equation:

$$E_{eq} = \frac{E}{1 + (wL)^2(T_i + T_F)AE/(24T_i^2T_f^2)}$$  (12.25)

where   $T_i$ = cable tension at start of load increment
$T_f$ = cable tension at end of load increment

Applying the above equation to the previous two conditions of $L = 400$ ft and $f_i = 60$ ksi, $f_f = 100$ ksi,

$$E_{eq} = 27.597 \times 10^6 \text{ psi}$$

This value is slightly greater than $27.50 \times 10^6$ psi, the average of $27.17 \times 10^6$ and $27.82 \times 10^6$.

Reference 10 states, "The analysis of cable-stay bridges shall include the effects of non-linearity arising from cable-sag, and the girder and pylon legs resulting from deflections of those components under the combined effects of bending moments and axial forces." From a practical standpoint, the nonlinear effects could be ignored using an equivalent $E$ for the cable. After all the details of the design are completed a nonlinear analysis could be performed using a selected possible overload.

## 12.5  FACTORS OF SAFETY

The ideal approach to design is to have the same factor of safety against collapse for all structural elements. This, of course, is not likely to be achieved in the design of a highway bridge, partly because of lack of knowledge of live-load history. In recent years the engineering profession has reached a high degree of sophistication in analysis that is not matched by knowledge of loads or effects of fabrication and erection imperfections. This sometimes leads to two misconceptions.The first misconception is the belief that since some phases of design and construction contain considerable approximation it is not necessary to be precise in anything. The other foolish idea is to fanatically believe in one's numbers, since elaborate and costly computer software and hardware were used in generating them. Good engineering design recognizes the nonprecise elements of design, fabrication, and erection and uses procedures based on sound judgment of the necessary precision in keeping with the unique nature of each project.

AASHTO Design Specifications provide limited guidance for the design of cable-stayed bridges. In the future there will most likely be some additions on this subject. However, since each design has its own unique conditions it is best to leave many aspects of the design to the best judgement of the design engineer. Well-qualified engineers only should be responsible for such bridge designs.

The AASHTO Specifications are the source of such items as allowable stress in the structural steel elements for some conditions of stress. For other conditions, such as buckling stresses and allowable cable stresses, the engineer has to rely on his own judgement or on material in technical journals.

Cables are usually designed with a breaking strength based on a requirement of the ASTM Specification or a requirement of the design specification. The limit of load in the cable due to working loads is then taken as the breaking strength divided by the selected factor of safety. The selected factor of safety should be based on accuracy of loading history, fabrication and erection control, accuracy of the mathematical model, and so forth. Factors of safety have been recommended in some publications (2, 10, 11) and vary from 2.2 to 3.0. The upper value of 3.0 is quite conservative unless there are uncertainties pertaining to the project. The value of 2.2 has been used only where the combination of dead, live, wind, and so on (Group II or III) was the loading.


## 12.6  AERODYNAMICS


All bridges are subjected to wind action of varying magnitudes. Cable-suspended structures are more flexible than conventional structures and, as such, require more careful consideration of response to wind or earthquake excited action. This is a complex subject and only general explanations can be given here along with some guidelines.

The cable-stayed bridge is much more aerodynamically stable than the conventional suspension bridge. The straight cable provides more restraint than the parabolic cable with vertical suspender ropes. Nevertheless, the cable-stayed bridge should be designed such that it will not be subject to any oscillations that will jeopardize the integrity of the structure or cause users of the bridge to feel motions during moderate air movements.

Experience with cable-stayed bridges has been very good. Of cable-stayed bridges built to date, only one structure has shown perceptible motions under wind after completion. The pylon tower can respond to vortex shedding effects under moderate or high winds before cables are attached. Precautions may be necessary during this period of construction. This tower is analogous to a tall chimney. Some such chimney structures in the past have had to add wind spoilers to alleviate an undesirable vibration problem. Attaching and tensioning cables should not be delayed unneccessarily. The presence of cables (the more the better) greatly reduces the possibility of the tower reaching any distressing amplitudes of vibration.

The response of any structure depends on the external shape of the structure, the turbulence of the approaching air flow, the Reynolds number, and the characteristics of vibration of the structure. When air flows around a structural shape projecting above the earth, such as a bridge tower, two types of loading are produced. One is a static force in the direction of the wind and the other is an alternating force perpendicular to the direction of the wind. This alternating force is produced by the shedding of vortices and induces vibration in the structure. These vortices produce periodic forces. The period is determined from the frequency of vortex shedding. The frequency can be determined by the equation

$$f = \frac{SV}{D} \tag{12.26}$$

where $f$ = frequency in cps
$S$ = nondimensional constant called the Strouhal number
$V$ = wind velocity, in./sec
$D$ = diameter of cross section, in.

For approximate values for a rectangular section, $D$ can be taken as the diagonal distance of the section.

The Strouhal number is a function of the Reynolds number, which is defined as

$$R = \frac{VD\rho}{\mu} \tag{12.27}$$

where $V$ = velocity of wind
$D$ = diameter of cross section
$\rho$ = mass density of air
$\mu$ = viscosity of air

For standard air at sea level

$$R = 44.516\,VD$$

where $V$ is in in./sec. and $D$ is in inches.

The Strouhal number varies with the Reynolds number, but for subcritical values of the Reynolds number (less than $3 \times 10^5$) the Strouhal number can be taken as a constant value of 0.20, which is the case for bridge towers.

The frequency of vortex shedding can be calculated for various wind velocities and the natural frequency of the tower can be calculated. The velocity of the wind can then be determined when the natural frequency of the tower (usually the first mode) matches the frequency of vortex shedding. If this velocity is within the realm of possibility (usually under 100 mph), then resonance in the tower and resulting large amplitudes of vibration are possible. For instance, if a tower has a natural frequency of 1 cps and a diameter of pylon of 60 in., then the velocity of wind that would produce vortices shedding at the same frequency of the tower is

$$V = \frac{fD}{S} = \frac{1.0(60)}{0.2} = 300 \text{ in/sec or } 17 \text{ mph}$$

This velocity is well within the possible environment. Additional cables attached to the pylon would increase the frequency considerably, as well as provide damping.

The frequency of the first mode of vibration of a cantilever beam can be calculated from the equation

$$f = 9.56 \sqrt{\frac{EI}{mL}} \tag{12.28}$$

where $m$ is the mass per unit length in lb-sec$^2$/in$^2$. It may be desirable to determine the vibration of the cables due to vortex shedding. Equation 12.26 can be used to check the vortex shedding frequency. Since $D$ is rather small and $S$ can also be taken as 0.2, the value of $f$ at various wind speeds is much larger than for the tower. The frequency of the cable can be determined from the equation

$$f_n = n\pi \sqrt{\frac{g}{8y}} \tag{12.29}$$

where $y$ is the sag of the cable.

The sag can be determined from the equation for maximum cable tension:

$$T = H\sqrt{1 + \left(\frac{h}{L} + \frac{4y'}{L}\right)^2} \tag{12.30}$$

where the terms in the equation are shown in Fig. 12.23. $H$ is given by

$$H = \frac{wL^2}{8y'} \tag{12.31}$$

where $y'$ is the maximum sag of the cable from the line connecting the ends of the cable. The easiest solution is to solve for $y'$ on a trial and error basis.

For a cable composed of 60 $\frac{1}{4}$ in. diameter wires, area of steel is 2.95 in$^2$. It is assumed that the cable is stressed to 100 ksi. Then

$$T = 100(2.95) = 295 \text{ k}$$

Taking the weight of the cable as 2.95(3.4) lb/ft. plus 10 lb/ft for protection material,

$$T = \frac{wS^2}{8y'}$$

where $S = (L^2 + h^2)^{1/2}$

For $L = 300$ ft and $h = 120$ ft, $S = 323.1$ ft. Then

$$295 = \frac{0.020 \ (323.1)^2}{8 \ (y')}$$

$$y' = 0.885 \text{ ft}$$

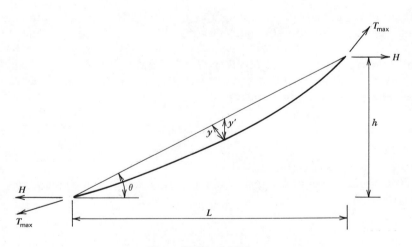

**Figure 12.23** Cable forces.

Using the exact equation

$$T_{max} = H\sqrt{1 + \left(\frac{h}{L} + \frac{4y'}{L}\right)^2}$$

where

$$H = \frac{wL^2}{8y'} = \frac{0.020\,(300)^2}{8\,(0.885)} = 254.2\,\text{k}$$

$$\frac{h}{L} = \frac{120}{300} = 0.40$$

$$\left(\frac{h}{L} + \frac{4y'}{L}\right)^2 = \left(0.40 + \frac{4(0.85)}{300}\right)^2 = 0.1696$$

$$T_{max} = 254.2\sqrt{1 + 0.1696} = 275 < 295\,\text{k}$$

$y' = 0.82$ ft (slightly less than 0.89 ft) is tried. Thus

$$H = 274.4\,\text{k}$$

$$\left(\frac{h}{L} + \frac{4y'}{L}\right)^2 = 0.1689$$

$$T_{max} = 27.4\sqrt{1 + 0.1689} = 296.6\,\text{k} \qquad \text{Use } y' = 0.82\,\text{ft}$$

However,

$$y = 0.82\cos\theta = 0.82\left(\frac{300}{323}\right) = 0.76$$

The frequency of the first mode is

$$f_n = 1(\pi)\sqrt{\frac{32.2}{8y}} = \pi\sqrt{\frac{32.2}{8(0.76)}} = 7.2\,\text{Hz}$$

The velocity of wind at which the vortex frequency matches the first-mode frequency now can be calculated, taking the Strouhal number as 0.2 and cable diameter (with protection sleeve) as 6.0 in.

$$V = \frac{Df_n}{S} = \frac{6(7.2)}{0.2} = 216\,\text{in/sec}$$

$$V = 216\,\text{in./sec} = \frac{216}{17.6} = 12.3\,\text{mph}$$

Checking the Reynolds number gives $R = 44.52(216)(6) = 57.7 \times 10^3 <$ $5 \times 10^5$.

The Strouhal number of 0.2 is correct.

The above calculations indicate that the frequency of vortex shedding matches the natural frequency of the first mode of vibration of the cable at 12 mph. This is without any damping. There is considerable elastic damping in the cable so that large amplitudes do not develop. In fact, large amplitudes cannot develop because of the high tension in the cable. Vibrations in the cable can develop bending stresses at the end fittings of the cable, and a breaking of wires as a result of fatigue may be possible. Special treatment of the cable to prevent unfavorable conditions caused by vibration may be necessary.

In most cable-stayed bridges a cable is composed of several individual cables lumped together. This increases the value of $D$ but not the value of $y$. This larger $D$ increases the required wind velocity necessary to develop any appreciable vibration. Such cable vibrations have not produced any problems in existing cable-stayed bridges; however, the bridge engineer should always investigate vibration possibilities.

Studies have shown that the most important single wind-stability factor for the bridge girder is its cross-sectional shape (12). Wind-tunnel tests of cable-suspended structures have shown that a streamlined box girder provides the most stable section. Wind-tunnel tests are usually conducted on models of major cable-suspended structures. The tests can be of two types, a full-scale model or a test of a model of a segment of the girder cross section. This latter test is much less expensive than wind-tunnel tests of a full model. In a test of section model only, the static coefficients $C_L$, $C_D$, and $C_M$ (lift, drag, and moment) can be evaluated. Dynamic flutter derivatives can also be obtained. With these factors determined, an analytical dynamic analysis can be made to obtain the response to wind of the real bridge.

The sectional model of about 1/50 scale can be varied in shape and tested in the wind tunnel to obtain the optimum shape of girder cross section. The model should be a good replica of possible girder shapes, including parapets.

A test of a full bridge model can have several drawbacks. Because of the limit in size of the available wind tunnels, the full bridge model has to be of such a small scale that an accurate dynamic modeling is almost impossible.

The data from the aerodynamic wind-tunnel section model tests can be used for the following analysis.

1. *Vortex shedding response*: In this analysis, the response of the bridge to various frequencies of vortices is made. In a cable-stayed bridge, the

frequency at which vibration amplitudes may be dangerous or even annoying should only be caused by a wind velocity greater than would ever be expected.

2. *Flutter.* This type of oscillation can be dangerous to a bridge structure but is, in most cases, only dangerous to the structure at very high wind velocities. The wind-tunnel test indicates if the shape of the cross section is likely to produce flutter in the girder, at possible velocities of the wind.

3. *Divergence.* The wind-tunnel test of a model of the girder cross section can determine, with analysis, at what wind velocities nonrestorable bridge uplift and twist can occur.

4. *Buffeting.* This response is caused by the turbulence of the wind as it makes contact with the bridge. The amplitudes that may occur at various locations along the structure can be predicted from the results of the sectional model tests and a dynamic analysis.

The cable-stayed bridge with many stays shows good resistance to wind-induced oscillations. The many cable forces disturb such oscillations very quickly. There is a considerable system damping in the structure when multiple cables are used. As each cable acts as a spring with a different value of stiffness, it is difficult to obtain and maintain the various modes of oscillation.

## 12.7 ERECTION

The cable-stayed bridge is favorably suited to difficult construction sites. Deep canyons, deep waterways, or waterways that have heavy ship traffic can be bridged without falsework in the central span of the cable-stayed bridge. The requirement of clear shipping lanes on the Rhine River led to the quick adoption of the cable-stayed bridge for this famous waterway.

In a two- or three-span bridge, the end span(s) normally requires some falsework. When the girder cross section is a closed box, the bending and torsional stiffness are adequate to permit cantilever erection of longer lengths. With cantilever erection, the cable stays are connected to the girder as the bridge projects out from the tower (Figure 12.24) The cables can be tensioned as needed to introduce the calculated prestress in the girder. Provision for final cable adjustment is usually required.

An alternate method that has been used on a few bridges is to build the complete girder on temporary piers and the tower piers. The girder is cantilevered between the piers (Fig. 12.25). The towers and girders can be erected simultaneously. The girder can be erected in a predetermined

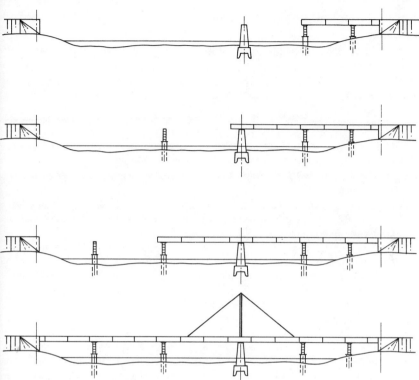

**Figure 12.25** Bridge erected entirely on temporary piers.

elevation so that as the cables are attached and the temporary piers removed, the final desired profile is obtained and the final cable tensions are achieved by a final tensioning of the cables or by jacking the tower saddles. This staging method is not practical when the water crossing has high-velocity stream flow, a deep channel, or heavy ship traffic.

The construction phase should be closely monitored to be sure that erection stresses are kept within the permissible limits. Erection methods can be suggested by the design engineers and specifications should always require that erection methods be submitted to the design engineers to have the stresses checked for their approval or rejection of the method.

Feasible erection schemes should always be developed during the design process. This is necessary before design details such as cable attachments can be finalized.

The bridge may be in a very vulnerable condition during erection with respect to the action of high winds. The response of the bridge to wind forces during all stages of erection should be checked by qualified engineers.

The cable-stayed bridge is suitable for direct and uncomplicated erection methods.

## REFERENCES

1   "Bibliography and Data on Cable-stayed Bridges," ASCE Committee on Long-span Steel Bridges, *Am. Soc. Civil Eng. STR Journal*, Vol. 103, No. ST 10, Oct 1977.

2   F. Leonhardt and W. Zellner, "Cable-stayed Bridge—Report on Latest Developments," *Proc. Can. Struc. Eng. Conf.*, 1970, Ontario, Canada.

3   W. Podolny, Jr., and J. B. Scalzi, *Construction and Design of Cable-Stayed Bridges*, Wiley, 1976.

4   W. L. Gute, "Design and Construction of the Sitka Harbor Bridge," American Society of Civil Engineers National Structural Engineering Meeting, April 1973, San Francisco, California, preprint 1957.

5   E. Dubrova, "On Economic Effectiveness of Application of Precast Reinforced Concrete and Steel for Large Bridges (USSR)," IABSE Bulletin 28, 1972.

6   B. E. Lazer, "Stiffness Analysis of Cable-stayed Bridges." *J. Am. Soc. Civil Eng., Struct. Div.*, Vol. 98, No. ST 7, July 1972.

7   Man-Chung Tang, "Analysis of Cable-stayed Girder Bridges," *Am. Soc. Civil Eng., Struct. Div.*, Vol. 97, No. ST 5, Paper 8116, May 1971.

8   B. S. Smith, "The Single Plane Cable-stayed Girder Bridges: A Method of Analysis Suitable for Computer Use," Institution of Civil Engineers, Paper No. 7011, Nov. 1967.

9   B. E. Lazer, M. S. Troitsky, and M. M. Douglass, "Load Balancing Analysis of Cable-stayed Bridges," *Am. Soc. Civil Eng., Struct. Div.*, Vol. 98, No. ST 8, Paper 9122, August 1972.

10   "Tentative Recommendations for Cable-stayed Bridge Structures," ASCE Task Committee on Cable-Suspended Structures, Structural Division.

11   Man-Chung Tang, "Design of Cable-stayed Girder Bridges," *J. Am. Soc. Civil Eng., Struct. Div.*, Vol. 98, No. ST 8, Paper 9151, August 1972.

12   R. H. Scanlan and R. H. Gade, "Motion of Suspended Bridge Spans Under Gusty Wind," *Am. Soc. Civil Eng. Struct. Div.*, Vol. 103, No. ST 9, Paper 13222, September 1977.

# Index

**459**

| DATE DUE | | | |
|---|---|---|---|
| | | | |
| | | | |
| | | | |
| | | | |
| | | | |
| | | | |
| | | | |
| | | | |
| | | | |
| | | | |
| | | | |